Critical Values of Student's *t* Distribution

One-tailed value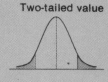

Two-tailed value

DEGREES OF FREEDOM	ONE-TAILED VALUE					
	0.25	0.10	0.05	0.025	0.01	0.005
	TWO-TAILED VALUE					
	0.50	0.20	0.10	0.05	0.02	0.01
1	1.000	3.078	6.314	12.706	31.821	63.657
2	0.816	1.886	2.920	4.303	6.965	9.925
3	.765	1.638	2.353	3.182	4.541	5.841
4	.741	1.533	2.132	2.776	3.747	4.604
5	.727	1.476	2.015	2.571	3.365	4.032
6	.718	1.440	1.943	2.447	3.143	3.707
7	.711	1.415	1.895	2.365	2.998	3.499
8	.706	1.397	1.860	2.306	2.896	3.355
9	.703	1.383	1.833	2.262	2.821	3.250
10	.700	1.372	1.812	2.228	2.764	3.169
11	.697	1.363	1.796	2.201	2.718	3.106
12	.695	1.356	1.782	2.179	2.681	3.055
13	.694	1.350	1.771	2.160	2.650	3.012
14	.692	1.345	1.761	2.145	2.626	2.977
15	.691	1.341	1.753	2.131	2.602	2.947
16	.690	1.337	1.746	2.120	2.583	2.921
17	.689	1.333	1.740	2.110	2.567	2.898
18	.688	1.330	1.734	2.101	2.552	2.878
19	.688	1.328	1.729	2.093	2.539	2.861
20	.687	1.325	1.725	2.086	2.528	2.845
21	.686	1.323	1.721	2.080	2.518	2.831
22	.686	1.321	1.717	2.074	2.508	2.819
23	.685	1.319	1.714	2.069	2.500	2.807
24	.685	1.318	1.711	2.064	2.492	2.797
25	.684	1.316	1.708	2.060	2.485	2.787
26	.684	1.315	1.706	2.056	2.479	2.779
27	.684	1.314	1.703	2.052	2.473	2.771
28	.683	1.313	1.701	2.048	2.467	2.763
29	.683	1.311	1.699	2.045	2.462	2.756
30	.683	1.310	1.697	2.042	2.457	2.750
35	.682	1.306	1.690	2.030	2.438	2.724
40	.681	1.303	1.684	2.021	2.423	2.704
45	.680	1.301	1.680	2.014	2.412	2.690
50	.680	1.299	1.676	2.008	2.403	2.678
55	.679	1.297	1.673	2.004	2.396	2.669
60	.679	1.296	1.671	2.000	2.390	2.660
70	.678	1.294	1.667	1.994	2.381	2.648
80	.678	1.293	1.665	1.989	2.374	2.638
90	.678	1.291	1.662	1.986	2.368	2.631
100	.677	1.290	1.661	1.982	2.364	2.625
120	.677	1.289	1.658	1.980	2.358	2.617
∞	.674	1.282	1.645	1.960	2.326	2.576

How large a sample
interval

t = one tailed
looking for probability

STATISTICS
An Introduction

Robert D. Mason

Douglas A. Lind

William G. Marchal

The University of Toledo

HARCOURT BRACE JOVANOVICH, PUBLISHERS

San Diego   New York   Chicago   Atlanta   Washington, D.C.

London   Sydney   Toronto

To Dorothy, Jane, and Andrea

ISBN: 0-15-583525-4
Library of Congress Catalog Card Number: 82-82707
Printed in the United States of America

Front endpapers: The Normal Probability Distribution table is from *CRC Standard Mathematical Tables*, Fifteenth Edition (West Palm Beach, Fla.: Copyright The Chemical Rubber Co., CRC Press, Inc.). Critical Values of Student's *t* Distribution table is from *The Ways and Means of Statistics* by Leonard J. Tashman and Kathleen R. Lamborn © 1979 by Harcourt Brace Jovanovich, Inc. Both are reproduced by permission.

Preface

This text has been designed for use in introductory statistics courses. Principles of both descriptive and inferential statistics are discussed, illustrated, and applied in situations close to most students' own experience. The mathematics can be handled easily by students with a limited background.

We have written and organized the material to provide the greatest possible flexibility of use in both content and format. The book's content is appropriate for any general statistics course. Illustrations and exercises are drawn from disciplines as varied as sociology, education, politics, demography, meteorology, sports, and mathematics. Considerable latitude has also been built into the organization of the chapter topics, so that the text can easily be adapted for use in one-semester, one-quarter, or two-quarter courses. A one-semester course, for example, might include Chapters 1–10, 13, and 14. Time permitting, Chapter 16 would be a logical addition.

A number of special features set this text apart in motivating and assisting students as they progress through the material:

- Each chapter begins with a set of performance objectives—what the reader should be able to do on completion of that chapter. The objectives serve the dual function of advance organizers and motivators.

- A brief chapter introduction reviews important concepts presented in the previous chapter and explains how they are linked to the present chapter. The introduction is followed by a chapter overview.

- The discussion of each concept is followed by a realistic statistical problem and its solution.

- A number of self-reviews are interspersed in each chapter. Each self-review is closely patterned after the chapter problems that preceded it. Answers and methods of solution are always shown in the margin. These self-reviews help students to monitor their own progress and provide them with constant reinforcement.

- Many interesting real-world exercises are incorporated in the body and at the end of each chapter. The answers and methods of solution for all even-numbered exercises are supplied in the book's appendix.

- Every important new term and formula is defined and placed in a box for easy reference.

- Both a summary and a chapter outline are included in every chapter. They are a valuable aid to students in pulling together a chapter's main ideas.

- An end-of-chapter achievement test covers all the material in that chapter, helping students to evaluate their own overall comprehension of the subject matter. Answers and methods of solution are provided in the appendix.

- A set of chapter highlights is included after each of six major groups of chapters. They include a brief review, key concepts, key terms, key symbols, review problems, and two ongoing cases (one dealing with a grocery store chain, and the other with a hospital) to be analyzed by the students.

- Throughout the book, a number of computer applications using the Statistical Package for the Social Sciences, MINI-TAB, and the BASIC programming language illustrate the computer's potential for problem solving.

- A complete glossary of terms and a set of standard statistical tables are also included in the book's appendix.

- The normal and t distributions are repeated inside the front cover for easy reference.

- In addition, a simple flowchart inside the back cover aids students in selecting the appropriate formula for computing the mean and the standard deviation.

A complete ancillary package accompanies the text. It includes a Study Guide, a Solutions Manual, and a set of suggested test questions for each chapter. The Study Guide is comprehensive, with an organization similar to that of the main text. Each chapter includes chapter objectives, an extensive summary, solved problems, student exercises with answers and methods of solution in the appendix, and tear-out chapter assignments to be graded by the instructor. Ample space is provided for computations. The Solutions Manual contains the full solutions to all exercises in the textbook and to all chapter assignments in the Study Guide.

Finally, we wish to acknowledge the valuable contribution made to this book by friends, students, colleagues, and collaborators throughout its various stages of development. In particular, we would like to express our thanks and appreciation to Richard J. Beres, John J. Bodner, Patricia S. DeJarnette, and Toni M. Somers at the University of Toledo, and Jerry Bergman, Spring Arbor College, for their helpful suggestions and assistance in class testing portions of the manuscript. Special thanks to Patricia E. Hetrick, who proofread the manuscript and checked the solutions, and to Dolores A. Lucitte and Jean K. Schaefer for typing some of the material.

We also wish to express our gratitude to John M. Rogers (California Polytechnic State University, San Luis Obispo), Bayard Baylis (The King's College, New York), William W. Lau (California State University, Fullerton), Kenneth R. Eberhard (Chabot College), and David Macky (San Diego State University), whose careful reviews made this a better text.

At Harcourt Brace Jovanovich, we are indebted to Terence W. Retchless, Susan Crosier, Jonathan Kroll, Sue Lasbury, Diane Polster, and Nancy A. Shehorn for their valuable assistance throughout the book's production. Special thanks to Liana Beckett for the outstanding job she did editing our manuscript.

We are grateful to the Literary Executor of the late Sir Ronald A. Fisher, F.R.S., to Dr. Frank Yates, F.R.S., and to Longman Group Ltd., London, for permission to reprint Table IV from their book *Statistical Tables for Biological, Agricultural and Medical Research* (6th edition, 1974).

ROBERT D. MASON

DOUGLAS A. LIND

WILLIAM G. MARCHAL

Contents

4   DESCRIPTIVE STATISTICS: MEASURES OF DISPERSION AND SKEWNESS

85

5   AN INTRODUCTION TO PROBABILITY

131

13 CORRELATION ANALYSIS   367

14 REGRESSION ANALYSIS   399

15 MULTIPLE REGRESSION AND CORRELATION ANALYSIS   441

16 ANALYSIS OF NOMINAL-LEVEL DATA: THE CHI-SQUARE DISTRIBUTION

473

17 NONPARAMETRIC METHODS: ANALYSIS OF RANKED DATA

501

ANSWERS TO CHAPTER ACHIEVEMENT TESTS

541

SOLUTIONS TO EVEN-NUMBERED EXERCISES

553

APPENDIX TABLES   601

GLOSSARY   618

INDEX   623

1
What Is Statistics?

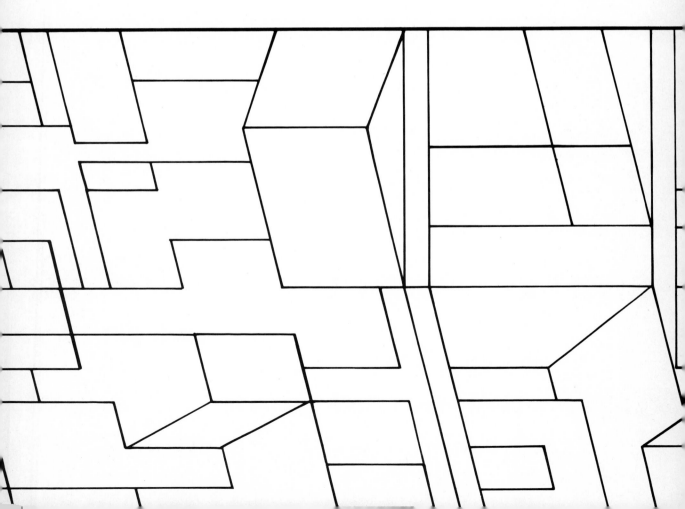

Introduction

Whenever we watch television, listen to the radio, or read a newspaper or magazine, all of us are exposed to—and sometimes overwhelmed by—assorted facts and figures commonly labelled "statistics." We may read or hear that

- The Consumer Price Index increased 0.8% last month.
- More than 97% of all households in the United States have television sets.
- One violent crime is committed every 27 seconds.
- The Dallas Cowboys defeated the Los Angeles Rams 29 to 17. The Cowboys had 24 first downs, the Rams 22. The passing yards were 275 for Dallas, 203 for Los Angeles. Dallas was penalized 108 yards, Los Angeles 78 yards.

As a result, you may envision a statistician as someone who sits in the press box and collects numerical data on passing yardage, punt returns, and so on and releases statistics like these:

Rams - Cowboys

October 18

Score by Periods

Los Angeles ..	0	10	7	0—17
Dallas	12	14	0	3—29

Scoring

Dallas—Springs 1 run (Septien kick).
Dallas—Field goal Septien 40.
Dallas—Safety, Haden tackled in end zone.
Dallas—Dorsett 44 run (Septien kick).
Los Angeles—Tyler 2 run (Corral kick).
Los Angeles—Field goal Corral 40.
Dallas—Hill 63 pass from D. White (Septien kick).
Los Angeles—D. Hill 43 pass from Haden (Corral kick).
Dallas—Field goal Septien 39.

Team Statistics

	Los Angeles	Dallas
First downs ..	22	24
Rushes - Yards ...	34 - 171	42 - 221

Individual Statistics

Rushing—Dallas, Dorsett 27 - 159, Springs 12 - 41, D. White 3 - 21; Los Angeles, Tyler 16 - 90, Bryant 14 - 53, Haden 2 - 17, Guman 2 - 11.

Passing—Dallas, D. White 15 - 33 - 2 —277; Los Angeles, Haden 13 - 30 - 3 —237.

Receiving—Dallas, Hill 4 - 97, Pearson 4 - 78, Saldi 2 - 17, Cosbie 1 - 28, Dorsett 1 - 22, DuPree 1 - 15, Johnson 1 - 15, Springs 1 - 5; Los Angeles, Miller 4 - 51, Moore 3 - 53, D. Hill 2 - 88, Bryant 2 - 13, Childs 1 - 19, Arnold 1 - 13.

Kickoff Returns—Los Angeles, D. Hill 6 - 95; Dallas, J. Jones 2 - 40, Fellows 1 - 14.

Punt Returns—Los Angeles, Irvin 3 - 13; Dallas, J. Jones 1 - 9, Fellows 2 - 6.

Interceptions—Los Angeles, Cromwell 2 - 0; Dallas, Walls 2 - 0, Thurman 1 - 0.

Punting—Los Angeles, Corral 5 - 48.0; Dallas, D. White 5 - 42.6.

Such sports data *are*, in fact, one variety of statistics—a collection of numerical data. However, statistics also has a broader meaning. The boxed definition that follows is the one we will consider throughout.

Statistics   The body of techniques used to facilitate the collection, organization, presentation, analysis, and interpretation of data for the purpose of making better decisions.

Types of Statistics

Descriptive Statistics

As stated in the definition, statistics involves, among other things, the collection, organization, and presentation of numerical data. Masses of unorganized data stored in a computer, or collected by Gallup-type polls with respect to the preference of voters, are usually of little value. Techniques are available, however, to organize such data into some meaningful form. These aids in organizing, analyzing, and describing (or summarizing) a large collection of numbers are collectively referred to as **descriptive statistics.**

Descriptive Statistics   Methods used to describe the data that have been collected.

We will discuss one such summary technique, called a frequency distribution, in Chapter 2. In that chapter, we will also examine how data can be presented graphically. For example, over 300 separate figures were needed to create a chart of a population's life expectancy at birth, by ethnic background and sex, since 1900 (see Figure 1-1). At a single glance, this chart reveals that the life expectancy for all groups has risen dramatically since 1900—with the life expectancy for blacks having increased over 100%.

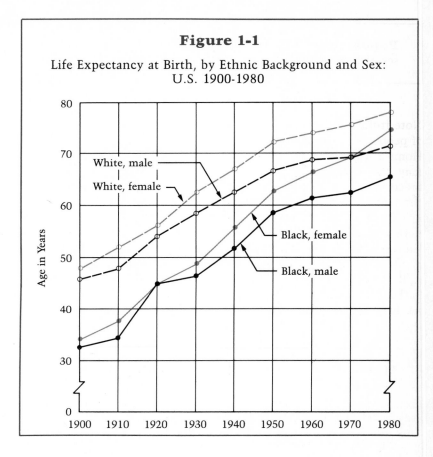

Figure 1-1

Life Expectancy at Birth, by Ethnic Background and Sex: U.S. 1900-1980

In Chapter 3 we will study the tendency of numerical data to cluster about a central value. An analysis of the spread or variation in the data using measures such as the standard deviation and the variance will be found in Chapter 4.

Inferential Statistics

Descriptive statistics is only one facet of the science of statistics. Another is **inferential statistics** or **inductive statistics.** Inferential statistical methods are concerned with finding out something about a **population.** Gallup, Harris, and other pollsters do just that when they are hired before an election to find out how voters (the population) plan to vote on election day.

> **Population** A collection of all possible members of a set of individuals, objects, or measurements.

Note from the definition that a population may consist of a group of people, such as all the senior citizens in the Vistful Vista Mobile Home Park or all the voters in Precinct 7. However, a population may also consist of a group of objects, such as all the loaves of bread baked by Blimpie's Bakery during a three-hour period or all the fish in a pond.

To find out how voters will react on election day, pollsters take a **sample** from the population. A sample is a portion, or part, of the population of interest. In the case of political polls, usually about 2,000 voters are sampled out of the population of about 80 million registered voters. Based on the sample results, an inference is then made about the reaction of *all* voters on election day. Thus, we may define inferential statistics as follows:

> **Inferential Statistics** A decision, estimate, prediction, or generalization about a population based on a sample.

Underlying statistical inference is probability. Probability and probability theory are discussed in Chapters 5, 6, and 7. Chapters 8 through 17 deal with methods of selecting a sample and making inferences about the population based on the sample results.

The Nature of Statistical Data

Statistical data occur whenever measurements are made or observations are classified. The data may be the heights of female adults, the weights of newborn babies, the number of robberies committed in various districts of the city, the preference for a certain design of a new car, the measurement of a personality trait, or the "yes" or "no" response to the question: "Have you ever been married?" Recall that in the definition of statistics we said it is the study of *data. Data are more than just numbers.* Yet, in order to make measurement simple, or "quantifiable," we must convert data into numerical form. How then can we transform a "yes" or "no" response into numerical data? We simply record, for example, a 1 for the "yes" response and a 0 for the "no" response. In this artificial way we can make this response into numerical data. Similarly, for the question: "What is your marital status?" we could record a 1, 2, 3, or 4 to indicate the responses "single," "married," "widowed," or "divorced." In these artificial ways, data may be converted to an artificial numerical form. However, certain mathematical operations would not make any sense. For example, why would you compute the average marital status of fifty people? Of course, it would be inappropriate.

Data may be classified into four general types, or levels, of measurement. The four levels are nominal, ordinal, interval, and ratio.

Nominal Level of Measurement

The information presented in Table 1-1 is an example of the nominal measurement scale. This level is the most "primitive," the "lowest," or the most limited type of measurement.

The words **nominal level of measurement,** or **nominal scaled** are used when dealing with this type of data, which can only be classified into categories. Strictly speaking, the information in Table 1-1 is simply a count, or tabulation, of the number of Protestants (78,952,000), the number of Roman Catholics (30,669,000), and so on.

Note that the arrangement of the religions could have been changed. Roman Catholic could have been listed first; Jewish, second; and so on. This indicates that for the nominal level of measurement *there is no particular order for the groupings.* Further, the categories are considered to be **mutually exclusive.**

Table 1-1

Religious Affiliation of the U.S. Population
14 Years Old and Over (self-reported)

Religion	Number
Protestant	78,952,000
Roman Catholic	30,669,000
Jewish	3,868,000
Other religion	1,545,000
No religion or religion not reported	3,195,000
Total	118,229,000

> **Mutually Exclusive** An individual or item that, by virtue of being included in one category, must be excluded from another.
>
> Ex. If your male you cant be female etc...

In this example, a person cannot be classified both as a Protestant and as having no religion.

In order to process data on religious preference, sex, type of crime, and other nominal scaled data, the categories of interest are often coded 1,2,3, . . . with, say, 1 representing Protestant, 2 representing Roman Catholic, 3 representing Jewish, and so on. This facilitates counting when computers or other data processing devices are used. Of course, it would not make sense to perform any arithmetic operations on such numbers! Also, note that in the data on religious preference the categories are **exhaustive.**

> **Exhaustive** Each person, object, or item must appear in at least one category.

If, for example, a person reported she is of the Greek Orthodox religion, she would be counted in the "Other religion" category. A person who refused to name his religion would be counted in the "No religion or religion not reported" category.

Ordinal Level of Measurement

The information in Table 1-2 is an illustration of **ordinal scaling**. Observe that here one category is ranked higher, or more outstanding,

Table 1-2
Ratings of 50 Field Social Workers by Supervisors

Rating	Number of Ratings
Superior	8
Good	20
Fair	11
Poor	9
No rating	2*

* Has been a field worker less than one month.

than the next one. That is, "superior" is a higher rating than a "good" rating, and a "good" rating is higher than a "fair" rating, and so on. When a category is higher than the preceding one, the information is considered to be of ordinal scale.

If we adopt a code where superior is coded 4, good is coded 3, and fair is coded 2, a 4 ranking is obviously higher than a 3 ranking, and a 3 ranking is higher than a 2 ranking. However, one could not say that a field worker rated 4 (superior) is twice as competent as a field worker rated 2 (fair). It can only be said that a rating of "superior" is better, or greater than, a rating of "good," and a "good" rating is better than a "fair" rating. This "greater than" concept can be illustrated by a diagram using a straight line. The one that follows also shows the uneven, or unequal, distances that may exist between various intervals.

Rating of worker	Superior	Good	Fair	Poor
Code number representing rating	4	3	2	1

To review: the major difference between a nominal and an ordinal level of measurement is the "greater than" relationship between ordinal level categories. Otherwise, the ordinal scale includes the same characteristics as the nominal scale—that is, the categories are mutually exclusive and exhaustive.

Interval Level of Measurement

The **interval level of measurement** is the next higher level. The length of service for several members of a hospital staff illustrates the concept of interval scaling:

Maria		Jennie	Rob	Grace
10		20	25	30

Length of Service (in years)

Jennie has been with the hospital ten years longer than Maria. Jennie has five years less service than Rob. And, the number of years of service between Jennie and Rob and Grace and Rob is the same. Thus, for *interval level of measurement, the distances between numbers are of a known constant size.* In the Length-of-Service graph, the length of the 10-year interval is exactly twice as long as that of the 5-year intervals.

In addition to the equidistant—or constant size—characteristic, interval scaled measurement has all the features of nominal and ordinal measurements. The lengths of service of the members of the hospital staff are mutually exclusive, that is, Jennie could not have exactly 10, 20, 25, and 30 years' service with the hospital at the same time. And, the "greater than" feature of ordinal data permits ranking the length of service from the longest (Grace) to the shortest (Maria). The interval level of measurement also assumes that the categories are exhaustive, that is, all the cases are included.

Ratio Level of Measurement

The **ratio level** is the highest level of measurement. Like the interval scale, the observations are ordered and the distance between successive observations is measured. In addition, the ratio scale uses the number zero to indicate the absence of the characteristic being measured. Money is an example of the ratio scale of measurement. The zero point is meaningful—that is, at zero you have none! To put it another way, ten dollars is worth twice as much as five dollars. Here the *ratio* of two to one has physical meaning. Temperatures on the Celsius scale, on the other hand, are *not* ratio data: 30°C is *not* twice as hot as 15°C. While all other characteristics of the interval level are in effect for the ratio scale, most hypothesis-testing situations described in this text will make use of the interval scale of measurement only.

A Word of Encouragement

If you are a freshman or sophomore in Arts and Letters or one of the social sciences, this may be your first college course with a quantitative orientation. Theory, and symbols such as Σ, σ and μ, are used extensively here. Do not be intimidated by them. They are merely a convenient shorthand that helps to condense the subject matter significantly. As a result, however, the material is slow reading, and only rarely will you feel you have a complete understanding of it unless you have gone over it more than once.

One of the ways in which you can verify your understanding of the material, as you progress through the book, is for you to work through each of several self-review problems that are included in every chapter. Checking your answers against those provided in the margin will allow you to find out immediately whether or not you understand previously covered subject matter. Taking the achievement test at the end of each chapter and comparing your answers with those in the book's Answer section will also be helpful. Solutions to all even-numbered exercises can also be found at the back of the book.

2
Summarizing Data: Frequency Distributions and Graphic Presentation

OBJECTIVES

When you have completed this chapter, you will be able to
- Construct a frequency distribution.
- Draw a histogram, a frequency polygon, and a cumulative frequency polygon.
- Design line graphs, bar graphs, and pie graphs.

Introduction

With this chapter we begin our study of **descriptive statistics.** Recall from Chapter 1 that the general purpose of descriptive statistics is to summarize data. Two important elements of descriptive statistics are the arrangement and the display of numerical information. Raw numbers alone provide little insight into the underlying pattern of the data from which conclusions are to be drawn. For example, we may believe that physicians generally have high incomes, or that young people have lower incomes than old people. However, even if we were given access to Internal Revenue Service files, the income data would be of little value unless arranged, sorted, and condensed. This chapter introduces various techniques that can be used in arranging, sorting, and depicting relevant data—specifically, the **frequency distribution** and various statistical charts.

Frequency Distribution

A frequency distribution is a useful statistical tool for organizing numerical information into a more meaningful form.

Frequency Distribution An arrangement of the data that shows the frequency of occurrence of the values of interest.

A frequency distribution is simply a table that shows how often each value (such as a wage, a height, or a test score) occurs. The idea of a frequency distribution is illustrated by the following problem and its solution.

Problem

Let us assume the federal government plans to publish guidelines for young newlywed couples, including suggested amounts to be budgeted for food. As a first step, the government has asked a small group of 100 newlyweds to keep for one month a detailed diary listing all their food expenditures. (This is a common way of collecting this type of information, with respondents usually paid a small sum of money for their cooperation.) The amount spent weekly on food by each of the 100 newly married young couples, rounded to the nearest dollar, is shown in Table 2-1.

Table 2-1
Weekly Amounts Spent on Food by 100 Young Newlywed Couples

$57	$66	$54	$52	$49	$57	$60	$55	$47	$59
50	46	64	52	58	52	68	61	63	55
58	58	56	57	62	61	57	52	63	45
55	52	57	41	55	67	58	68	57	57
64	59	47	59	60	53	42	56	67	52
48	47	56	50	58	57	57	66	53	59
60	58	53	61	51	54	62	53	67	46
51	50	60	56	63	62	47	57	61	74
56	46	57	53	48	64	65	65	63	54
69	70	40	72	71	53	58	61	44	67

low   high

Raw Data. The numerical information in Table 2-1 is usually called the **raw data.** As such, the mass of raw data has very little meaning—that is, it reveals little about the expenditure pattern for food. How can the raw data be reorganized to describe the weekly expenditures for food by newlyweds in the most useful way?

Solution

In constructing a frequency distribution the usual practice is to:

1. Decide on the groupings. To condense the large number of observations, all the values that occur within a particular interval are grouped. These groups, or intervals, are called **classes:**

$40 - $44
45 - 49

2. Tally the raw data into the classes.

3. Count the number of tallies in each class.

In Table 2-1 the smallest weekly amount spent on food by any newlywed is $40. Using an interval of $5 between each class, the classes would appear as:

$40 - $44
45 - 49
50 - 54
55 - 59
60 - 64
65 - 69
70 - 74

Note that there is no overlapping—that is, there is no question where the amount $45 is tallied. It is in the $45 - $49 class.

As noted, the second step in organizing the newlywed data is to tally the weekly food amounts into the appropriate classes. For example, the amount in the upper left-hand corner of Table 2-1 is $57. It is tallied in the $55 - $59 class. A tally mark (/) is placed opposite the $55 - $59 class. The same procedure is followed for each of the remaining weekly amounts spent on food. The resulting frequency distribution is shown in Table 2-2.

Table 2-2
Frequency Distribution: Weekly Amounts Spent
on Food by 100 Newlywed Couples

Weekly Amounts Spent on Food	Tallies	Number of Newlyweds
$40 - $44	////	4
45 - 49	ᵗʰʰ ᵗʰʰ /	11
50 - 54	ᵗʰʰ ᵗʰʰ ᵗʰʰ ᵗʰʰ	20
55 - 59	ᵗʰʰ ᵗʰʰ ᵗʰʰ ᵗʰʰ ᵗʰʰ ᵗʰʰ /	31
60 - 64	ᵗʰʰ ᵗʰʰ ᵗʰʰ ////	19
65 - 69	ᵗʰʰ ᵗʰʰ /	11
70 - 74	////	4
	Total	100

The number of tallies that occur in each class is called the **class frequency.** For example, 4 is the class frequency for the class of $40 - $44.

True Limits. Recall that the newlywed data has been rounded to the nearest dollar. Using conventional rounding rules, a weekly amount of $39.50 would be rounded *up* to $40. And, any amount over $44 but under $44.50 would be rounded *down* to $44. Therefore, the *true* class limits of the $40 - $44 class are $39.50 - $44.50. Using the true class limits of $39.50 - $44.50 and $44.50 - $49.50, and so on, each interval is exactly $5.00.

Midpoints. The midpoint divides a class into two equal parts. It can be determined two ways: (1) It is halfway between the *true* class limits. Halfway between the true limits of $39.50 and $44.50 is $42. (2) Another way to find the midpoint is to use the *stated* class limits. For the first class, that would be halfway between $40 and $44, which is also $42. Later in the chapter we will use midpoints to draw frequency polygons, while true limits will be useful in computing certain averages in Chapter 3, and measures of dispersion in Chapter 4.

Describing the Data

Now that the raw data is organized into a frequency distribution, the pattern of weekly food expenditures by the newlyweds can be described. Table 2-2 shows that the smallest weekly amount spent is about $40, the largest about $74. Moreover, most of the newlyweds spend between $50 and $65 with the largest cluster in the $55 - $59 class. The single value of $57 (the midpoint of the $55 - $59 class) is considered a "typical" weekly amount spent for food.

Incidentally, in the process of grouping the food expenditures into a frequency distribution some accuracy is often lost. For example, while we have just estimated that the typical, or most frequent, amount spent on food is $57, this figure might not be identical to one computed using the raw data. However, the difference is usually not great. The advantages of grouping data into a frequency distribution generally outweigh the disadvantage of working with unwieldy raw data.

Note: The following is the first of several self-review problems that will appear in each chapter following discussion of a major topic. The problems allow you to test immediately your comprehension of the preceding text material. First, cover the answers in the margin. Then complete the self-review and check your own answers against those given.

Self-Review 2-1

a. Age   Number

15 - 24	///	3
25 - 34	////	4
35 - 44	//	2
45 - 54	/	1
	Total	10

b. Raw data.

c. Based on the frequency distribution, the ages range from 15 to 54. The largest concentration is in the 25 - 34 class.

A reminder: cover the answers in the margin.

Police files reveal these to be the ages of persons arrested for purse snatching: 16, 41, 25, 21, 30, 17, 29, 50, 30, and 39.

a. Using 15 years as the beginning number for the first class and an interval of 10 years, organize the age data into a frequency distribution.
b. What are the numbers 16, 41, 25, . . . , called?
c. Describe the age distribution of the purse snatchers.

Chapter Exercises

1. The following data represent the amounts spent on textbooks for the current quarter by a sample of 35 students at an urban university:

$57	$34	$27	$41	$25	$18	$39
33	37	39	38	47	31	42
60	58	31	47	37	16	64
34	41	43	50	46	63	51
41	30	42	37	48	28	34

Starting with $15 as the beginning number for the first class, and using an interval of $10, organize the data into a frequency distribution. Briefly describe the distribution.

2. A survey of 50 people included the question: "How many times did you visit relatives during the past month?" The number of visits recorded were

5	3	3	1	4
4	5	3	4	2
6	6	6	7	0
14	0	2	4	4
4	5	2	3	5
3	4	3	6	8
4	7	6	5	9
11	3	12	4	7
3	5	15	0	10
1	8	1	2	12

Starting with 0 as the beginning number for the first class, and using an interval of 3, organize the data into a frequency distribution. Briefly describe the distribution.

A reminder: the solutions to all even-numbered exercises are at the back of the book.

Suggestions for Constructing a Frequency Distribution

The following are some suggestions for the construction of frequency distributions:

1. **Overlapping Classes.** Overlapping classes, such as $40 - $45 and $45 - $50 for the newlywed example, should be avoided. Otherwise, it would be unclear where to tally $45.

2. **Equal-Sized Classes.** If possible, intervals between classes should be equal. The use of equal intervals allows computation of certain averages and measures of dispersion that will be discussed in the next two chapters. Classes of unequal size are often necessary, however, in order to accommodate all the data, as Table 2-3 illustrates. Had the interval been kept at a constant size of $1,000, the frequency distribution would include at least 1,000 classes—obviously too many to make any meaningful analysis. (Note that the Internal Revenue Service actually uses the

---

Table 2-3
Adjusted Gross Income for Individuals Filing Income
Tax Returns

Adjusted Gross Income Class	Number of Returns (in thousands)
Under $2,000	135
2,000 - 2,999	3,399
3,000 - 4,999	8,175
5,000 - 9,999	19,740
10,000 - 14,999	15,539
15,000 - 24,999	14,944
25,000 - 49,999	4,451
50,000 - 99,999	699
100,000 - 499,999	162
500,000 - 999,999	3
$1,000,000 and over	1

frequency distribution in Table 2-3 to show the adjusted gross income, before taxes, for over 67 million persons who filed income tax forms during the year.)

3. **Open-Ended Classes.** If possible, open-ended classes should be avoided. Table 2-3 has two such classes, namely, "Under $2,000" and "$1,000,000 and over." Strictly speaking, the class "Under $2,000" is not open-ended, since it has an implied lower limit of zero. However, usually we avoid classes like this because the class midpoint ($1,000) is not representative of all the values in the class. If a frequency distribution has an open end, a commonly used measure of central tendency called the arithmetic mean cannot be used. (More about the arithmetic mean in Chapter 3.)

4. **Number of Classes.** No fewer than 5 and no more than 20 classes should be used in the construction of a frequency distribution. Too few, or too many, classes give little insight into the distribution of the data. For example, the following age distribution contains only two classes. Very little can be said about the distribution of the ages of the inmates in Social Prison.

Age of Prisoners	Number of Prisoners
20 - 39	432
40 - 59	431

5. **Class Size.** It is common practice to use intervals such as 5; 10; 20; 100; 1,000; and so on. If it has been decided to have classes of equal size, the width of the interval can be approximated by subtracting the lowest value from the highest value and dividing by the number of classes, that is

$$\text{Width of class interval} = \frac{\text{highest value} - \text{lowest value}}{\text{number of classes}}$$

For example, suppose it had been established that the newlywed food budget data in Table 2-1 should be placed into 7 classes. What would the width of the interval be?

$$\text{Width of interval} = \frac{\$74 - \$40}{7}$$

$$= \$4.86 \text{ (only a guideline)}$$

An interval of $4.86 is cumbersome to use. As shown in Table 2-2, a rounded-off figure of $5 is a more convenient interval.

Portraying the Frequency Distribution Graphically

The popularity of illustrated magazines such as *Better Homes & Gardens*, *Time*, *Ebony*, and *Omni* is a tribute to the Chinese proverb that "a picture is worth a thousand words." Pictures of a special variety are also used extensively to help hospital administrators, business executives, and consumers get a quick grasp of statistical reports. Such pictures, of course, are what we call *graphs* or *charts*. The three graphic forms commonly employed to portray a frequency distribution are the **histogram,** the **frequency polygon,** and the **cumulative frequency polygon.**

Histogram

A **histogram** is one of the most easily interpreted charts. We will illustrate its construction by using the distribution of the weekly amounts spent on food by the newlyweds.

The class frequencies (number of newlyweds) are plotted on the vertical axis (*Y*-axis). The variable (weekly amounts spent on food) is scaled on the horizontal axis (*X*-axis). It is customary to use the true class limits, as shown in Table 2-4. Note that 4 couples spent between $39.50 and $44.50. The first step in constructing a histogram is to draw vertical lines from both $39.50 and $44.50 on the *X*-axis to points opposite 4 on the *Y*-axis. Then, as shown in Figure 2-1, the tops of these two lines are connected to form a bar. (The area of the bar represents the number of newlyweds in that class.)

Table 2-4

Weekly Amounts Spent on Food
by 100 Newlywed Couples

Weekly Amounts Spent on Food Stated Limits	True Limits	Number of Newlyweds
$40 - $44	$39.50 - $44.50	4
45 - 49	44.50 - 49.50	11
50 - 54	49.50 - 54.50	20
55 - 59	54.50 - 59.50	31
60 - 64	59.50 - 64.50	19
65 - 69	64.50 - 69.50	11
70 - 74	69.50 - 74.50	4
		100

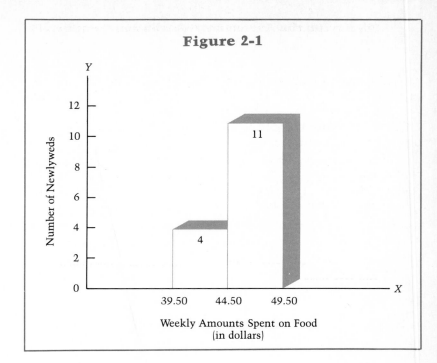

Figure 2-1

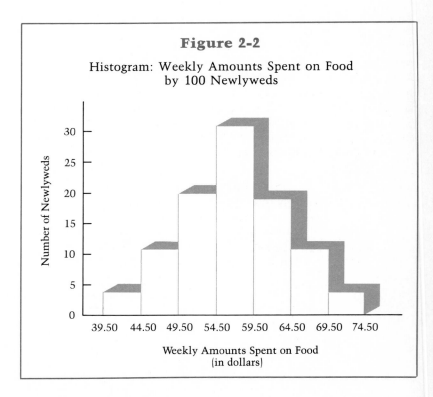

Figure 2-2

Histogram: Weekly Amounts Spent on Food
by 100 Newlyweds

Next, lines are drawn vertically from $44.50 and $49.50 to 11 on the *Y*-axis. Then, the tops of these two lines are connected to form a bar.

This procedure is continued for each of the classes. Figure 2-2 is the completed histogram.

As noted, the *X*-axis in the previous histogram shows the true limits. However, it would also be acceptable to show the rounded figures—$40, $45, $50, and so on. To construct a histogram, the interval level of measurement should be used.

The age distribution of guards hired by the Corrections Department within the last year is

Age	Number
20 - 29	2
30 - 39	13
40 - 49	20
50 - 59	12
60 - 69	3

Construct a histogram of these data.

Self-Review 2-2

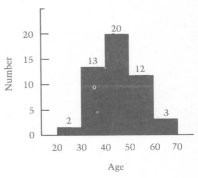

Frequency Polygon

To illustrate the construction of a **frequency polygon,** the frequency distribution listing the weekly amounts spent on food by 100 newlyweds is reintroduced (see Table 2-5). This time, a column

Table 2-5
Frequency Distribution: Weekly Amounts Spent
on Food by 100 Newlywed Couples

Weekly Amounts Spent on Food	Class Midpoints	Number of Newlyweds
$40 - $44	$42	4
45 - 49	47	11
50 - 54	52	20
55 - 59	57	31
60 - 64	62	19
65 - 69	67	11
70 - 74	72	4

showing the class midpoints is included; midpoints are one of the components needed to graph frequency polygons.

The class frequencies (the number of newlyweds, in this problem) are plotted on the vertical scale (*Y*-axis). Note that in Figure 2-3 the numbers on the *Y*-axis increase in magnitude from 0 to 35, with each division representing an interval of 5.

The classes are represented on the horizontal axis (*X*-axis). In this illustration, the *X*-axis shows the weekly amounts spent on food. Each class is represented by its midpoint. For example, the midpoint of the $40 - $44 class is $42, the midpoint of the $45 - $49 class is $47, and so on.

Referring to Table 2-5, there are 4 newlyweds in the first class of $40 - $44. Therefore, the coordinates of the first plot, or graph segment, are *X* = $42, *Y* = 4. As shown, a dot is placed at the coordinates $42 and 4. The coordinates of the next plot are $47 and 11. The dots are connected in order by straight line segments. The point representing the first class is joined to the one representing the second class, and so on.

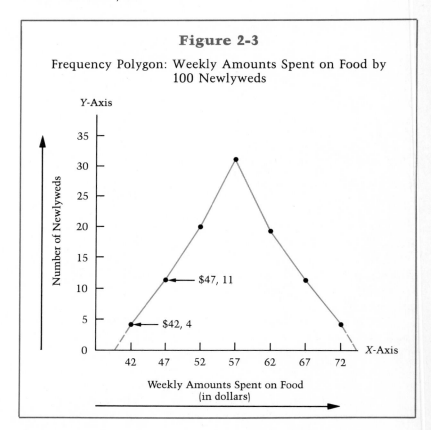

Figure 2-3

Frequency Polygon: Weekly Amounts Spent on Food by 100 Newlyweds

A technical note: the usual practice is to extend the two extremes of the frequency polygon to the X-axis. By convention, this is accomplished by using dashed, rather than solid, lines for the extension. In Figure 2-3, the left end of the polygon extends to the midpoint of a class below the lowest class of interest. The upper end of the polygon is treated in a similar way.

Refer back to Self-Review 2-2. Portray the ages of the guards in the form of a frequency polygon.

Self-Review 2-3

Chapter Exercises

3. In Exercise 1 you constructed the following frequency distribution:

Amounts Spent on Textbooks	Number
$15 - $24	2
25 - 34	10
35 - 44	12
45 - 54	6
55 - 64	5

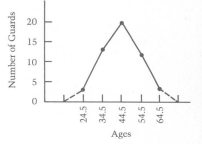

a. Portray the distribution in the form of a histogram.
b. Portray the distribution in the form of a frequency polygon.

4. Refer back to Exercise 2. Draw a histogram and a frequency polygon using the distribution of the number of visits to relatives.

The frequency polygon is an ideal way to compare two or more frequency distributions. As an illustration, recall that the group of 100 newlyweds consisted of only young couples. Suppose the federal government now wished to compare the amounts spent on food by young newlyweds with the expenditures by newly married couples over 65.

The frequency polygons in Figure 2-4 represent the expenditures for these two groups. It is readily apparent that the newlyweds over 65 spend more per week on food. The typical (most frequent) weekly amount spent by the newlyweds over 65 is about $67, compared with about $57 for the younger newlywed couples.

It should be noted again that in order to construct a frequency polygon the level of measurement must be interval level.

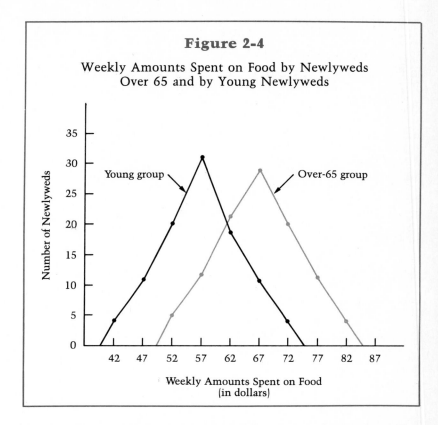

Figure 2-4

Weekly Amounts Spent on Food by Newlyweds
Over 65 and by Young Newlyweds

Percent Frequency Distribution and Percent Frequency Polygon

The number of young newlyweds and the number of newlyweds over 65 were about equal. This made the comparative analysis of the two distributions relatively easy. It is difficult to make comparisons, however, if the size of one group is much larger than the other one. For example, 50 persons were surveyed on controversial subjects such as abortion and strip mining. Four years later, a much larger group was asked the same questions. The research report gave the age distributions for both groups (see Table 2-6).

Because the number in the 1978 survey is small relative to the 1982 survey, the accompanying frequency polygon is difficult to interpret. The 1978 population is compressed at the bottom of Figure 2-5.

A better approach is to convert the class frequencies to percentages of the total. For example, the class frequency in the 20 - 29 class (1978) can be converted to a percent by dividing 2 by the total number of persons (2/50 = 0.04 = 4%). The class frequency

Table 2-6

Age Distributions for the 1978 and 1982 Surveys

Age (in years)	Number of Persons	
	1978	1982
20 - 29	2	74
30 - 39	13	400
40 - 49	20	980
50 - 59	12	460
60 - 69	3	86
Totals	50	2,000

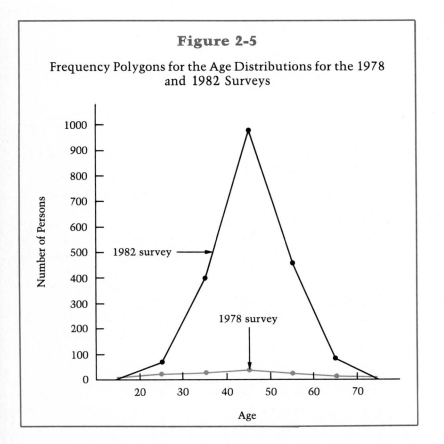

Figure 2-5

Frequency Polygons for the Age Distributions for the 1978 and 1982 Surveys

in the 50 - 59 class (1982) is changed to a percent of the total by computing $460/2{,}000 = 0.23 = 23\%$.

The percents of the total are shown in columns 4 and 5 in Table 2-7. An analysis of the table and the accompanying chart (Figure 2-6) reveals that, when using this method of conversion, the shapes of both age distributions are nearly identical. In Figure 2-6, the percentages of the total are plotted in the form of frequency polygons.

Table 2-7

Age Distributions for the 1978 and 1982 Surveys

Age (1)	Number of Persons		Percent of the Total	
	1978 (2)	1982 (3)	1978 (4)	1982 (5)
20 - 29	2	74	4.0	3.7
30 - 39	13	400	26.0	20.0
40 - 49	20	980	40.0	49.0
50 - 59	12	460	24.0	23.0
60 - 69	3	86	6.0	4.3
Total	50	2,000	100.0	100.0

Note: the ages were rounded to the nearest year, that is, the age of a person 29 years 6 months was rounded up to 30 years. The age 29 years 5 months was rounded down to 29.

Cumulative Frequency Distribution and Cumulative Frequency Polygon

We may be interested in knowing the percent of observations that are less than a particular value. Or, the question might be: What percent of the observations is greater than a particular value? For example, what percent of the newlyweds spend more than $58 a week on food? A **cumulative frequency distribution** or a **cumulative frequency polygon** can be used to answer these questions.

As the name implies, a cumulative frequency distribution requires cumulative class frequencies. There are two cumulative frequency distributions—a "less-than" cumulative frequency distribution and a "more-than" cumulative frequency distribution. In order to show the construction of both types, the expenditures on food of young newlyweds are repeated in Table 2-8.

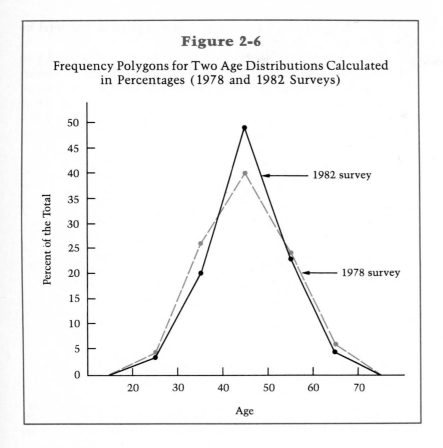

Figure 2-6

Frequency Polygons for Two Age Distributions Calculated
in Percentages (1978 and 1982 Surveys)

1982 survey

1978 survey

Table 2-8

Weekly Amounts Spent on Food
by 100 Newlywed Couples

Stated Limits	True Limits	Number of Newlyweds f
$40 - $44	$39.50 - $44.50	4
45 - 49	44.50 - 49.50	11
50 - 54	49.50 - 54.50	20
55 - 59	54.50 - 59.50	31
60 - 64	59.50 - 64.50	19
65 - 69	64.50 - 69.50	11
70 - 74	69.50 - 74.50	4

A less-than cumulative frequency distribution is constructed by adding the frequencies from the *lowest to the highest class.* In Table 2-9, note that for the lowest class 4 newlywed couples spent less than the upper true class limit of $44.50. Those 4, plus the 11 in the next-lowest class, or 15, spent less than $49.50. Then 4 + 11 + 20, or 35 newlyweds, spent less than the upper true class limit of $54.50. This adding process is continued for all the frequencies. Table 2-9 gives the complete set of cumulative frequencies.

Table 2-9

Less-Than Cumulative Frequency Distribution for the Amounts Spent on Food by Newlyweds

Stated Limits	True Limits	Frequency	Cumulative Frequency
$40 - $44	$39.50 - $44.50	4	4
45 - 49	44.50 - 49.50	11	15
50 - 54	49.50 - 54.50	20	35
55 - 59	54.50 - 59.50	31	66
60 - 64	59.50 - 64.50	19	85
65 - 69	64.50 - 69.50	11	96
70 - 74	69.50 - 74.50	4	100

The *upper true class limits* and the cumulative frequencies are plotted to portray the weekly amounts spent on food in the form of a **less-than cumulative frequency polygon** (Table 2-10). The first three plots are $44.50 and 4, $49.50 and 15, and $54.50 and 35. These are illustrated in Figure 2-7. As shown, one practice is to scale the cumulative frequencies on the left side of the chart and the cumulative percents on the right.

Table 2-10

Upper True Class Limit	Cumulative Frequency
$44.50	4
49.50	15
54.50	35
59.50	66
64.50	85
69.50	96
74.50	100

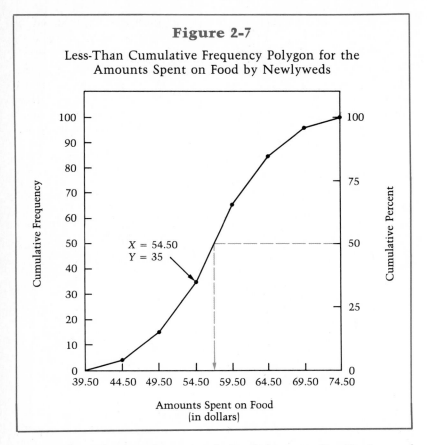

Figure 2-7

Less-Than Cumulative Frequency Polygon for the
Amounts Spent on Food by Newlyweds

Based on the less-than cumulative frequency distribution and the accompanying less-than cumulative polygon, statements can be made such as "about half (50%) of the newlyweds spend less than $57 a week on food." The $57 is found by drawing a horizontal line from the 50% mark on the right-hand margin of the vertical axis to the polygon slope, and then moving straight down to the X-axis to read the figure $57 (approximately). The less-than frequency polygon does not simply stop somewhere in space above the X-axis. Since no values occur before the smallest class, the true lower limit for that class is zero. No (0) newlyweds spent less than $39.50.

A technical note: the upper true class limits were scaled on the X-axis mainly to show the construction of the cumulative polygon. However, in order to make the chart easier to read, rounded amounts such as $40, $45, and $50 are often substituted for true limits like $39.50, $44.50, and $49.50 in labeling the X-axis. The actual plots do not change—that is, they are still $44.50 and 4, $49.50 and 15, and so on.

Table 2-11

Upper True Class Limits	Cumulative Frequency
$44.50	100
49.50	96
54.50	85
59.50	65
64.50	34
69.50	15
74.50	4

The **more-than cumulative frequency distribution** is constructed by adding the frequencies from the *highest class to the lowest class* (Table 2-11). Next, the *lower true class limits* and the corresponding cumulative frequencies are plotted. (See Figure 2-8.)

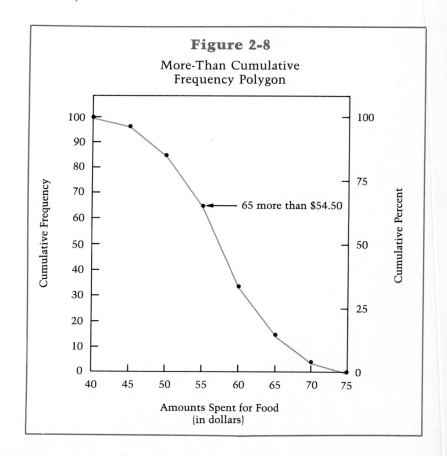

Figure 2-8

More-Than Cumulative
Frequency Polygon

Returning to the age distribution of guards in Self-Review 2-2:

Age	Number
20 - 29	2
30 - 39	13
40 - 49	20
50 - 59	12
60 - 69	3

a. Construct a less-than cumulative frequency distribution.
b. Draw a less-than cumulative frequency polygon.
c. Based on the graph, about half the guards hired by the Corrections Department were less than what age?

Self-Review 2-4

a.

Age	Cumulative Number
20 - 29	2
30 - 39	15
40 - 49	35
50 - 59	47
60 - 69	50

b.

Upper True Class Limits	Number Less-Than
19.5	0
29.5	2
39.5	15
49.5	35
59.5	47
69.5	50

Chapter Exercises

5. A Comprehensive Example. A study by Laurette Looney of the grade point averages of a sample of 120 undergraduate students with GPAs of 1.80 or higher revealed the following information:

2.30	2.81	2.71	2.95	2.25	3.15	2.19	2.22	3.07	3.70
2.18	2.00	2.17	2.16	2.54	2.41	2.87	2.45	3.25	3.79
2.28	2.20	2.39	1.99	2.71	2.22	2.11	2.63	3.49	2.21
3.97	2.41	2.59	2.59	3.18	2.18	2.30	1.81	2.01	2.49
2.45	2.60	2.73	2.72	2.48	2.89	2.49	3.27	2.82	2.00
2.78	3.00	3.89	3.81	2.26	2.12	2.65	3.51	2.21	2.19
2.55	3.40	2.51	2.47	1.89	2.24	3.08	2.01	2.42	2.17
2.32	3.31	2.31	2.47	2.90	2.50	3.33	1.80	3.77	3.43
2.31	3.62	2.24	2.29	2.12	1.88	3.51	2.07	2.61	2.57
1.99	2.51	2.58	2.22	2.36	3.12	2.06	2.22	3.01	1.91
2.18	2.59	2.37	2.97	2.52	3.37	2.85	2.44	3.23	2.76
2.38	4.00	1.97	2.14	2.65	2.22	2.09	2.61	3.42	2.40

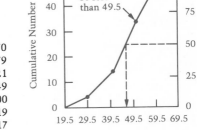

c. About 47 years.

a. Organize the grade point averages into a frequency distribution.
b. Draw a histogram.
c. Draw a frequency polygon.
d. Organize the data into a less-than cumulative frequency distribution.
e. Draw a less-than cumulative frequency polygon.
f. Organize the data into a more-than cumulative frequency distribution.

g. Draw a more-than cumulative frequency polygon.
h. Write a brief paragraph summarizing your findings.

Other Graphic Techniques

The frequency polygon, the cumulative frequency polygon, and the histogram have strong visual appeal. There are other graphs, however, that are commonly used in government reports, research reports, newspapers, and journals. Several are presented in this section. Others, such as the scatter diagram, will be introduced in later chapters.

Line Charts.   A **line chart** is particularly useful in portraying data over a period of time. We will use the registrations of motorcycles since 1940 as an example. Time (years, in this case) is *always* plotted on the horizontal axis. The number of motorcycles registered is scaled on the Y-axis. The first plot is $X = 1940$, $Y = 166,000$;

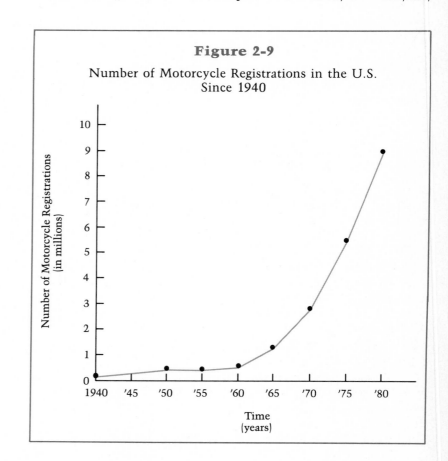

Figure 2-9

Number of Motorcycle Registrations in the U.S.
Since 1940

the second plot is 1950 and 454,000; and the last plot is 1980 and 9,000,000. The line chart is very easy to interpret, namely, motorcycle registration has increased rapidly, especially since 1965 (see Figure 2-9).

Year	Motorcycle Registrations (000)
1940	166
1950	454
1955	412
1960	574
1965	1,382
1970	2,824
1975	5,494
1980	9,000

The farm population of the United States since 1940 is

Self-Review 2-5

Year	Farm Population (000,000)
1940	30.5
1950	23.0
1960	15.6
1970	9.7
1980	7.9

a. Plot the farm population in the form of a line chart.
b. Interpret your chart.

a.

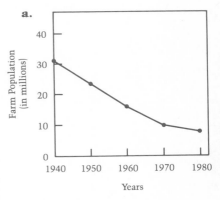

b. The number of persons on farms has been decreasing rapidly since 1940.

Bar Charts. A **bar chart** can be used to portray any one of three levels of measurement—nominal, ordinal, or interval. The religion reported by the population of the United States 14 years old and over is used again to illustrate the construction of a bar chart. The data in Figures 2-10 and 2-11 are nominal level. The chart may be organized so that the bars are either horizontal as in Figure 2-10, or vertical as in Figure 2-11.

Technical notes: as shown in these two bar charts, a small space usually separates each bar. Color is often used—especially in the annual reports of businesses, and in magazines such as *Time*. Actual numbers may be placed on the graph, such as 79 million for Protestant, 31 million for Roman Catholic, and so on.

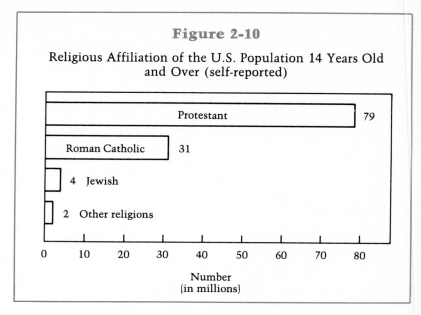

Figure 2-10

Religious Affiliation of the U.S. Population 14 Years Old and Over (self-reported)

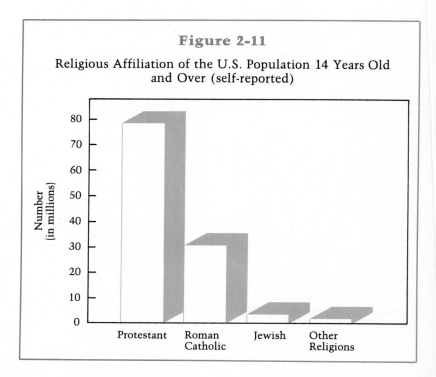

Figure 2-11

Religious Affiliation of the U.S. Population 14 Years Old and Over (self-reported)

In Self-Review 2-5, we listed the farm population of the United States since 1940 as

Year	Farm Population (000,000)
1940	30.5
1950	23.0
1960	15.6
1970	9.7
1980	7.9

a. Draw a bar chart placing the bars horizontally.
b. Draw a bar chart placing the bars vertically.

a.

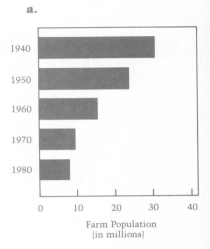

Farm Population (in millions)

Pie Charts. A very popular graph used to portray parts of the total is called a **pie chart.** As an example, suppose that students living on the campus of Solid State University had reported their annual expenditures on books, tuition, and so on. The figures were grouped and typical amounts for each category determined (refer to Table 2-12).

First we divide a circle (the pie) into equal "slices," as shown by the notches along the perimeter of the circle in Figure 2-12. Here the slices represent 5% each (the total area of the pie is 100%).

In order to represent tuition and other expenditures, each category must first be converted to percents of the total. Tuition, for example, accounts for 40% of the total, found by computing $\$3,200/\$8,000 = 0.40 = 40\%$ (see Table 2-12).

b.

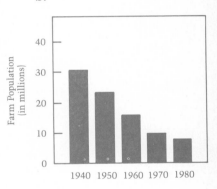

Table 2-12
Typical Amounts Spent on Tuition
and Other Expenditures at
Solid State University

Expenditure	Amount	Percent of Total
Tuition	$3,200	40%
Room and board	2,400	30
Transportation	1,200	15
Books	400	5
Recreation	800	10
Total	$8,000	100

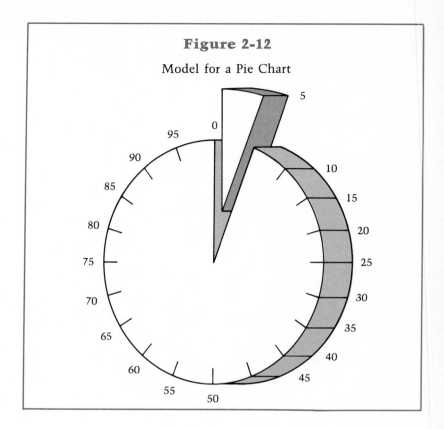

Figure 2-12

Model for a Pie Chart

In constructing this pie chart, it was arbitrarily decided to plot the 5% spent on books first. A line was drawn from 0 to the center of the pie and another line from the center to 5 on the edge of the circle. The enclosed piece of the pie represents expenditures on books.

Expenditures for recreation (10%) are plotted next by adding the 10% to the 5% for books, for a total of 15%. A line is drawn from 15% to the center of the pie. The area between 5% and 15% represents the typical amount spent on recreation. This procedure is continued for all remaining expenditures. The completed pie chart is shown in Figure 2-13.

From Figure 2-13 it is readily apparent that the largest annual expenditure is for tuition, followed by the expenditure for room and board (30%), and so on.

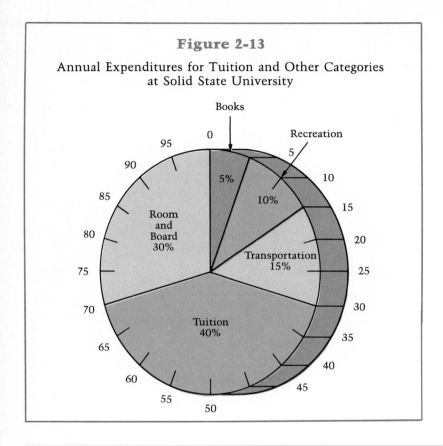

Figure 2-13

Annual Expenditures for Tuition and Other Categories
at Solid State University

The annual expenditures of Carefree Township during last year were

Expenditure	Amount
Roads	$1,000,000
Schools	600,000
Administration	300,000
Graft	100,000

Portray the expenditures in the form of a pie chart.

Self-Review 2-7

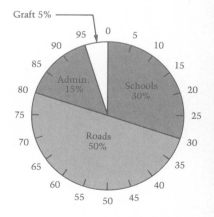

Chapter Exercises

6. In its *Uniform Crime Reports for the United States*, the Federal Bureau of Investigation reported the number of murders:

Year	Number (000)
1968	13.8
1969	14.8
1970	16.0
1971	17.8
1972	18.7
1973	19.6
1974	20.7
1975	20.5
1976	21.0
1977	22.1
1978	22.9
1979	23.6
1980*	24.0

* Estimated

Draw a line chart that portrays the number of murders in the United States since 1968.

7. According to the Bureau of the Census, the population of the State of Arizona for the years 1910 - 1990 is

Year	Number
1910	204,000
1930	436,000
1950	750,000
1970	1,792,000
1990*	2,724,000

* Estimated

Plot the population data in the form of a bar chart.

8. The Council of Environmental Quality estimated the amounts to be spent by various groups on water pollution between 1975 and 1984:

Group	Amount (in billions of dollars)
Federal government	$ 2.5
State and local governments	39.8
Industry	57.0
Utilities	11.5

Portray the data by drawing a pie chart.

A frequency distribution is a table that shows how often each value occurs. To construct a frequency distribution the raw data may be tallied into predetermined classes. The tallies are counted to arrive at the frequency for each class. The hourly wages of 248 employees grouped into five wage classes might appear as

Summary

Hourly Wage	Number
$ 3.00 - $ 4.99	21
5.00 - 6.99	62
7.00 - 8.99	90
9.00 - 10.99	58
11.00 - 12.99	17

A frequency polygon uses class frequencies and midpoints in its construction. A percent frequency polygon is based on class frequencies converted to percents of the total and on class midpoints. Cumulative frequency polygons are based on the cumulative class frequencies and on the true class limits. The histogram is a set of rectangles whose heights are the class frequencies and whose widths are the class intervals.

Other graphic techniques include line charts, bar charts, and pie charts. All have reader appeal.

Summarizing Data: Frequency Distributions and Graphic Presentations

Chapter Outline

Frequency Distributions

I. Objective. To condense raw data into a form that can be readily analyzed.

II. Definition. A frequency distribution is a table showing the number of observations for each class.

III. Procedure
   A. Select the lower limit of the first class, and the class interval.
   B. Tally the raw data into the classes.
   C. Count the number of tallies in each class.

IV. Suggestions
   A. Set up classes that avoid overlapping. Example: 200 - 299, 300 - 399; avoid 200 - 300, 300 - 400.
   B. All class intervals should be of equal width.
   C. Avoid open-ended classes.
   D. Generally, frequency distributions should contain no fewer than 5 and no more than 20 classes.

Graphic Presentation of Frequency Distributions

I. Objective. To capture reader attention, and to visually portray the data in a form that is easy to read and analyze.

II. *Frequency polygon.* Plot each class frequency and corresponding midpoint. Example:

Hourly Wages	Class Frequency
$3.50 - $4.49	5
4.50 - 5.49	8
5.50 - 6.49	2

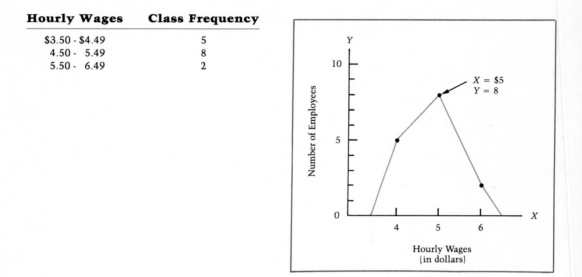

III. *Cumulative Frequency Polygon.* Plot cumulative frequencies and corresponding true class limits.

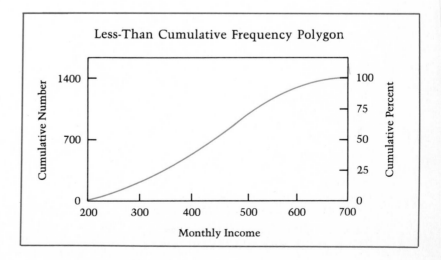

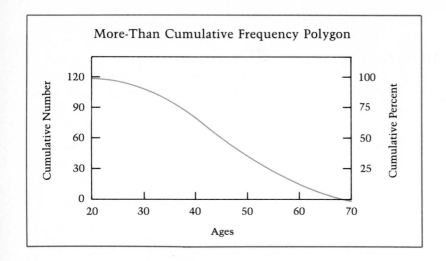

IV. *Histogram.* Plot class frequency and upper and lower true limits of the class to form a bar. Illustration:

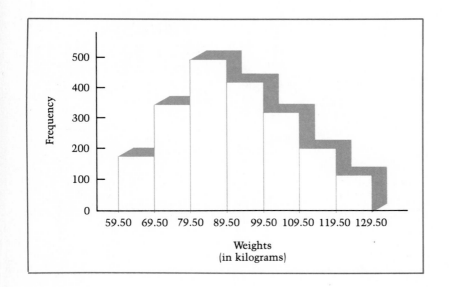

Other Graphic Techniques

I. *Line chart.* Ideal for portraying data over a period of time. Plot time on the horizontal axis. Illustration:

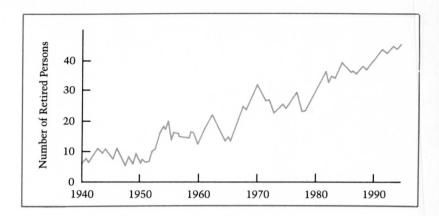

II. *Bar chart.* Can be used to present nominal, ordinal, or interval data. Bars may be vertical or horizontal. Illustrations:

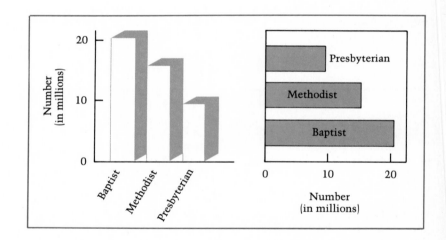

III. *Pie chart.* Convert the number in each category to a percent of the total. Plot the percents. Illustration:

Political Affiliation	Number in Survey	Percent of Total
Democrat	1,100	55
Republican	500	25
Independent	300	15
All others	100	5
	2,000	100

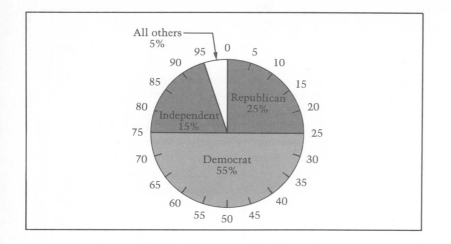

Chapter Exercises

9. The following are the number of births per 1,000 population for 60 countries:

34	24	10	15	22	15
17	22	10	17	15	31
25	32	15	20	31	18
37	12	15	18	28	27
19	13	20	19	40	35
19	16	22	13	43	35
27	18	16	13	31	20
19	14	10	13	44	32
45	12	17	18	34	38
24	16	14	30	24	32

   a. Organize the data into a frequency distribution. Start with a lower limit of 10 and use an interval of 5.

   b. Draw a histogram.

   c. Draw a frequency polygon.

   d. Draw a more-than cumulative frequency polygon.

   e. The United States has a birth rate of 15 per 1,000 population. What percent of the countries have a birth rate equal to, or greater than, the United States?

10. Citizens' Trust lists the end-of-month checking-account balances for forty of its customers:
$203, 37, 141, 43, 55, 303, 252, 758, 321, 123, 425, 27, 72, 87, 215, 358, 521, 863, 284, 279, 608, 302, 703, 68, 149, 327, 127, 125, 489, 234, 498, 968, 350, 57, 75, 503, 712, 440, 185, 404.

    a. Tally the data into a frequency distribution using $100 class intervals and $0 as the starting point.

    b. Draw a less-than and a more-than cumulative frequency polygon.

    c. The bank considers a "preferred" customer to be one with an account balance of $500 or more. What percent of all customers are preferred customers?

    d. The bank is planning to offer free checking accounts to about 30% of its customers. Customers with the larger balances may participate. What account balance would be the cutoff point for the free checking account?

11. The following are attendance figures for the Detroit Tigers in games played at Tiger Stadium:

48,612	13,252	21,659	14,419	39,884
10,754	24,856	47,855	19,079	32,951
8,862	9,397	21,350	35,395	10,303
4,612	19,909	51,032	26,447	9,896
7,088	14,923	14,454	14,963	8,204
8,818	16,038	51,041	15,028	16,410
8,317	17,894	27,630	36,523	20,371
10,059	12,503	21,764	51,822	12,156
11,802	9,283	14,835	10,104	8,949
12,250	10,866	45,905	8,141	5,328
14,092	36,377	30,110	17,385	7,147
14,583	24,038	23,841	34,760	10,522
51,650	24,824	44,068	14,979	12,814
11,757	16,095	22,553	13,426	
10,642	9,321	14,924	9,503	

    a. Organize the data into a frequency distribution.

    b. Draw a histogram.

    c. Using relative (percent) frequencies, draw a frequency polygon.

    d. Draw a less-than cumulative frequency polygon.

    e. Draw a more-than cumulative frequency polygon.

    f. Interpret your findings.

12. The total foreign investment in the United States since 1973 is shown in the table that follows.

Year	Investment (in billions)
1973	174.9
1974	197.4
1975	220.5
1976	263.4
1977	307.8
1978	370.4
1979	413.9
1980	480.9

a. Portray the foreign investment in the form of a line chart.
b. Portray the foreign investment in the form of a bar chart.

13. The following graph shows a less-than cumulative frequency polygon of the ages of a group of automobiles. The plotted values are (3, 35), (6, 60), (9, 75), (12, 78), and (15, 80). Develop a table of the corresponding frequency distribution.

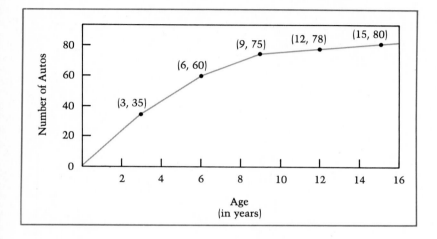

14. The sales manager for Ed Smith Auto Mart wishes to place radio ads for the super-new Belchfire 4 on one of two local stations. Belchfire 4 is a sports car aimed toward a younger buyer. Surveys of listener audiences indicate the following distribution of ages for the two stations.

| | **Station** | |
Ages	**KSOP**	**KROK**
15 - 24	10	58
25 - 34	22	52
35 - 44	53	35
45 - 54	45	30
55 - 64	60	25
65 - 74	10	0
	200	200

Which station would you recommend he select and why?

This is the first in a series of Chapter Achievement Tests. Their purpose is to allow you to evaluate your understanding of the material. Do all the problems. Then check your answers against those given in the Answer section at the back of the book.

Chapter Achievement Test

I. Multiple Choice (4 points each).

1. When arranging data into classes it is suggested that there be
   a. fewer than 5 classes
   b. between 5 and 20 classes
   c. more than 20 classes
   d. between 10 and 40 classes

2. When constructing a line chart, "time" is plotted along the
   a. vertical axis
   b. horizontal axis
   c. either axis
   d. neither axis

3. A frequency distribution requires data to be of what scale?
   a. nominal
   b. interval
   c. ordinal
   d. none of these are correct

4. The class midpoint is
   a. the number of observations in a class
   b. the center of the class
   c. the upper limit of the class
   d. the width of the class

The following frequency distribution records the number of empty seats on flights from Cleveland to Tampa:

Number of Empty Seats	Frequency
0 - 4	3
5 - 9	8
10 - 14	15
15 - 19	18
20 - 24	12
25 - 29	6

5. The midpoint of the 0 - 4 class is
   a. 2
   b. 4
   c. 2.5
   d. 0

6. The true lower limit of the 0 - 4 class is
   a. 0
   b. −0.5
   c. 2.0
   d. 2.5

7. The size of the class interval is
   a. 5
   b. 4
   c. 4.5
   d. 3

8. What percent of the flights had fewer than 19 empty seats?
   a. 15%
   b. about 29%
   c. about 71%
   d. cannot be determined from grouped data

The following less-than cumulative frequency polygon was developed for the distance commuting students travel from home to a college campus:

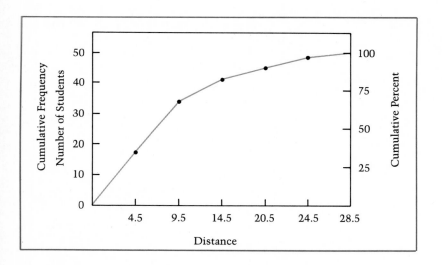

9. What percent of the students travel less than 7 miles to campus?
   a. about 50%
   b. all
   c. about 36%
   d. cannot be determined

10. Seventy-five percent of the students travel how many miles or less to campus?
    a. about 5
    b. about 7
    c. about 12
    d. about 20

II. Problems (20 points each).

11. A hospital administrator has commissioned a study of the length of time a patient must spend waiting in the emergency room before receiving treatment. For a typical day, the information obtained is as follows:

12	6	5	23	10
25	19	11	25	18
10	14	12	10	16
5	19	17	17	14
2	21	9	6	21

   Construct a frequency distribution using an interval of 5.

12. The Bureau of Motor Vehicles of the State of Nevada made a study of the tread depth of tires. Sixty passenger cars that had stopped at a highway rest station were tested. The tread depth, measured to the nearest $1/32''$, is reported in the table below. (a) Draw a less-than cumulative frequency polygon. (b) What percent of the tires have less than $7/32''$ tread? (c) Forty percent of the tires have less than how much tread?

Tread Depth (1/32″)	Frequency
0 - 1 (bald)	1
2 - 3	3
4 - 5	7
6 - 7	8
8 - 9	10
10 - 11	15
12 - 13	9
14 - 15	5
16 - 17	2

13. A family's monthly expenses were listed as follows: housing $400, utilities $140, medical $25, food $190, transportation $150, clothing $50, savings $50, miscellaneous $50. Draw a pie chart showing this information in terms of percents.

3

Descriptive Statistics: Measures of Central Tendency

OBJECTIVES

When you have completed this chapter, you will be able to
- Compute the mean, median, and mode.
- Describe the characteristics of the mean, median, and mode.
- Describe the position of the mean, median, and mode for both symmetric and skewed distributions.

Introduction

In Chapter 2, the weekly amounts spent on food by a group of newlyweds were organized into a table. Next, graphs such as histograms or frequency polygons were drawn to help describe the data. In this chapter, the data will be further summarized by computing several **measures of central tendency.** As the name implies, these are measures that locate the center of a set of observations. Recall that the newlywed data tended to cluster about $55 - $59. Virtually all data tend to "mound up" in the middle around some value. In summarizing them, the goal is to determine a number that is either typical of the set of data or that best describes it. This number is often referred to as an **average.**

What Is an Average?

One may think of an average as a gauge of what is considered to be typical or "normal." An annual family income of $1,500, for example, would be considered well-below average when compared with all families in the United States. The weight of a 450-pound man is thought of as being well above the typical weight of men.

> **Average**   A single value that is representative of the set of data considered as a whole. An average locates the "center" of a set of data.

Examples of Averages

- The average life expectancy at birth in the United States is 69.0 years for males and 76.7 years for females.
- The median family income in the United States in 1980 was $20,079.
- The average cost of drilling an oil well is $1,943,000.
- The Surgeon General reported that 54 million Americans smoke, on the average, 31.2 cigarettes a day.
- In 1950, the median age of the population of the United States was 30.2 years.
- The projected median age for the year 2000 is 35.5 years.

There are many averages, each having distinct characteristics, advantages, and disadvantages. For the same set of data all the averages might have different values. Three commonly used averages will be examined in this chapter—namely, the **arithmetic mean,** the **median,** and the **mode.**

The Sample Mean

Among the different measures of central tendency the one used most frequently is the **arithmetic mean,** or simply the mean.

> **Arithmetic Mean**   The sum of the values divided by the number of values.

In everyday language, the mean is often called the "average." However, as you will soon see, there are several kinds of averages, so it is well to develop the habit of speaking precisely regarding the various averages.

Because the arithmetic mean of many different sets of data will be calculated, it is convenient to use a simple formula in which the components of the arithmetic mean are represented by a few general symbols. For example, if X_1, X_2, . . . , X_n represent the values of n items or observations in a sample, such as test scores, the arithmetic mean of these sample items, written $\overline{X}$, is defined as

$$\overline{X} = \frac{X_1 + X_2 + \cdots + X_n}{n}$$

[which is a shorthand way of saying that the sample mean of all scores ($\overline{X}$) is equal to the sum of all scores (X_1, X_2, and so on) divided by the number of scores]

For simplicity, instead of writing X_1, X_2, . . . , X_n for all the observations, we use the symbol Σ (which is the Greek capital letter *sigma* equivalent to our S), which here represents a summation, or addition. Hence, the formula for computing the sample mean of n items is

$$\text{Mean} = \frac{\text{sum of the values}}{\text{number of values}}$$

or, in symbolic notation:

$$\overline{X} = \frac{\Sigma X}{n}$$

where

$\overline{X}$ (read X-bar, or bar X) is the designation for the arithmetic mean of a sample.

Σ is the Greek capital letter *sigma,* which is the symbol for addition. In this case, it directs one to sum all the X values.

X refers to the individual values in the sample.

Problem

The estimated ages of five rocks uncovered at an excavation site were 5, 4, 9, 2, and 10 million years What is the mean age of this sample of five rocks?

Solution

Using the above notation, n refers to the number of observations in the sample, which is 5. The values for the various Xs are: $X_1 = 5$, $X_2 = 4$, $X_3 = 9$, $X_4 = 2$, and $X_5 = 10$. Hence, the arithmetic mean of the sample of the five rocks is computed as follows:

$$\overline{X} = \frac{\Sigma X}{n} = \frac{X_1 + X_2 + X_3 + X_4 + X_5}{n}$$

$$= \frac{5 + 4 + 9 + 2 + 10}{5} = 6 \text{ (million years)}$$

A reminder: cover the answers in the margin.

A sample of six school buses in the Carlton District travel the following distances each day: 14.2, 16.1, 7.9, 10.6, 11.2, and 12.0 kilometers.

a. Give the formula for the arithmetic mean.
b. Insert the appropriate figures and compute the arithmetic-mean distance traveled by the buses in the sample.

a. $\bar{X} = \dfrac{\Sigma X}{n}$

b. $\bar{X} = \dfrac{72}{6}$

$\quad = 12.0$ kilometers.

Chapter Exercises

1. The agility test scores for a sample of eight retarded children are 95, 62, 42, 96, 90, 70, 99, and 70. What is the arithmetic-mean score for the group?

2. The monthly incomes (in thousands of dollars) for a sample of nine minority executives are $2, $2, $2, $2, $10, $10, $10, $25, and $100. Find the arithmetic-mean monthly income.

The Population Mean

As shown in the previous section, it is standard statistical practice to denote the mean of a sample by the symbol $\bar{X}$. The number of observations in the sample is denoted by the lower-case n. A number such as $\bar{X}$, that is, a number computed from a sample, is called a **statistic.**

> $\times$
>
> **Statistic**   One measurable characteristic of a sample.

Usually a statistic is computed for the purpose of estimating an unknown population **parameter.** Hence, the sample mean $\bar{X}$ is a statistic that may be thought of as an estimate of the population mean from which the sample was drawn. The population mean is a parameter.

> $\times$
>
> **Parameter**   One measurable characteristic of a population.

In the previous example about the ages of rocks dug up at an excavation site, the sample of five rocks represented *all* the rocks at the site; the sample mean of 6 million years estimates the population mean. To put it another way, the 6 represents an estimate of the mean age that would be obtained if every single rock at the site were checked for age and the arithmetic mean of all the rocks determined. Clearly, obtaining the population mean can be very tedious and time-consuming.

The computation of the population mean is the same as that of a sample, except that the symbols are slightly different. By convention, the population mean is denoted by the Greek letter μ (*mu*). The number of observations in a population is denoted by the capital letter N. Hence, the mean of a population is computed from the following formula:

$$\mu = \frac{X_1 + X_2 + \cdots + X_N}{N} = \frac{\Sigma X}{N}$$

Problem

For the four years of Governor Lamb's administration, the number of pieces of legislation enacted by the state legislature was 310, 780, 960, and 842. What is the arithmetic-mean number of pieces of legislation?

Solution

Notice that this is a population because it covers all four years of the Lamb administration. The population consists of four observations and the population mean is

$$\mu = \frac{\Sigma X}{N} = \frac{310 + 780 + 960 + 842}{4} = \frac{2892}{4} = 723.0$$

The Properties of the Arithmetic Mean

The popularity of the mean as a measure of central tendency is not accidental. Not only is it simple, familiar, and easy to calculate, but it also has the following desirable properties:

1. It can be calculated for any set of interval-level data, so it always exists.

2. A set of data has only one arithmetic mean, so it is a unique value.

3. It is quite reliable. (This property is discussed in connection with the chapter on sampling.)

4. All the data items are used in its calculation.

The arithmetic mean has one additional important characteristic, namely, *the sum of the deviations from the mean of each value will always be zero.* This concept can be illustrated by a seesaw like those found on most playgrounds. Suppose that three children of about equal weight approach the seesaw with the balance point set at 5. One child sits at one end of the board and another next to him. The third child sits at point 8, but the board will not balance (see Figure 3-1). Finally, she moves to point 12 and the weights balance; now the children can have fun seesawing back and forth (see Figure 3-2). The mean of 5 can be considered the center of gravity of the weights.

Note in Figure 3-2 that the distance (deviation) from the balance point to the child sitting on the right end of the board (+7) is counterbalanced by the deviations to the left of the balance point (−3 and −4). The balance point of 5 is the arithmetic mean of the distances. This can be shown algebraically by letting each observation be represented by X and the mean by $\overline{X}$. In the seesaw problem $\overline{X} = 5$. The sum of the deviations from the mean is 0. That is, $\Sigma(X - \overline{X}) = (1 - 5) + (2 - 5) + (12 - 5) = 0$. The mean is the only average where this is always true.

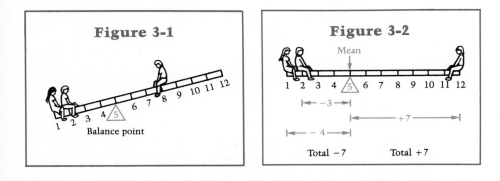

Figure 3-1

Balance point

Figure 3-2

Mean

Total −7   Total +7

___

As shown on the following seesaw, one child sits at point 1, another at 2, another at 3, and a fourth at 10.

Self-Review 3-2

a. $\overline{X} = \dfrac{\Sigma X}{n}$

$\quad = 16/4$

$\quad = 4$

b. Arithmetic mean.

c. $\Sigma(X - \overline{X}) = (1 - 4) + (2 - 4) +$
$(3 - 4) + (10 - 4) = 0$

a. Where must the balance point be placed to make the seesaw balance?

b. What is this balance point called?

c. Show that the sum of the deviations from the mean equals 0.

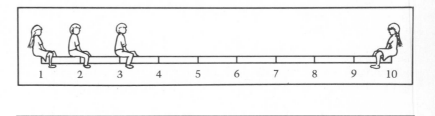

Suppose it had been reported that the ages of the senior citizens on the Happy Boys slow-pitch softball team were 60, 70, 80, and 90. It might be concluded that the arithmetic-mean age is 75, found by $(60 + 70 + 80 + 90)/4$. This is true only if the *same* number of senior citizens are aged 60, 70, 80, and 90. However, suppose one is 60, one 70, one 80, and nine are 90 years. To find the mean, 60 is **weighted** by 1, 70 is weighted by 1, 80 is weighted by 1, and 90 by 9. The resulting average is aptly called the **weighted arithmetic mean.**

In general, the weighted mean of a group of numbers designated $X_1, X_2, X_3, \ldots, X_n$ with corresponding weights $w_1, w_2, w_3, \ldots, w_n$ is computed by

$$\overline{X}_w = \frac{w_1 X_1 + w_2 X_2 + w_3 X_3 + \cdots + w_n X_n}{w_1 + w_2 + w_3 + \cdots + w_n}$$

or shortened to

$$\overline{X}_w = \frac{\Sigma w \cdot X}{\Sigma w}$$

Problem

Fifteen secretaries have 20 years of service, three have 30 years of service, and two have 50 years of service. What is the weighted-mean length of service?

Solution

$$\overline{X}_w = \frac{15(20) + 3(30) + 2(50)}{15 + 3 + 2}$$

$$= \frac{490}{20}$$

$$= 24.5$$

The weighted arithmetic-mean length of service is 24.5 years.

The Median

If the data contain an observation that is either very large or very small compared to the other values, that value may make the arithmetic mean unrepresentative. To avoid this possibility, sometimes we describe the "center" of the data with other measures of central value. One measure used to designate the central value of a set of data is the **median.**

> **Median**   The midpoint of the values after they have been arranged from the smallest to the largest (or the largest to the smallest). There will be as many values above the median as below the median.

For an *odd* number of values: the median is the middle value after the data have been ordered from low to high.

Problems

A. The test scores of a sample of five students are 92, 86, 2, 96, and 90. What is the median test score?

B. The incomes of a sample of seven federal prison wardens are $31,500; $33,900; $37,100; $18,600; $22,500; $34,200; and $34,200. What is the median income?

Solutions

A. Arrange the test scores from low to high. Then select the middle value.

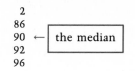

2
86
90 ← the median
92
96

The median score is 90.

B. Arrange the incomes of seven prison wardens from low to high. Then select the middle value.

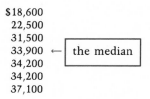

$18,600
22,500
31,500
33,900 ← the median
34,200
34,200
37,100

The median income is $33,900. Note that there are the same number of test scores (or incomes) below the median as above it.

Self-Review 3-3

a. Arranged from low to high:

18
22
(34) median
40
75

b. 2, 2

The ages of prisoners in cell block D are 22, 75, 18, 40, and 34.

a. Determine the median age.
b. How many prisoners are above the median? Below it?

For an *even* number of values: first arrange the values from low to high. It is common practice to determine the median by finding the arithmetic mean of the two center values.

Problems

A. A park ranger recorded the lengths of a sample of several rainbow trout caught in the Yellowstone River. The lengths (in inches) were $24\frac{1}{2}$, 15, $10\frac{1}{2}$, $12\frac{1}{2}$, 21, and 26. What was the median length?

B. The weights of eight collegiate football players are 198, 240, 230, 240, 210, 250, 225, and 188 pounds. What is the median weight?

Solutions

A. Arranging the lengths from low to high:

$10\frac{1}{2}$
$12\frac{1}{2}$
15
21
$24\frac{1}{2}$
26

The median is 18, which is halfway between 15 and 21.

B. Arranging the weights from low to high:

188
198
210
225
230
240
240
250

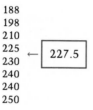

The median weight is 227.5.

---

a. In their last eight games the Rams football team scored the following number of points: 13, 34, 0, 0, 9, 42, 21, and 7. What is the median number of points scored?

b. The number of telephones in use in the United States for four recent years were: 149,010,000; 138,290,000; 155,170,000; and 143,970,000. Determine the median number in use.

Self-Review 3-4

a. 11 points—halfway between 9 and 13.

b. 146,490,000 telephones—halfway between 143,970,000 and 149,010,000.

Chapter Exercises

3. The agility test scores for the retarded children from Exercise 1 are 95, 62, 42, 96, 90, 70, 99, and 70. What is the median score?

4. The monthly incomes (in thousands of dollars) for several minority executives from Exercise 2 are $2, $10, $2, $25, $2, $10, $100, $2, and $10. What is the median income?

Like the mean, the median is unique for any set of data. It is easy to calculate, after the data have been arranged according to size. Unlike the mean, however, the median is not affected by extremely large or extremely small observations. Also, while the arithmetic mean requires the interval scale, the median merely requires that the data be at least of ordinal scale.

Mode

The **mode** is yet another value that may be used to describe the central tendency of data.

Mode The value of the observation that appears most often.

Value that occurs most often

The mode has two major advantages. First, it requires no calculation and second, it can be computed for both nominal and ordinal data.

Problem

The length of confinement (in days) for a sample of nine patients in Ward C are 17, 19, 19, 4, 19, 26, 3, 21, and 19. What is the modal length of confinement?

Solution

The mode is 19 days because that value appears most frequently. There is no mode for these hourly incomes: $4, $9, $7, $16, and $10. There are two modes (76 and 81) for these test scores: 81, 39, 100, 81, 69, 76, 42, and 76.

Choosing an Average

The question often arises: "What average should I employ—the mean, the median, or the mode?" We will now consider some of the factors influencing a decision.

The mean is probably the most frequently used measure of central tendency. It is the measure most people think of when an "average" is mentioned. The mean is considered the most reliable, or precise, average because the means of several samples taken from a population will not fluctuate as widely as either the median or the mode. (We will return to this discussion later in the book.)

The selection of an average depends in part on the level of data. As noted in Chapter 1, there are four such levels: nominal, ordinal, interval, and ratio. The arithmetic mean requires interval-level data (see Table 3-1). The mean of ordinal data (Table 3-2), or of nominal-level data (Table 3-3) would be a meaningless figure.

Table 3-1	Table 3-2		Table 3-3	
Interval-Level Data	Ordinal-Level Data		Nominal-Scale Data	
Weight (in pounds)	**Rank**	**Number**	**Party**	**Number**
	Lieutenant	44	Democratic	2,561
2	Captain	104	Socialist	732
1	Major	63	Republican	1,602
3	Colonel	51	Independent	1,814
12	General	6		
3				
3				

$\overline{X} = 4$ pounds

Median = 3 pounds Median and modal Only modal category can be lo-

Mode = 3 pounds categories = captain cated.

Modal category = Democratic

How, for example, can one possibly add lieutenant, captain, major, colonel, and general to compute the mean? Nor can we order the four political party designations in Table 3-3 (Democratic, Socialist, Republican, Independent) from low to high to arrive at the median.

Computation of the Mean from Grouped Data

Data on ages, incomes, education, and the like are often released by the Bureau of the Census and others in the form of a frequency distribution. In order to estimate the mean of a frequency distribution, it is assumed that *the values are spread evenly throughout each class.* Logically, the mean of all the values in a class is its **midpoint.** The midpoint of a class, therefore, is used to represent the class. These computations are similar to those for finding a weighted mean.

The arithmetic mean of sample data organized in a frequency distribution is estimated by

$$\overline{X} = \frac{\Sigma fX}{n}$$

where

$\overline{X}$ is the designation for the sample mean.

f is the frequency in each class.

X is the midpoint of each class.

fX is the frequency of each class times its midpoint.

ΣfX directs one to add these products.

n is the total number of observations in the sample.

Problem

Table 3-4 repeats the weekly amounts spent on food by the sample of 100 newlyweds introduced earlier. What is the arithmetic mean amount spent on food by the newlywed couples?

Solution

It is assumed that the 4 newlyweds in the $40 - $44 class spent $42 a week (the class midpoint). Thus, the 4 newlyweds spent an approximate total of 4 × $42, or $168. Next, the class midpoint of $47 is used to represent the $45 - $49 class. The total amount spent by the 11 newlyweds in that class is 11 × $47 = $517. This process is continued for all the classes. The total amount spent by the 100 newlyweds is $5,695 (see Table 3-5).

Solving for the arithmetic mean:

$$\overline{X} = \frac{\Sigma fX}{n}$$

$$= \frac{\$5,695}{100}$$

$$= \$56.95$$

Table 3-4
Weekly Amounts Spent on Food by 100 Newlywed Couples

Class Interval	Frequency
$40 - $44	4
45 - 49	11
50 - 54	20
55 - 59	31
60 - 64	19
65 - 69	11
70 - 74	4

Table 3-5
Weekly Amounts Spent on Food by 100 Newlywed Couples

Class Interval	Frequency f	Midpoint X	Frequency × Midpoint fX
$40 - $44	4	$42	$ 168
45 - 49	11	47	517
50 - 54	20	52	1,040
55 - 59	31	57	1,767
60 - 64	19	62	1,178
65 - 69	11	67	737
70 - 74	4	72	288
	$\Sigma f = n = 100$		$\Sigma fX = \$5,695$

Hence, the mean amount spent on food for the sample of 100 newlyweds is $56.95.

When the class intervals are all equal, and only then, the calculations can be greatly simplified by using the **coded method.** With this method the scale of X is "transformed" so that a class midpoint near an estimate of the mean is coded 0. Class midpoints smaller than the estimated mean are coded $-1, -2, -3$, and so on. Midpoints greater than the estimated mean are coded 1, 2, 3, and so on. Therefore, the distance between two classes is always one unit. The estimated mean is usually near the concentration of the data. Often, the class with the largest number of observations is selected as the estimated mean, although any class may be chosen.

The formula for computing the arithmetic mean of grouped data is

$$\mu = \overline{X}_0 + i\left(\frac{\Sigma fd}{N}\right) \text{ for populations}$$

$$\text{and}$$

$$\overline{X} = \overline{X}_0 + i\left(\frac{\Sigma fd}{n}\right) \text{ for samples}$$

where

$\overline{X}_0$ is the estimated mean.

i is the class interval.

f is the class frequency.

d is the unit deviation from the estimated mean.

Problem

Calculate the arithmetic-mean amount spent on food by the newly-wed couples. Use the coded method.

Solution

The data in Table 3-4 are concentrated in the $55 - $59 class (it has the largest number of observations) and the midpoint of $57 is selected as the estimated mean. The midpoint of the class before that of the estimated mean is $52.00, which is one unit below the estimated mean; therefore, it is coded as -1. Similarly, the midpoint of the $70 - $74 class is $72, which is 3 units above the estimated mean and coded 3. The coded midpoints are shown in column 3 of Table 3-6.

Table 3-6

Weekly Amounts Spent by Newlyweds on Food

Class Interval	Frequency f	Coded Value d	Frequency × Coded Value $f \cdot d$	
42 $40 - $44	4	× −3	−12	
47 45 - 49	11	× −2	−22	−54
52 50 - 54	20	× −1	−20	
57 55 - 59	31	× 0	0	add
62 60 - 64	19	× 1	19	
67 65 - 69	11	× 2	22	53
72 70 - 74	4	× 3	12	
	100		− 1	

The fourth column of Table 3-6, labeled $f \cdot d$, is obtained by multiplying the deviation from the estimated mean (d) by the number of frequencies in that class (f). This multiplication is carried out as directed in the Σfd portion of the formula. The result is -1.0.

The remaining values of $i = 5$, the class interval, and $n = 100$, the sample size, are inserted in the formula for samples and the mean computed:

$$\overline{X} = \overline{X}_0 + i\left(\frac{\Sigma fd}{n}\right)$$

midpoint *interval*

$$= \$57 + \$5\left(\frac{-1}{100}\right) = \$57 - \$0.05 = \$56.95$$

By using the coded method, the same result is obtained as when the actual midpoints are used, but the calculations are simpler. <u>This is so because coded values are smaller than actual midpoints.</u>

A sample of television viewers were asked to rate a new soap opera entitled "Mother Knows Best." The highest possible score is 74. A score near 0 indicates that the viewer thinks this is a very poor program. A score near 74 indicates it is an outstanding soap opera. The scores were organized into a frequency distribution:

Rating	Number of Ratings
0 - 14 7	2 —2
15 - 29 22	6 —1
30 - 44 37	10 0
45 - 59 52	8 1
60 - 74 67	4 2
	30

Determine the mean rating using (a) the actual class midpoints, and (b) the coded method.

n = Sample Size

Chapter Exercises

5. A group of vehicles traveling on Interstate 75 were clocked by radar, and the following frequency distribution was constructed:

ifd = 5⁴

Speeds (mph)	Number of Vehicles		
40 - 44	42 × 9 -2	378	
45 - 49	47 × 15 -1	705	
50 - 54	52 × 30 0	1560	
55 - 59	57 × 17 1	969	
60 - 64	62 × 17 2	1054	
65 - 69	67 × 12 3	804	
		5470	

$\overline{X} = \frac{\Sigma fx}{n}$

$\overline{X} = \frac{5470}{100}$

$\overline{X} = 54.7$

a. What is the total number of frequencies? 100
b. Determine the midpoint for the class 50 - 54. 52

$$\frac{\Sigma mp \times f}{100}$$

Self-Review 3-5

a. Actual Midpoints

fX

14
132
370
416
268

1,200

$$\overline{X} = \frac{\Sigma fX}{n}$$

$$= \frac{1200}{30}$$

$$= 40$$

b. Coded Midpoints

f	d	fd
2	-2	-4
6	-1	-6
10	0	0
8	1	8
4	2	8
		6

$$\overline{X} = \overline{X}_0 + i\left(\frac{\Sigma fd}{n}\right)$$

$$= 37 + 15\left(\frac{6}{30}\right)$$

$$= 37 + 3$$

$$= 40$$

c. Find the arithmetic-mean speed using the class midpoints.
d. Find the arithmetic-mean speed using the coded method. 54.7

6. A study of the mileage per gallon of 35 compact cars of the same make and model revealed:

Miles per Gallon	Number of Cars
20 - 24	2
25 - 29	7
30 - 34	15
35 - 39	8
40 - 44	3

a. What is the total number of frequencies? 35
b. What is the midpoint of the 30 - 34 class? 32
c. Compute the arithmetic-mean mileage using the class midpoints.
d. Compute the arithmetic-mean mileage using the coded method.

In conclusion, two points can be made regarding the computation of the arithmetic mean for grouped data. First, if the grouped data represent a population, and not a sample, the symbols are altered slightly but the computations are the same. The formula is

$$\mu = \frac{\Sigma fX}{N}$$

where

μ is the population mean.

ΣfX is the sum of the products of the class midpoints times the number of frequencies in each class.

N is the total number of observations in the population.

Secondly, it is unlikely that the mean of grouped data organized in a frequency distribution will agree exactly with the mean of the raw data. Consequently, the mean based on data in a frequency distribution can be considered only an *estimate* of the actual mean. In the newlywed problem, the mean using the frequency distribution is $56.95; the mean amount based on the raw data presented in Chapter 2 is $56.93. In most practical situations, this loss of exactness is not important.

Handwritten notes in left margin:

6d –

$\bar{X} = \bar{X}_o + i \left(\frac{\Sigma fd}{n} \right)$

$\bar{X} = 32 + 5 \left(\frac{3}{35} \right)$

$\bar{X} = 32 + .4$

$\bar{X} = 32.4$

Computation of the Median from Grouped Data

Median — middle most value

The median of data organized into a frequency distribution can be estimated by (1) locating the class in which the median lies and (2) interpolating within that class to arrive at the median.

Problem

The weekly amounts spent by newlyweds on food are reintroduced to illustrate the procedure for approximating the median (see Table 3-7). The cumulative frequencies in the right-hand column will be helpful in locating the class in which the median lies. What is the median amount spent on food by the newlyweds?

Solution

Note that this table includes 100 newlyweds. The middle observation is determined by $n/2$, or $100/2 = 50$. (Note: in order to be consistent with the way the median was located earlier in the chapter, the expenditure of the $(n + 1)/2$, or 50.5th case should be located. However, it is common practice to use the more convenient $n/2$ value. The error involved is very slight.) The class interval containing the 50th newlywed is found by referring to the cumulative frequency column. Note that 35 newlyweds spent $54 or less, and that 66 newlyweds spent $59 or less. Logically, the 50th amount is in the $55 - $59 class, which contains 31 members ($66 - 35$).

Recall from Chapter 2 that the lower limit of the $55 - $59 class is really $54.50 and the upper limit is $59.50. These were

Table 3-7

Weekly Amounts Spent on Food by 100 Newlyweds

Class	Frequency f	Cumulative Frequency CF
42 $40 - $44	4	4
47 45 - 49	11	15
52 50 - 54 54.5	20	35
57 55 - 59 59.5	31	66
62 60 - 64	19	85
67 65 - 69	11	96
72 70 - 74	4	100

median in this class →

100

add 1 to the value, divide by 2.

$$\frac{n+1}{2} \quad \frac{100+1}{2} = \frac{101}{2} = 50.5$$

66,
-35
31

referred to as the *true limits* of that class. Thus, the median weekly amount spent on food is between $54.50 and $59.50. In order to compute the median, again, the assumption is made that the amounts spent on food by the 31 newlyweds in the median class are evenly distributed between the lower and upper limits of that class. Shown in a diagram:

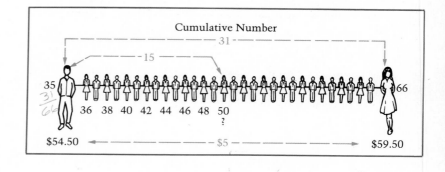

The next step is to interpolate and determine the median from the diagram. There are 15 newlyweds between the 35th and the 50th newlywed (50 − 35). There are 31 newlyweds in the class containing the median. Thus, the median amount spent is 15/31 of the distance between $54.50 and $59.50. Since the distance between those two amounts is $5.00, then 15/31 of $5.00 is $2.42. By adding $2.42 to $54.50 we obtain the estimated median of $56.92.

$$\text{Median} = \$54.50 + \frac{15}{31} \text{ of } \$5.00$$

$$= \$56.92$$

For grouped data, then, the median may be computed by using the formula:

$$\text{Median} = L + \frac{\frac{n}{2} - CF}{f} (i)$$

where

L is the true lower limit of the class containing the median (the median is in the $55 - $59 class and $54.50 is the true lower limit of that class).

n is the total number of frequencies (in this problem it is 100).

CF is the cumulative number of frequencies in all the classes immediately preceding the class containing the median (the class containing the median is $55 - $59, and the cumulative number of newlyweds prior to that class is 35).

f is the frequency in the class containing the median (there are 31 newlyweds in that class).

i is the width of the class in which the median lies (the class limits are $54.50 and $59.50 and its width is $5.00, found by $59.50 − $54.50).

Inserting these values into the formula:

$$\text{Median} = L + \frac{\frac{n}{2} - CF}{f} \ (i)$$

$$= \$54.50 + \frac{\frac{100}{2} - 35}{31} \ (\$5)$$

$$= \$54.50 + \frac{15}{31} \times \$5$$

$$= \$56.92$$

This is the same median as computed previously.

Problem

The hourly wages for a sample of apprentice plumbers were grouped into the following frequency distribution. Estimate the median hourly wage of the apprentice plumbers.

Hourly Wages	Number f
$ 8.00 - $ 8.49	3
8.50 - 8.99	6
9.00 - 9.49	12
9.50 - 9.99	10
10.00 - 10.49	7
10.50 - 10.99	2

Solution

The median class is located by dividing the total number of observations by 2: $n/2 = 40/2 = 20$. The class containing the 20th apprentice plumber can be located by referring to the cumulative frequency column in the following table. Notice that 9 plumbers earn \$8.99 or less and 21 earn \$9.49 or less. Hence, the median is in the \$9.00 - \$9.49 class. The true lower limit of that class is \$8.995; CF is 9, the cumulative number of frequencies in the class preceding the median class; f is 12, the number of frequencies in the median class; and $i = \$0.50$, the width of the median class. Solving:

$$\text{Median} = L + \frac{\frac{n}{2} - CF}{f}\,(i) = \$8.995 + \frac{\frac{40}{2} - 9}{12}\,(\$0.50) = \$9.45$$

$$= 8.995 + .475$$
$$= 9.47$$

	Hourly Wages	Frequency (f)	Cumulative Frequency (CF)
	\$ 8.00 - \$ 8.49	3	3
	8.50 - 8.99	6	9 12
median class →	9.00 - 9.49	12	21 33
	9.50 - 9.99	10	31 64
	10.00 - 10.49	7	38 102
	10.50 - 10.99	2	40 142
	Total	40	

Self-Review 3-6

$$\frac{50}{2} - 13$$

$$\text{Median} = 39.5 + \frac{50 - 13}{20}\,(10)$$

$$= 39.5 + \frac{12}{20}\,(10)$$

$$= 45.5 \text{ years}$$

On arriving in Hawaii, a sample of vacationers were asked their ages by the tourist bureau. This information was organized into the following frequency distribution (the true limits and the cumulative frequencies are given for clarification):

Ages	True Limits	Number of Vacationers	Cumulative Number
20 - 29	19.5 - 29.5	4	4
30 - 39	29.5 - 39.5 L	9	13 = CF
40 - 49	39.5 - 49.5 upper	20	33
50 - 59	49.5 - 59.5	8	41
60 - 69	59.5 - 69.5	5	46
70 - 79	69.5 - 79.5	4	50 = n
		50	

median class

Compute the median age.

take the total # 50 add 1 divide by 2

$$\frac{50 + 1}{2}$$

The Mode from Grouped Data

The mode is the *midpoint* of the class interval with the largest frequency. Refer back to Table 3-6 for an example. The largest frequency in a class is 31. Thus, the mode is the midpoint of the $55 - $59 class, or $57. This indicates that the greatest number of newlyweds spent $57 a week on food.

Refer to Self-Review 3-6 for the ages of the Hawaii vacationers.

a. What is the mode of the age distribution of vacationers arriving in Hawaii? This question also could be stated as: What is the *modal* age of the vacationers arriving in Hawaii?

b. Explain what the mode indicates.

Self-Review 3-7

a. 44.5 years.

b. The largest number of vacationers were 44.5 years of age.

Chapter Exercises

7. Refer back to the frequency distribution of speeds on I-75 in Exercise 5.
a. Which class interval contains the median value?
b. What is the true lower limit of the class containing the median?
c. How wide are the classes?
d. Determine the cumulative number of frequencies in all classes smaller than the median class.
e. Compute the median speed.
f. What is the mode of the distribution of speeds?
g. Discuss which measure of central tendency you would use to "average" this data.

8. A sociologist is studying the impact of television on the family. Of particular interest is the number of hours of television school-aged children watch each day. A random sample of 50 homes reveals the following number of hours watched on a Wednesday between 4 P.M. and 11 P.M.:

Number of Hours	Frequency
0	9
1	14
2	10
3	9
4	5
5 or more	3
	50

(a.) Compute the median number of hours of TV viewing.

b. Determine the modal number of hours of TV watched.

9. An advertising campaign is being designed for a local car dealer. One relevant piece of information, the ages of recent purchasers of cars sold by the dealership, is shown in the following frequency distribution:

Age	Frequency
Under 20	7
20 - 29	13
30 - 39	26
40 - 49	15
50 - 59	6
60 or over	3
Number of cars sold	70

a. Compute the median age.

b. What is the modal age?

Choosing an Average for Data in a Frequency Distribution

If the data being organized into a frequency distribution contain extremely high or extremely low values, the resulting distribution is said to be **skewed.** The frequency distribution in Figure 3-3 is **positively skewed.** It can be identified by the long "tail" on the right.

Note the relative positions of the three averages. The arithmetic mean ($\overline{X}$), is being pulled upward toward the tail by a few very highly paid employees. It is the highest of the three averages. If the skewness is very pronounced, the mean is not an appropriate average to describe the central tendency of the data.

The median (M) is the next higher measure of central tendency for a positively skewed distribution. It, of course, divides the incomes into two parts and may be more typical of the data. The mode (MO) is the smallest of the three measures. It occurs at the peak, or apex, of the curve and indicates that the largest number of employees earns $200 a week.

The distribution of the ratings in Figure 3-4 is **negatively skewed.** The arithmetic mean is being pulled down by a few very low ratings and once again, therefore, it is not the best average to

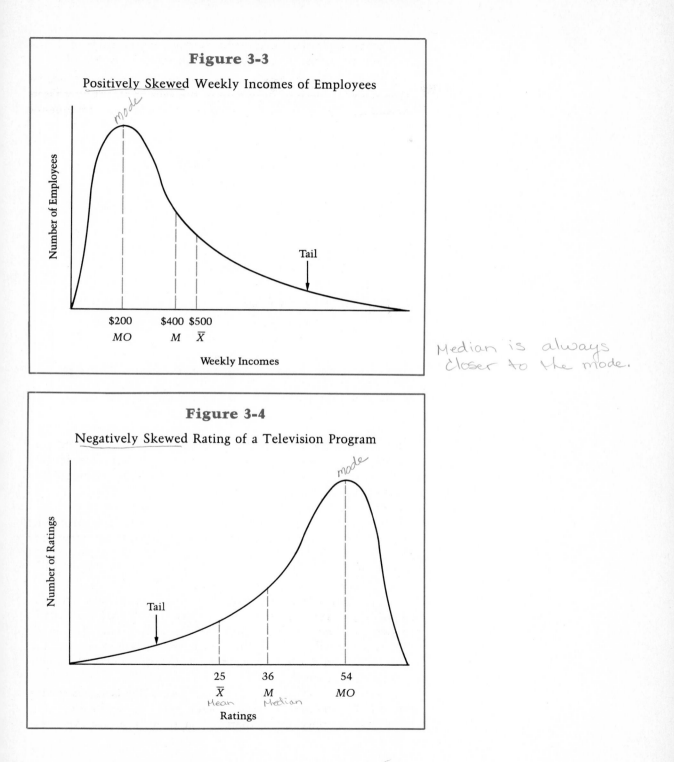

Figure 3-3

Positively Skewed Weekly Incomes of Employees

Number of Employees

mode

Tail

| $200 | $400 | $500 |
| MO | M | X̄ |

Weekly Incomes

Median is always closer to the mode.

Figure 3-4

Negatively Skewed Rating of a Television Program

Number of Ratings

mode

Tail

25	36	54
X̄	M	MO
Mean	Median	

Ratings

use. The median is the next higher average, with the mode of 54 the highest of the three averages.

By contrast, Figure 3-5 reveals that there is no skewness in the distribution of the annual incomes of the social workers. This type of distribution is called symmetrical. A **symmetric distribution** has the same shape on either side of the median. That is, if the distribution graph were folded in half at the median value, the two halves would be identical. Therefore, with such a distribution all three values—mode, median, and mean—are the same. In Figure 3-5 the value for all three is $19,000.

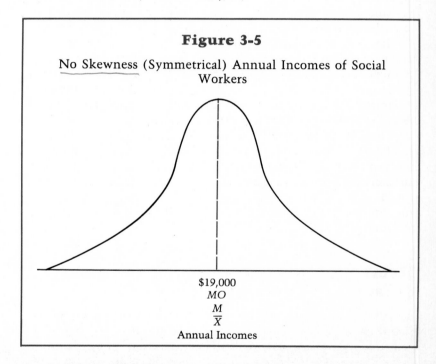

Figure 3-5

No Skewness (Symmetrical) Annual Incomes of Social Workers

$19,000
MO
M
$\overline{X}$

Annual Incomes

Open-ended frequency distributions pose a problem for the computation of the arithmetic mean. Unless the midpoint of the open-ended class can be approximated, the mean cannot be computed. In the example below, unless the midpoint of the "$30,000 and over" class can be approximated, the mean of this open-ended distribution cannot be calculated.

Income	Number
$10,000 - $19,999	42
20,000 - 29,999	86
30,000 and over	19

The median or the mode, however, can be used to represent the central value of the income distribution.

Summary

This chapter dealt with three measures of central tendency used to describe the typical value of a set of data. The most commonly used average is the arithmetic mean. For raw data, it is computed by summing the values and dividing the total by the number of values. For data grouped into a frequency distribution, each class frequency is multiplied by its midpoint and the products summed. This sum is then divided by the number of observations to obtain the arithmetic mean.

The median divides the data into two equal parts. For raw data, the values are arranged from low to high and the center value is the median. For data in a frequency distribution, first the median class is identified from the cumulative frequency distribution, and interpolation is used next to determine the median. The mode is the value of the item that appears most often.

A number of characteristics make the mean a very reliable average. However, because all the values are included in its computation, the mean may not be representative of the data where there are extremely high or low values. The median or the mode are preferred in such skewed samples. The mode can be used for all levels of data, the median requires at least ordinal data, and the mean at least interval scale.

Descriptive Statistics: Measures of Central Tendency

Chapter Outline

Objective. To determine a single value considered representative of a set of values.

I. Arithmetic Mean
A. Definition

$$\text{Mean} = \frac{\text{sum of the values}}{\text{number of values}}$$

1. The formula for the mean of a sample of raw data is

$$\bar{X} = \frac{\Sigma X}{n} \qquad \left[\text{Example: 3, 2, 7,} \qquad \bar{X} = \frac{\Sigma X}{n} = \frac{12}{3} = 4 \right]$$

2. The formula for the mean of a population of *raw* data is

$$\mu = \frac{\Sigma X}{N}$$

3. The mean of a sample of *grouped* data is

$$\overline{X} = \frac{\Sigma fX}{n}$$

Example:

Ages	Number f	Midpoint X	fX
2 - 6	3	4	12
7 - 11	5	9	45
12 - 16	2	14	28
	10		85

$$\overline{X} = \frac{\Sigma fX}{n} = \frac{85}{10} = 8.5 \text{ years}$$

4. The mean of a population of *grouped* data is

$$\mu = \frac{\Sigma fX}{N}$$

B. Characteristics of the Arithmetic Mean
   1. The arithmetic mean is appropriate for interval-level measurements such as incomes, ages, or weights if the data are not highly skewed.
   2. It should not be used for ordinal- or nominal-level data.
   3. The mean of raw data and the mean of the same data organized in a frequency distribution may be slightly different. This loss of exactness, however, is usually very minor.
   4. Likewise, the mean of a positively or negatively skewed distribution might not be representative. The median might be a better choice.
   5. For a positively skewed distribution, the mean is the largest of the three averages. For a negatively skewed

distribution it is the smallest. The three averages are equal for symmetrical distributions. Examples:

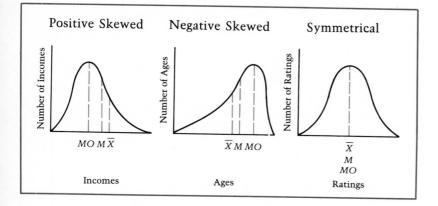

II. Median

    A. Definition. The median is the value of the middle observation after all observations have been arranged from low to high.

    B. Computation

        1. Raw-data, odd-numbered set of observations. Arrange the observations from low to high and determine the middle value.

        Example: 3, 12, 2. Rearranged, this gives 2, ③, 12; the median is 3.

        2. Raw-data, even-numbered set of observations. Arrange the observations from low to high. The median is halfway between the two middle values.

        Example: 11, 5, 1, 7. Rearranged, this gives 1, 5, 7, 11; the median is ⑥, which is halfway between 5 and 7.

        3. Frequency distribution.

$$\text{Median} = L + \frac{\frac{n}{2} - CF}{f} \ (i)$$

    Example:

Ratings	Number f	CF
2 - 6	3	3
7 - 11	5	8
12 - 16	10	18
17 - 21	2	20

$$\text{Median} = L + \frac{\frac{n}{2} - CF}{f} \quad (i)$$

$$= 11.5 + \frac{\frac{20}{2} - 8}{10} \quad (5)$$

$$= 12.5$$

C. Characteristics
1. Appropriate for the ordinal, interval, and ratio levels of data.
2. Should be used when the data in a frequency distribution are highly skewed. For that reason, the government usually reports the median family income instead of the mean family income.
3. It can be computed even if the distribution is open-ended.
4. If the distribution is skewed, the median is the middle of the three averages (mean, median, mode), as shown by the previous diagrams.

III. Mode
A. Definition. The mode is the value of the item that appears most frequently.
B. Computation
1. Raw data. Determine the value of the observation or observations that appear most often. There may be one mode, several modes, or no mode for a set of raw data.
2. Frequency distribution. Locate the class containing the highest frequency. The mode is the midpoint of that class.
C. Characteristics
1. Can be determined for all levels of data.
2. It is the lowest of the three averages for a positively skewed distribution; it is the highest for a negatively skewed distribution.

Chapter Exercises

10. The U.S. Law Enforcement Assistance Administration reported that on December 31 there were, for each of the past five years, 213,000, 211,000, 196,000, 241,000, and 263,000 prisoners in federal and state institutions. Consider this information to be a population.
a. Find the arithmetic-mean number of prisoners.
b. Find the median number of prisoners.

11. The U.S. National Oceanic & Atmospheric Administration reported the following monthly breakdown of normal daily mean temperatures for Juneau, Alaska, and San Juan, Puerto Rico:

Month	Temperature (Fahrenheit) Juneau	San Juan
January	23.5	75.4
February	28.0	75.3
March	31.9	76.3
April	38.9	77.5
May	46.8	79.2
June	53.2	80.5
July	55.7	80.9
August	54.3	81.3
September	49.2	81.1
October	41.8	80.6
November	32.5	78.7
December	27.3	76.8

a. Is the information a sample or population?
b. Find the mean annual temperature for both Juneau and San Juan.
c. Find the median annual temperature for both Juneau and San Juan.

12. The National Science Foundation reported that the median time candidates in the social sciences take to earn a doctorate degree, after receiving a bachelor's degree, is 8.5 years. Explain what the 8.5 years indicates.

13. The Surgeon General released these figures for the U.S.: 54 million smokers smoked 615 billion cigarettes in one year. What is the average annual number of cigarettes smoked by persons who smoke? What is the average daily number?

14. A report on the literacy rate for a sample of countries revealed the following percents:

Country	Literacy Rate (percent)
Turkey	55
Tunisia	32
Tonga	93
Benin	20
Bangladesh	25
Colombia	47
Iceland	99
Iran	37

a. Find the mean percent.
b. Find the median percent.

15. According to the Department of Commerce, the 1980 incomes of white families and those of all other families were as follows:

Income	Family Income White (percent)	Other (percent)
Under $5,000	4.9	15.2
$ 5,000 to $ 7,499	5.3	12.3
$ 7,500 to $ 9,999	6.0	9.8
$10,000 to $14,999	13.9	16.1
$15,000 and over	69.9	46.6

a. Find the median income for each group.
b. Comment on whether the mean or the mode would be an appropriate average.

16. In the June 9, 1980, issue, *U.S. News & World Report* reported their findings on executive pay. The survey included 1,052 top officials in 396 of the largest U.S. firms.

12 earned less than $100,000
213 earned $100,000 to $200,000
300 earned $200,000 to $300,000
241 earned $300,000 to $400,000
135 earned $400,000 to $500,000
76 earned $500,000 to $600,000
34 earned $600,000 to $700,000
41 earned $700,000 to $1,600,000

Source: "Survey's Findings on Executive Pay," *U.S. News & World Report,* June 9, 1980. © 1980.

a. Compute the median income. Explain what this figure indicates.
b. What problems would you encounter in determining the mean income?
c. What is the modal income?

17. Are the baseball batting averages in the American League higher than those in the National League? Consider these data to be a sample:

Batting Averages	American League	National League
350 - 374	2	0
325 - 349	10	5
300 - 324	17	9
275 - 299	28	25
250 - 274	36	27
225 - 249	22	17
200 - 224	12	10
175 - 199	1	4
150 - 174	4	0

a. Compare the mean batting averages for the two leagues.
b. Compare the median averages.
c. Compute the mode for each league.
d. Does this show that one league is better than the other?

18. A sample of seven independently owned service stations in the San Francisco metropolitan area revealed the following cost per gallon of regular gasoline:

Andy's Amoco	$1.21
Rossi's Gastown	1.30
Kernie's Mobil	1.27
Mark's Exxon	1.25
Yamamoto's Sohio	1.27
Ali's Sunoco	1.25
Deckers' Union 76	1.27

a. Calculate the mean, median, and modal cost per gallon.
b. Discuss which average is most representative of the information or numbers.

Do all the problems. Then check your answers against those given in the Answer section of the book.

Chapter Achievement Test

I. Matching Problems. Select the letter corresponding to the correct answer from the list appearing in the answer column. Each answer may be used more than once (5 points each).

Questions

1. Which measure of central value is determined by arranging the values from low to high and selecting the middle one?

Answers

A. Mode
B. Median
C. Arithmetic Mean

D. $\bar{X} = \dfrac{\Sigma X}{n}$

E. $\bar{X} = \dfrac{\Sigma fX}{n}$

F. $L + \dfrac{\dfrac{n}{2} - CF}{f}\ (i)$

G.

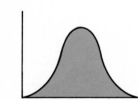

H.

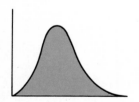

I.

2. Which measure of central value cannot be determined if the distribution has an open end?
3. Which formula is used for computing the arithmetic mean of raw data?
4. Which formula is used for computing the median of data grouped into a frequency distribution?
5. Which chart shows a symmetrical distribution?
6. Which chart shows a positively skewed distribution?
7. Which chart shows a distribution for which the mean and median are equal?
8. In a positively skewed distribution, which average is the lowest?

Questions 9 and 10 are based on the following frequency polygon:

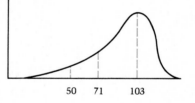

9. Which average is represented by the number 50?
10. Which average is represented by the number 71?

II. Problems (10 points each).
    11. The weekly incomes for a sample of four registered nurses are: $420, $445, $600, and $395.
        a. What is the mean weekly income?
        b. What is the median weekly income?
    12. The amounts spent in one day on outpatient care by a sample of five persons were $22.50, $85.90, $18.60, $140.20, and $76.10.
        a. What is the median amount spent?

Questions 13 - 15 are based on the following frequency distribution of a sample of the weights of king crab caught in Alaskan waters:

Weights (pounds)	Number of Crabs
0.0 to 0.4	3
0.5 to 0.9	8
1.0 to 1.4	20
1.5 to 1.9	10
2.0 to 2.4	6
2.5 to 2.9	3

13. What is the arithmetic-mean weight (to the nearest hundredth of a pound)?
14. What is the median weight (to the nearest hundredth of a pound)?
15. What is the modal weight (to the nearest hundredth of a pound)?

4

Descriptive Statistics: Measures of Dispersion and Skewness

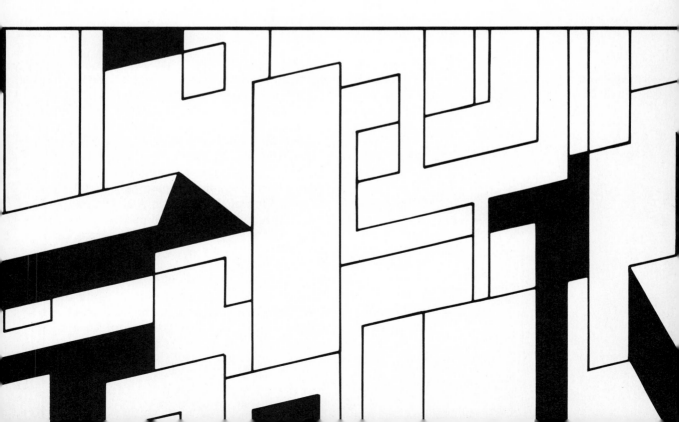

OBJECTIVES

When you have completed this chapter, you will be able to
- Describe the spread or dispersion in a set of data.
- Calculate and interpret the range, interquartile range, quartile deviation, mean deviation, variance, and standard deviation.
- Compute the coefficient of variation and the degree of skewness.

Introduction

In Chapter 2, we noted that the usual first step in summarizing raw data is to organize them into a frequency distribution. A histogram or a frequency polygon may be drawn to portray such distribution in graphic form. Chapter 3 described a summary measure called an "average." Three kinds of averages were examined, namely, the arithmetic mean, the median, and the mode. An average is a central value about which the data tend to cluster.

A direct comparison of the central value of two or more distributions may be misleading. For example, suppose that a study of the lengths of imprisonment for armed robbery in Alabama had shown the arithmetic-mean length to be 10 years. Suppose further that a similar study of the lengths of imprisonment in neighboring Georgia had also shown a mean length of 10 years. Based on the two means, one might conclude that the distributions of the lenghts of imprisonment were about the same in both states. Reference to Figure 4-1, however, reveals this conclusion to be incorrect. Despite the fact that the two means are equal, the lengths of imprisonment in Georgia are spread out, or dispersed, significantly more than those in Alabama.

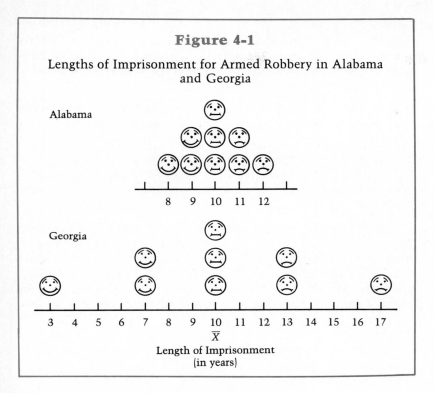

Figure 4-1

Lengths of Imprisonment for Armed Robbery in Alabama and Georgia

Length of Imprisonment
(in years)

Measures of Dispersion

As Figure 4-1 suggests, the next step in analyzing a set of data is to develop measures that describe the spread, or **dispersion,** of that set.

The Range

Perhaps the simplest measure of dispersion is the **range.** The range is the difference between the highest and lowest values in a set of data.

> **Range**   The result of subtracting the lowest value from the highest value.

Problem

Calculate the range of the length-of-imprisonment data in Figure 4-1 for both Alabama and Georgia. Compare the two ranges.

Solution

For the Alabama population, the longest stretch in prison is 12 years, the shortest is 8 years. The range is 4 years (12 − 8). The range for the Georgia prisoners is 14 years, or 17 − 3. A comparison of the two ranges indicates the following:

1. There is more spread in the Georgia data because the range of 14 years for Georgia is greater than the range of 4 years for Alabama.

2. The lengths of imprisonment for the Alabama inmates are clustered more closely about the mean of 10 years than those for the Georgia inmates, because the range of 4 years is less than the range of 14 years.

Obviously, the range is easy to compute. It does have a serious disadvantage, however. It is based on only two values—the two extreme observations. The range, therefore, does not reflect the variation in data that lie between the high and low observations. Further, one extremely high (or low) value might give a misleading picture of the dispersion. For example, suppose that the ages of a group at a rock concert were 18, 20, 19, 20, 19, 17, 18, 18, and 78. Reporting the range of 61 years, found by subtracting 17 from 78, would give the impression that there is considerable variation in the ages of the audience. Except for one person, however, the distance between the youngest (17 years) and the oldest (20 years) is only 3 years! For that reason, the range can be considered only a rough index of the variation in a set of data.

Self-Review 4-1

A reminder: cover the answers in the margin.

A radar check of several automobiles traveling on the Sunshine Highway recorded these speeds in kilometers per hour: 86, 70, 91, 110, and 89. Speeds recorded at about the same time on a nearby state highway were 89, 87, 89, 92, 85, and 92.

a. Sunshine: $\overline{X} = \dfrac{446}{5} = 89.2$

State: $\overline{X} = \dfrac{534}{6} = 89.0$

b. Sunshine: 40 km per hour (110 − 70).
State: 7 km per hour (92 − 85).

c. More dispersion in speeds on Sunshine because 40 is greater than 7. Speeds on state highway clustered closer to mean because 7 is less than 40.

a. Compute the mean speed for each set of data.
b. Compute the range for each set of data.
c. Compare the dispersion for the two sets of speed checks.

The Interquartile Range and the Quartile Deviation

Recall that half of the ordered values are above the median and half below it. The lower half of the ordered values can be further subdivided into two parts, so that one-fourth of all the ordered values are smaller than a particular value. The point where this subdivision occurs is called the **first quartile** and is designated Q_1. Similarly, the upper half of the ordered distribution can be divided into two equal parts. The **third quartile,** designated Q_3, is the point below which lie three-fourths of the ranked observations. The median, of course, is the same as the second quartile.

The **interquartile range** is the distance between the first quartile and the third quartile.

Interquartile Range   The result obtained by subtracting the first quartile from the third quartile ($Q_3 - Q_1$).

Figure 4-2 illustrates this concept; note that the third quartile is 20 days, the first quartile is 12 days. The interquartile range is 8 days, found by subtracting 12 from 20. Thus, the interquartile range reports the difference between the two values that encompass the middle 50% of the observations.

Another measure of dispersion is the **quartile deviation,** designated QD, which is one-half of the interquartile range.

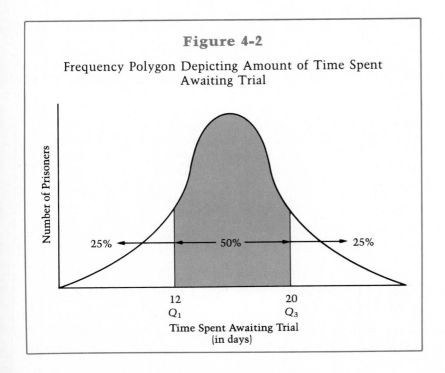

Figure 4-2

Frequency Polygon Depicting Amount of Time Spent Awaiting Trial

Number of Prisoners

25% ← → 50% ← → 25%

12   20
Q_1 Q_3

Time Spent Awaiting Trial
(in days)

$$\text{Quartile deviation} = \frac{\text{third quartile} - \text{first quartile}}{2}$$

$$QD = \frac{Q_3 - Q_1}{2}$$

As mentioned earlier, one of the objections to the range as a measure of spread is that its value is based on the two extreme observations of a set of data. An unusually large or small extreme value might give a distorted picture of the spread in the data. The interquartile range and the quartile deviation offset this objection because they are not based on the extreme, or end, values. The first and third quartiles may be more representative of the data. Since the QD is based on the relative positions of Q_1 and Q_3 in the frequency distribution, it may be computed not only for interval scaled data, but also for ordinal scaled data. Later in the chapter you will learn to compute the interquartile range and the quartile deviation for data organized into a frequency distribution.

Mean Deviation

One of the disadvantages of the range, the interquartile deviation, and the quartile deviation is that they do not make use of all the observations. However, several measures that incorporate all values do exist; they are based on the deviation of each observation from its mean. These measures include the **mean deviation** (often called average deviation), the **variance,** and the **standard deviation.**

The **mean deviation,** abbreviated *MD,* is computed by first finding the difference between each observation and the mean. Next, these deviations are summed using their absolute values—that is, disregarding their algebraic signs (+ and −). Finally, the sum obtained is divided by the number of observations to arrive at the mean deviation.

$$MD = \frac{\Sigma|X - \overline{X}|}{n}$$

where

X is the value of each observation.

$\overline{X}$ is the mean.

n is the number of observations.

(The symbol | | indicates the absolute, or unsigned, value of a number.)

Problem

The test kitchen of Cellu-Lite, a large producer of cake mixes, constantly monitors the weight, moisture content, and flavor of its cakes. The weights (in grams) of a sample of five peach upside-down cakes are 498, 500, 503, 498, and 501. What is the mean deviation? How is it interpreted?

Solution

First, we find the mean weight of the cakes, which is 500 grams, (498 + 500 + 503 + 498 + 501)/5 = 2500/5 = 500. Next, we find the deviation from the mean for each cake. For example, the first cake selected, weighing 498 grams, deviates 2 from the mean of 500. Calculations for the mean deviation follow (see Table 4-1).

Interpreting the result: on the average, the weights of the five peach upside-down cakes deviate 1.6 grams from the mean of 500 grams.

Table 4-1

Calculations Needed for the Mean Deviation

Diagram Showing Deviations from the Mean

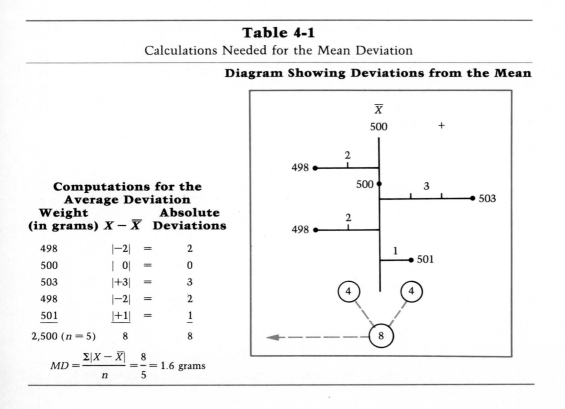

Computations for the Average Deviation

Weight (in grams)	$X - \overline{X}$		Absolute Deviations
498	$\|-2\|$	=	2
500	$\| \ 0\|$	=	0
503	$\|+3\|$	=	3
498	$\|-2\|$	=	2
501	$\|+1\|$	=	1
2,500 ($n = 5$)	8		8

$$MD = \frac{\Sigma|X - \overline{X}|}{n} = \frac{8}{5} = 1.6 \text{ grams}$$

Self Review 4-2

a. $\bar{X} = \dfrac{3,000}{6} = 500$

X	$\|X - \bar{X}\|$
484	$\|-16\|$
503	$\|+3\|$
496	$\|-4\|$
510	$\|+10\|$
491	$\|-9\|$
516	$\|+16\|$
3,000	58

$MD = \dfrac{58}{6} = 9.7$ grams

b. On the average, blueberry cakes deviate 9.7 grams from the mean of 500 grams.

c. Greater variation in weights of blueberry cakes because 9.7 > 1.6. Peach cakes clustered closer to the mean weight.

The Cellu-Lite test kitchen now wants to compare the variation in the weights of blueberry cakes with that of the peach upside-down cakes studied in the previous case. The weights of the blueberry cakes are 484, 503, 496, 510, 491, and 516 grams.

a. Compute the mean and the mean deviation.
b. Interpret the mean deviation.
c. Compare the variation in the weights of the peach upside-down cakes with the variation in the weights of the blueberry cakes.

Chapter Exercises

1. The amount (in gallons) of fuel oil consumed during the winter heating season for a sample of newer homes is 600, 590, 605, 600, 603, 610, 597, and 595. A second group of older homes used the following amounts: 810, 750, 790, 800, and 850.
   a. What is the range for the newer homes? For the older homes?
   b. What is the mean deviation for the newer homes and for the older homes?
   c. Compare the dispersion in the data for the new homes and older homes.

2. The Department of Fisheries stocked two ponds with trout fingerlings. The fish in one pond (pond A) were fed the traditional food. Those in the other pond (pond B) were fed an experimental food. Consider this as sample data. After four years the lengths of the fish were

Lengths in inches	
Pond A	Pond B
12.5	18.0
14.0	20.0
13.5	12.0
14.5	14.5
15.0	19.0
16.5	13.5
16.0	14.5
12.0	13.0
14.0	20.5
14.0	14.0

   a. Compute the arithmetic-mean lengths for both ponds.
   b. Compute the range for both sets of data. Interpret.
   c. Compute the mean deviation for each set. Interpret.

The Variance and the Standard Deviation

The mean deviation has two distinct advantages: it uses all available data in its computation, and it is easy to understand—that is, the mean deviation is the average amount by which the observations differ from the mean. As a measure of variation, however, it does have a major flaw: the absolute values used in its computation are difficult to manipulate algebraically.

Two other measures of dispersion—the **variance** and the **standard deviation**—are more versatile than the mean deviation. As a result, they will be used extensively in later chapters.

Like the mean deviation, the variance and the standard deviation are based on the deviations from the mean. However, unlike the mean deviation, the sign (+ or −) of the deviation from the mean is *not* ignored in their computation. To compute the variance and the standard deviation, the difference between each value and the mean is *squared*. This squaring eliminates the possibility of negative numbers (since multiplication of two negative numbers yields a positive number). The squared deviations are then totaled, and the total is divided by the number of values.

> **Variance**   The arithmetic mean of the squared deviations from the mean.

> **Standard Deviation**   The square root of the variance.

A measure of total variability

Because of a difference in computation, it is necessary to distinguish between the variance of a *population,* designated by σ^2 (the lowercase Greek *sigma*), and the variance of a *sample,* which is designated s^2. Recall that in Chapter 3 we made a distinction between a parameter, which is a measure of a population, and a statistic, which is a measure of a sample. The computational formulas for the variance of a population and the variance of a sample are

$$
\text{Variance of a population}
$$
$$
\sigma^2 = \frac{\Sigma(X - \mu)^2}{N}
$$

where

  μ is the population mean.

X is the value of each observation in the population.

N is the number of observation in the population.

Variance of a sample

$$s^2 = \frac{\Sigma(X - \overline{X})^2}{n - 1}$$

---

where

X is the value of each observation in the sample.

$\overline{X}$ is the sample mean.

n is the number of observations in the sample.

Notice that the sample variance is computed by dividing the sum of the squared deviations by $n - 1$ rather than by n. The use of the sample mean $\overline{X}$ instead of the population mean μ causes the numerator to be too small. So we "underestimate" the denominator as well, to compensate. Stated differently, s^2 would underestimate σ^2 if $\Sigma(X - \overline{X})^2$ were divided by n. For a practical illustration, see the problem and solution that follow.

Problem

Refer to the weights for a sample of five peach upside-down cakes in Table 4-1. What are the sample variance and the standard deviation?

Solution

The data needed to calculate the variance and the standard deviation are included in Table 4-2. The sample-mean weight, $\overline{X}$, is 500 grams.

To determine the sample variance, the sum of the squared deviations (18) is divided by the number of observations minus one.

$$s^2 = \frac{\Sigma(X - \overline{X})^2}{n - 1}$$

$$= \frac{18}{5 - 1}$$

$$= 4.5$$

The variance, therefore, is 4.5. The standard deviation is then determined by taking the square root of the variance. Here, the standard deviation (s) is 2.12 grams.

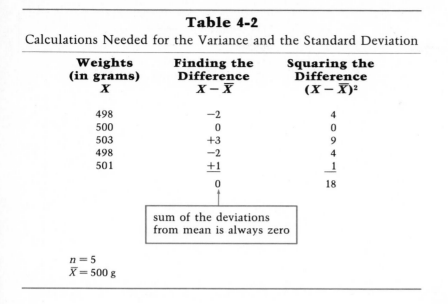

Table 4-2

Calculations Needed for the Variance and the Standard Deviation

Weights (in grams) X	Finding the Difference $X - \bar{X}$	Squaring the Difference $(X - \bar{X})^2$
498	−2	4
500	0	0
503	+3	9
498	−2	4
501	+1	1
	0	18

sum of the deviations from mean is always zero

$n = 5$
$\bar{X} = 500$ g

The test kitchen wants to compare the variation in the weights of the blueberry cakes in the previous self-review with the peach cakes—using the variance and the standard deviation. The weights of a sample of blueberry cakes were 484, 503, 496, 510, 491, and 516 grams.

a. Compute the variance and the standard deviation.
b. Compare the variation in the weights of the peach cakes and of the blueberry cakes.

Chapter Exercises

3. The amounts of fuel oil consumed for a sample of homes in Exercise 1 are repeated below (amounts shown are in gallons).
Newer homes: 600, 590, 605, 600, 603, 610, 597, and 595.
Older homes: 810, 750, 790, 800, and 850.
   a. What are the variance and the standard deviation for the newer homes?
   b. What are the variance and the standard deviation for the older homes?
   c. Compare the dispersion in the consumption of fuel oil by newer and older homes.

Self-Review 4-3

a. $\bar{X} = 3{,}000/6 = 500$ grams

X	$X - \bar{X}$	$(X - \bar{X})^2$
484	−16	256
503	+3	9
496	−4	16
510	+10	100
491	−9	81
516	+16	256
3,000	0	718

$$s^2 = \frac{718}{6 - 1} = 143.6$$

The variance is 143.6.

$$s = \sqrt{143.6} = 11.98$$

The standard deviation is 11.98 grams.

b. There is greater variation in blueberry cakes because variance of 11.98 > 4.5.

4. The lengths of trout fingerlings stocked in two ponds are repeated here from Exercise 2. Consider them as sample data.

Lengths in inches	
Pond A	Pond B
12.5	18.0
14.0	20.0
13.5	12.0
14.5	14.5
15.0	19.0
16.5	13.5
16.0	14.5
12.0	13.0
14.0	20.5
14.0	14.0

a. Determine the variance for both sets.
b. Determine the standard deviation for both sets.
c. Write an analysis of your findings with respect to the variation in the lengths of the fish in the two ponds.

Unless the mean is an integer—which often is not the case—subtracting each observation from the mean and then squaring the differences to arrive at the variance tends to be rather laborious. Rounding errors can also become a problem. A more convenient computational formula for both variance and standard deviation is based directly on the raw data. The variance formula is

Population   Sample

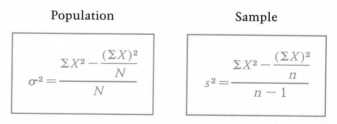

$$\sigma^2 = \frac{\Sigma X^2 - \dfrac{(\Sigma X)^2}{N}}{N}$$

$$s^2 = \frac{\Sigma X^2 - \dfrac{(\Sigma X)^2}{n}}{n-1}$$

The standard deviation (σ for a population, s for a sample) is still derived, as before, by computing the square root of the variance.

Note: ΣX^2 directs one to first square each observation and then sum the squares. This will not yield the same result as $(\Sigma X)^2$, which directs one to sum the numbers first and then square their total.

Problem

The number of days the patients currently in St. Luke Hospital's North Wing have been confined are 6, 4, 5, 3, and 4 days. Compute the standard deviation of the lengths of confinement using both methods of calculation described.

Solution

First, note that the five observations are a population, because they include all patients currently in the North Wing. The formulas for populations will therefore be used.

<table>
<tr><th colspan="3">**Method Using
Deviations from the Mean**
($\mu = 22/5 = 4.4$)</th><th colspan="2">**Method Using
Raw Data
Only**</th></tr>
<tr><th>X</th><th>$X - \mu$</th><th>$(X - \mu)^2$</th><th>X</th><th>X^2</th></tr>
<tr><td>6</td><td>1.6</td><td>2.56</td><td>6</td><td>36</td></tr>
<tr><td>4</td><td>-0.4</td><td>0.16</td><td>4</td><td>16</td></tr>
<tr><td>5</td><td>0.6</td><td>0.36</td><td>5</td><td>25</td></tr>
<tr><td>3</td><td>-1.4</td><td>1.96</td><td>3</td><td>9</td></tr>
<tr><td>4</td><td>-0.4</td><td>0.16</td><td>4</td><td>16</td></tr>
<tr><td>22</td><td>0.0</td><td>5.20</td><td>22</td><td>102</td></tr>
</table>

$$\sigma^2 = \frac{\Sigma(X - \mu)^2}{N}$$

$$= \frac{5.20}{5}$$

$$= 1.04$$

$$\sigma = \sqrt{1.04} = 1.0198 \text{ days}$$

$$\sigma^2 = \frac{\Sigma X^2 - \frac{(\Sigma X)^2}{N}}{N}$$

$$= \frac{102 - \frac{(22)^2}{5}}{5}$$

$$= \frac{102 - 96.8}{5} = 1.04$$

$$\sigma = \sqrt{1.04} = 1.0198 \text{ days}$$

Self-Review 4-4

Six families live on Merrimac Circle. The number of children in each family is 1, 2, 3, 5, 3, and 4.

a. Is this a sample or population?

b. Use the appropriate formula to compute the variance and the standard deviation.

a. Since there are only 6 families and all are being studied, this is a population.

b.

X	X^2
1	1
2	4
3	9
5	25
3	9
4	16
18	64

$$\sigma^2 = \frac{64 - \dfrac{(18)^2}{6}}{6} = \frac{64 - 54}{6}$$

$$= 1.6667$$

$$\sigma = \sqrt{1.6667} = 1.29 \text{ children}$$

Chapter Exercises

5. A study was made of the length of time it takes ambulances to travel from the scene of the accident to the hospital when responding to emergency calls in Zone A. The times in minutes were 6.1, 5.9, 4.8, 10.2, 9.6, and 6.1.
   a. Using the raw data only (not the deviations from the mean), determine the variance. Assume the data represent a population.
   b. Find the standard deviation.

6. In Exercise 4 you computed the variance and the standard deviation of the fingerling trout data using deviations from the mean. Verify your calculations using the raw data only. Consider these to be sample data.

Because the variance and the standard deviation use every observation in their computation, they are more reliable measures of spread than the range. Two or more variances (or standard deviations) can be used to compare the variability around their respective means. If the observations are clustered close to the mean, the standard deviation will be small. As the observations become more dispersed from the mean, the variance and standard deviation become larger.

It should be noted that if the raw data have a few extreme values, the variance and standard deviation may not be representative of the dispersion in the data. In such cases the deviations from the mean will be large, and squaring these deviations will result in an unusually large variance and standard deviation.

The standard deviation is especially valuable in sampling theory and statistical inference, topics that will be discussed starting with Chapter 8. The variance will be used extensively in Chapter 12.

Measures of Dispersion for Data in a Frequency Distribution

Range

The **range** for data in a frequency distribution may be determined either by

Method 1. Finding the difference between the midpoints of the highest class and the lowest class, or

Method 2. Finding the difference between the highest value and the lowest value

Note that if there are any open-ended classes in the frequency distribution, the range cannot be computed.

Problem

The weekly amounts spent on food by the sample of young newlyweds (from Chapter 3) is reintroduced in Table 4-3. What is the range for the amounts spent on food?

Table 4-3
Weekly Amounts Spent on Food by a Group of Newlyweds

Amount Spent	Number
$40 - $44	4
45 - 49	11
50 - 54	20
55 - 59	31
60 - 64	19
65 - 69	11
70 - 74	4

Solution

Method 1. The midpoint of the highest class is $72; the midpoint of the lowest class is $42. The range is $30, found by subtracting $42 from $72.

Method 2. The highest value is $74; the lowest is $40. The range is $34, found by subtracting $40 from $74.

Self-Review 4-5

A fast-food chain received several complaints with respect to the weight of their new hamburger, "The 125-Gramburger." A check of a sample of hamburgers revealed these weights:

Weight (in grams)	Number
116 - 118	7
119 - 121	19
122 - 124	28
125 - 127	16
128 - 130	5

Determine the range for the weights using both methods.

Method 1, using midpoints. Range is 12 grams, found by 129 − 117.

Method 2, using high and low values. Range is 14 grams, found by 130 − 116.

Interquartile Range and the Quartile Deviation

Recall from a previous section that the **interquartile range** and the **quartile deviation** are based on the distance between the third quartile (Q_3) and the first quartile (Q_1).

$$\text{Interquartile range} = Q_3 - Q_1$$

$$\text{Quartile deviation} = \frac{Q_3 - Q_1}{2}$$

For a symmetric distribution, the mean plus and minus the quartile deviation includes the middle 50% of the observations. Like other measures of dispersion, it can be employed to compare the dispersion of several sets of data.

Problem

Table 4-4 shows an expanded version of the familiar newlywed food budget. The true limits and the cumulative frequencies will be used shortly. What are the interquartile range and the quartile deviation?

Table 4-4

Weekly Amounts Spent on Food by a Group of Newlyweds

	Stated Limits	True Class Limits	Number	Cumulative Number
	$40 - $44	$39.50 - $44.50	4	4
	45 - 49	44.50 - 49.50	11	15
first quartile class →	50 - 54	49.50 - 54.50	20	35
	55 - 59	54.50 - 59.50	31	66
	60 - 64	59.50 - 64.50	19	85
	65 - 69	64.50 - 69.50	11	96
	70 - 74	69.50 - 74.50	4	100
			100	

Solution

When the data have been grouped into a frequency distribution, the computations for the first and third quartiles are similar to those for the median (which is the second quartile, Q_2), as discussed in

Chapter 3. The procedure for finding the first quartile, Q_1 is as follows:

Step 1. Determine how many observations are in the first quarter. Since there are 100 newlyweds, one fourth, or 25, fall in the first quarter.

Step 2. Locate the class in which the first quartile lies. Note that in the cumulative number column there are 15 weekly food amounts in the first two classes and 35 amounts in the first three classes. Clearly, the first quartile is in the third class, because 25 is larger than 15 but smaller than 35.

Step 3. Determine the distance one must move into the $49.50-$54.50 class to locate the first quartile. Notice that we are using the true limits. Q_1 is 10 observations into the $49.50-$54.50 class, which contains 20 observations. So the first quartile is 10/20 of the way between $49.50 and $54.50—or, $52.00:

$$Q_1 = \$49.50 + \frac{10}{20} \text{ of } \$5$$

$$= \$52.00$$

Shown in a diagram:

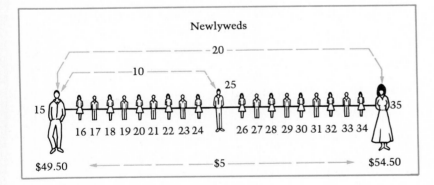

The first quartile may be found using the formula

$$Q_1 = L + \frac{\frac{n}{4} - CF}{f} \quad (i)$$

where

L is the lower true limit of the class containing the first quartile. Q_1 is in the \$49.50-\$54.50 class and \$49.50 is the lower true limit of that class.

n is the total number of frequencies. There are 100 newlywed couples.

CF is the cumulative number of frequencies in all of the classes preceding the class in which the first quartile lies. There are 15 cumulative frequencies in the class preceding the \$49.50-\$54.50 class.

f is the frequency in the first quartile class. There are 20 newlywed couples in that class.

i is the width of the class in which the first quartile falls. The width of the \$49.50-\$54.50 class is \$5.

Solving for Q_1:

$$Q_1 = L + \frac{\dfrac{n}{4} - CF}{f} \ (i)$$

$$= \$49.50 + \frac{\dfrac{100}{4} - 15}{20} \ (\$5)$$

$$= \$52.00 \text{ (same answer as before)}$$

The formula for the *third quartile, Q_3,* is

$$Q_3 = L + \frac{\dfrac{3n}{4} - CF}{f} \ (i)$$

The procedure is the same as for the first quartile, except that L, CF, f, and i refer to the values needed for the third quartile.

Problem
Compute Q_3 for the weekly amounts spent on food by newlyweds. Use the data in Table 4-4.

Solution

The third quartile class, or the 75th person, is located in the $60-$64 class. The true limits of the class are $59.50-$64.50, and this class contains 19 (f) observations. A total of 66 observations (CF) have accrued prior to the class containing the third quartile. Hence, the 75th observation is 9 observations into this class, found by $75 - 66$. The interval is $5 ($i$). The true lower limit of this class is $59.50 ($L$). Substituting these values for the symbols in the formula, we obtain

$$Q_3 = L + \frac{\frac{3}{4}n - CF}{f}\,(i)$$

$$= \$59.5 + \frac{75 - 66}{19}\,(\$5)$$

$$= \$59.5 + \$2.37$$

$$= \$61.87$$

Thus, the third quartile is $61.87.

In Self-Review 4-5 a fast-food chain had received complaints about the weight of "The 125-Gramburger." A check of 75 hamburgers had revealed these weights:

Weight (in grams)	Number
116 - 118	7
119 - 121	19
122 - 124	28
125 - 127	16
128 - 130	5

a. Determine the first quartile (Q_1).
b. Find the third quartile (Q_3).
c. What are the interquartile range and quartile deviation (QD)?

Chapter Exercises

7. The following is the distribution of per capita state taxes for our 50 states during a recent year:

Self-Review 4-6

Cumulative Frequencies

7
26
54
70
75

a. $Q_1 = 118.5 + \dfrac{\frac{1}{4}(75) - 7}{19}\,(3)$

$= 120.4$ grams

b. $Q_3 = 124.5 + \dfrac{\frac{3}{4}(75) - 54}{16}\,(3)$

$= 124.9$ grams

c. $Q_3 - Q_1 = 4.5$ grams

$QD = \dfrac{4.5}{2} = 2.25$ grams

Per Capita Tax	Frequency
$375 - $449	6
450 - 524	15
525 - 599	10
600 - 674	6
675 - 749	9
750 - 824	4
	50

a. Determine the first quartile. Explain what the first quartile indicates.
b. Determine the third quartile. Explain what the third quartile indicates.
c. What is the interquartile range?
d. What is the quartile deviation?

8. A study was made by a hospital administrator to compare the number of emergency admissions on a given Monday with those on the Friday of the same week:

	Number	
Emergency Admissions	Monday	Friday
4 - 7	1	1
8 - 11	4	4
12 - 15	15	21
16 - 19	26	22
20 - 23	16	13
24 - 27	7	3
28 - 31	3	0
	72	64

a. Calculate the quartile deviation for each set of data.
b. Compare the dispersion in the Monday and Friday admissions.

The Standard Deviation

If the data are grouped into a frequency distribution, the standard deviation for both samples and populations may be computed as follows:

Population   Sample

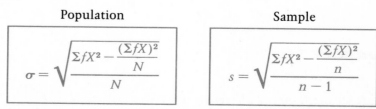

$$\sigma = \sqrt{\dfrac{\Sigma fX^2 - \dfrac{(\Sigma fX)^2}{N}}{N}}$$

$$s = \sqrt{\dfrac{\Sigma fX^2 - \dfrac{(\Sigma fX)^2}{n}}{n-1}}$$

where X represents the midpoint of each class and f the number of observations in each class. As usual, n is the sample size and N the population size.

Problem
Compute the standard deviation for the distribution of weekly amounts spent on food by the sample of young newlyweds.

Solution
The calculations essential for computing the standard deviation are shown in Table 4-5. Computing the standard deviation of the sample:

$$s = \sqrt{\frac{\Sigma fX^2 - \dfrac{(\Sigma fX)^2}{n}}{n-1}}$$

$$= \sqrt{\frac{329,305 - \dfrac{(5,695)^2}{100}}{100-1}}$$

$$= \sqrt{50.25}$$

$$= \$7.0887, \text{ or } \$7.09 \text{ rounded}$$

The variance, s^2, is 50.25, found by $(\$7.0887)^2$.

Table 4-5
Calculations Needed for the Standard Deviation

Weekly Amounts Spent on Food	Frequency f	Class Midpoint X	Frequency × Class Midpoint fX	$fX \cdot X$ or fX^2	
$40 - $44	4	$42	$ 168	7,056	$42 × $168
45 - 49	11	47	517	24,299	
50 - 54	20	52	1,040	54,080	$52 × $1,040
55 - 59	31	57	1,767	100,719	
60 - 64	19	62	1,178	73,036	
65 - 69	11	67	737	49,379	
70 - 74	4	72	288	20,736	
	100		$5,695	329,305	

In Chapter 3 we described the "coded method" for computing the mean. This method eased the computational burden for the mean by coding the midpoints of the various classes. Recall that this coding consisted of assigning the value of 0 to one of the class midpoints near the center of the frequency distribution. Class midpoints below the 0 class were coded -1, -2, and so on, whereas those above were coded 1, 2, 3, and so on. This method can also be employed to compute the standard deviation. The formulas are

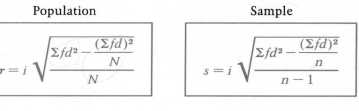

Population

$$\sigma = i \sqrt{\dfrac{\Sigma fd^2 - \dfrac{(\Sigma fd)^2}{N}}{N}}$$

Sample

$$s = i \sqrt{\dfrac{\Sigma fd^2 - \dfrac{(\Sigma fd)^2}{n}}{n-1}}$$

where

Σfd^2 is the sum of the class frequency (f) times the deviation squared (d^2).

Σfd is the sum of the products of frequency (f) and the deviation (d).

N is the population size.

n is the sample size.

i is the class interval.

Problem

Compute the standard deviation for the distribution of the weekly amounts spent on food by the sample of young newlyweds.

Solution

The essential calculations are shown in Table 4-6. Computing the standard deviation:

$$s = i \sqrt{\dfrac{\Sigma fd^2 - \dfrac{(\Sigma fd)^2}{n}}{n-1}}$$

$$= (\$5) \sqrt{\dfrac{199 - \dfrac{(-1)^2}{100}}{100 - 1}}$$

$$= \$5\sqrt{2.01}$$

$$= \$7.0887, \text{ or } \$7.09 \text{ rounded}$$

As expected, this result is exactly the same as that obtained earlier without coding.

The hourly wages, including tips, of a sample of waiters and waitresses are

Hourly Wages	Number
$ 1 - $ 3	4
4 - 6	10
7 - 9	20
10 - 12	11
13 - 15	5

a. Determine the standard deviation using both the long method and the coded method.
b. Compute the variance.

a. Long method:

f	X	fX	fX²
4	$ 2	$ 8	16
10	5	50	250
20	8	160	1,280
11	11	121	1,331
5	14	70	980
50		$409	$3,857

$$s = \sqrt{\frac{3857 - \frac{(409)^2}{50}}{50 - 1}}$$

$$s = \sqrt{\frac{511.38}{49}} = \sqrt{10.4363}$$

$$s = \$3.23$$

Coded method:

f	d	fd	fd²
4	−2	−8	16
10	−1	−10	10
20	0	0	0
11	1	11	11
5	2	10	20
50		3	57

$$s = \$3 \sqrt{\frac{57 - \frac{(3)^2}{50}}{50 - 1}}$$

$$s = \$3\sqrt{1.1596} = \$3.23$$

b. The variance is 10.43, found by ($3.23)².

Table 4-6

Calculations Needed for the Standard Deviation

Weekly Amounts Spent on Food	Frequency f	Coded Midpoint d	fd	fd²
$40 - $44	4	−3	−12	36
45 - 49	11	−2	−22	44
50 - 54	20	−1	−20	20
55 - 59	31	0	0	0
60 - 64	19	1	19	19
65 - 69	11	2	22	44
70 - 74	4	3	12	36
	100		−1	199

$$S = (5)\sqrt{\frac{199 - \frac{(-1)^2}{100}}{N = 100 - 1}}$$

Chapter Exercises

9. The per capita tax for the 50 states is repeated from Exercise 7:

Per Capita Tax	Frequency
$375 - $449	6
450 - 524	15
525 - 599	10
600 - 674	6
675 - 749	9
750 - 824	4
	50

a. Is this a sample or a population?
b. Compute the standard deviation using both methods.
c. Determine the variance.

10. The number of Monday and Friday emergency hospital admissions is repeated from Exercise 8:

Emergency Admissions		Number Monday	Friday
4 - 7	5.5	1	1
8 - 11	9.5	4	4
12 - 15	13.5	15	21
16 - 19	17.5	26	22
20 - 23	21.5	16	13
24 - 27	25.5	7	3
28 - 31	29.5	3	0
		72	64

a. Calculate both the standard deviation and the variance for each set of data.   For mon + Fri
b. Compare the dispersion in the Monday and Friday emergency admissions.

Relative Dispersion

A *direct* comparison of two or more measures of dispersion may result in an incorrect conclusion regarding the variation in the data. Also, if one distribution, for example, were in dollars and the other in meters, clearly it would be impossible to compare them. Therefore, it is preferable to use a *relative* measure of dispersion when either:

1. The means of the distributions being compared are far apart, or
2. The data are in different units.

One such relative measure of dispersion is known as the **coefficient of variation,** abbreviated *CV*.

✗ **Coefficient of Variation**   The standard deviation divided by the mean.

In order to express the coefficient of variation as a percent, $\frac{s}{\overline{X}}$ (for a sample) is multiplied by 100.

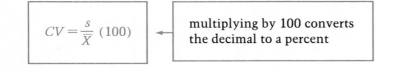

$$CV = \frac{s}{\overline{X}} \, (100)$$ ◄──   multiplying by 100 converts the decimal to a percent

Problem

The mean income of a sample of homeowners in Precinct 12 is $40,000, and the standard deviation is $4,000. In Precinct 9, the mean income of a sample of homeowners is $12,000, and the standard deviation is $1,200. Note that the means are far apart. Compare and interpret the relative dispersion in the two groups of incomes.

Solution

The first impulse is to say that there is more dispersion in the incomes in Precinct 12, because the standard deviation associated with that group ($4,000) is greater than the $1,200 for the Precinct 9 incomes. However, converting the two sets of measurement to relative terms, and using the coefficient of variation, the conclusion is that the relative dispersion is the same! Here are the calculations for the coefficients of variation:

Precinct 12

$$CV = \frac{s}{\overline{X}} \, (100) = \frac{\$4,000}{\$40,000} \, (100) = 10\%$$

Precinct 9

$$CV = \frac{s}{\overline{X}} \, (100) = \frac{\$1,200}{\$12,000} \, (100) = 10\%$$

Interpreting the results: the incomes in both precincts are spread out 10% from their respective means.

Problem

In order to illustrate the use of the coefficient of variation when two or more distributions are in different units, we will now compare the dispersion in the ages of the homeowners in Precinct 12 with the dispersion in their incomes. The coefficient of variation allows each unlike set of data to be converted to a common denominator (a percent).

The mean age of the sample of homeowners is 40 years, the standard deviation is 10 years. Recall that for their incomes $\overline{X} =$ $40,000, and the standard deviation is $4,000. Compare the dispersion in their ages and incomes.

Solution

$$\text{Income: } CV = \frac{s}{\overline{X}} (100) = \frac{\$4,000}{\$40,000} (100) = 10\%$$

$$\text{Age: } \quad CV = \frac{s}{\overline{X}} (100) = \quad \frac{10}{40} (100) \quad = 25\%$$

There is greater relative dispersion in the ages of the homeowners in Precinct 12 than in their incomes (because 25% is greater than 10%).

Self-Review 4-8

Mechanical: $\frac{30}{200} (100) = 15\%$

Social work: $\frac{40}{500} (100) = 8\%$

Larger relative dispersion in mechanical aptitude (15%) compared with social work aptitude (8%).

A sample of Harber High School seniors have recently completed two aptitude tests. One dealt with mechanical aptitude, the other with aptitude for social work. The results are

a. Mechanical aptitude: mean 200, standard deviation 30.
b. Aptitude for social work: mean 500, standard deviation 40.

Compare the relative dispersion in the test results.

Chapter Exercises

11. An investor is considering the purchase of one of two stocks. The yield of Venture Electronics has averaged $105 per share over the past 10 years with a standard deviation of $15 per share. Aerospace Ltd. has yielded an average of $330 per share during the same period with a standard deviation of $40. Compare the relative dispersion of the two stocks.

12. A recent study of Ohio College of Business faculty revealed that the arithmetic mean salary for nine months is $21,000 and the standard deviation of the sample is $3,000. The study also showed that the faculty had been employed an average (arithmetic mean) of 15 years with a standard deviation of 4 years. How does the relative dispersion in the distribution of salaries compare with that of the lengths of service?

The Coefficient of Skewness

An average describes the central tendency of a set of observations, while a measure of dispersion describes the variation in the data. The degree of **skewness** in a distribution can also be measured. Recall from Chapter 3 that if the mean, median, and mode are equal, there is no skewness (see Figure 4-3). If the mean is larger than the median and the mode, the distribution is said to have **positive skewness** (see Figure 4-4). If the mean is the smallest of the three averages, the distribution is **negatively skewed** (see Figure 4-5).

The degree of skewness is measured by the **coefficient of skewness,** abbreviated *sk*.

$$sk = \frac{3(\text{mean} - \text{median})}{\text{standard deviation}}$$

Usually its value ranges from −3 to +3.

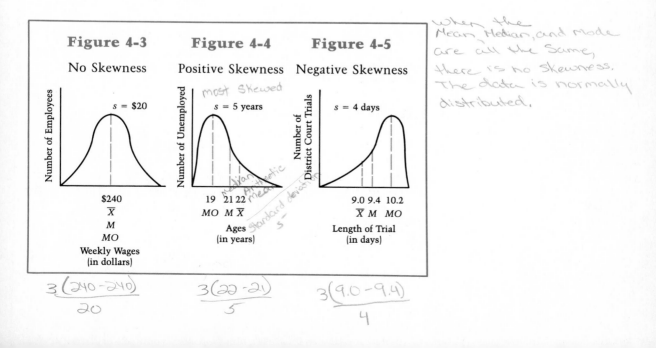

Figure 4-3 — No Skewness

Figure 4-4 — Positive Skewness

Figure 4-5 — Negative Skewness

(handwritten notes) When the Mean, Median, and Mode are all the same, there is no skewness. The data is normally distributed.

(handwritten) most skewed

(handwritten calculations) $\frac{3(240-240)}{20}$ $\frac{3(22-21)}{5}$ $\frac{3(9.0-9.4)}{4}$

Problem

What is the coefficient of skewness for the distribution of ages in Figure 4-4?

Solution

$$sk = \frac{3(\text{mean} - \text{median})}{\text{standard deviation}}$$

$$= \frac{3(22 - 21)}{5}$$

$$= +0.6$$

This indicates that there is a slight positive skewness in the age distribution.

Self-Review 4-9

a. −0.3, found by

$$\frac{3(9.0 - 9.4)}{4} = -0.3$$

b. 0, found by

$$\frac{3(\$240 - \$240)}{\$20} = 0$$

c. Figure 4-5: slight negative skewness. Figure 4-3: no skewness.

a. Determine the coefficient of skewness for Figure 4-5.
b. Determine the coefficient of skewness for Figure 4-3.
c. Interpret the two coefficients.

Chapter Exercises

13. The Flightdeck, a restaurant overlooking the Venice Airport, computed the arithmetic-mean dinner check for two persons to be $54.00, the median $50.50. The standard deviation was $3.75. What is the coefficient of skewness?

14. The research director of a large oil company conducted a study of buying consumer habits with respect to the amount of gasoline purchased at full-service pumps. The arithmetic-mean amount was 11.5 gallons, and the median amount was 11.95 gallons. The standard deviation of the sample was 4.5 gallons. Is this distribution negatively skewed, positively skewed, or symmetrical?

A Computer Example

The work of tallying raw data into a frequency distribution (Chapter 2), computing measures of central tendency (Chapter 3), and measures of dispersion (this chapter), can be accomplished quickly by a computer. A number of computer "packages" (special-purpose sets of programs) are used extensively in the social sciences, business, education, and other areas. They require very little knowledge of computer programming. Illustrations from two such packages will be used in this text: the Statistical Package for the Social Sciences (SPSS), and MINITAB. Check with your computing center to determine what statistical packages are available for your use. Computer packages are generally easy to use and will save you all the tedious manual calculations.

SPSS was used to generate the following output for the weekly amounts spent on food by the group of young newlyweds.

```
NEWLYWED FOOD STUDY

FILE    LIND


XAMT
```

CATEGORY LABEL	CODE	ABSOLUTE FREQ	RELATIVE FREQ (PCT)	ADJUSTED FREQ (PCT)	CUM FREQ (PCT)
40 - 44	42.	4	4.0	4.0	4.0
45 - 49	47.	11	11.0	11.0	15.0
50 - 54	52.	20	20.0	20.0	35.0
55 - 59	57.	31	31.0	31.0	66.0
60 - 69	62.	19	19.0	19.0	85.0
65 - 69	67.	11	11.0	11.0	96.0
70 - 74	72.	4	4.0	4.0	100.0
TOTAL		100	100.0	100.0	

```
MEAN         56.950    MEDIAN       56.919    STD DEV        7.089
VARIANCE     50.250    MINIMUM      42.000    MAXIMUM       72.000
VALID CASES     100    MISSING CASES      0
```

An average pinpoints the central tendency of a set of data. To describe the variability in the data the range, interquartile range, quartile deviation, mean deviation, variance, or standard deviation may be used.

The range is the difference between the highest and lowest values. Caution should be exercised in applying the range, because one extremely high (or low) observation may cause it to be an unreliable measure of spread. The interquartile range and the quartile

Summary

deviation are based on the third quartile and the first quartile. One of the weaknesses of these two measures, and of the range, is that they do not include all the observations. By contrast, the mean deviation, the variance, and the standard deviation all use every available value to measure the scatter, or dispersion, about the mean. In its computation, the mean deviation ignores the algebraic signs of the deviation about the mean. Both the variance and the standard deviation use the square of each difference in their computation. The standard deviation is especially useful in statistical inference, to be discussed in the forthcoming chapters.

A direct comparison of the dispersion in distributions having means that are either far apart, or that are measured in different units, may be misleading. To correct for this, the standard deviations are divided by their respective means, resulting in a relative measure of dispersion called the coefficient of variation. Since its outcome is in percents, even dissimilar dispersions can be compared to assess the variation.

The coefficient of skewness measures the degree to which the distribution is not symmetrical. A coefficient of zero indicates that the frequency distribution is symmetrical—that is, it is neither negatively nor positively skewed. The coefficient usually varies between −3 (negative skewness) and +3 (positive skewness).

Chapter Outline

Descriptive Statistics:
Measures of Dispersion and Skewness

I. Measures of Dispersion
Objective. To compare the reliability of measures of central value and/or to describe the spread in the data.

A. **Range**
  1. Defined as the difference between the highest and lowest values.
  2. Computed for raw data by subtracting the lowest value from the highest value. For a frequency distribution: either subtract the lower limit of the smallest class from the upper limit of the largest class, or the midpoint of the smallest class from the midpoint of the largest class.

B. **Interquartile range and quartile deviation**
  1. Interquartile range is defined as the distance between the first and third quartiles. Quartile deviation is half that distance.
  2. Computations. Interquartile range $= Q_3 - Q_1$; and quartile deviation:

$$\frac{Q_3 - Q_1}{2}$$

where

$$Q_3 = L + \frac{\frac{3n}{4} - CF}{f} \quad (i) \quad \text{and} \quad Q_1 = L + \frac{\frac{n}{4} - CF}{f} \quad (i)$$

C. **Mean deviation**
 1. Defined as the mean of the absolute deviations from the mean.
 2. Computed by:

$$MD = \frac{\Sigma |X - \overline{X}|}{N}$$

D. **Variance and standard deviation**
 1. Definitions. Variance is the arithmetic mean of the squared deviations from the mean. Standard deviation is the square root of the variance.
 2. Computations.
 For raw data:

Population	Sample

$$\sigma^2 = \frac{\Sigma X^2 - \frac{(\Sigma X)^2}{N}}{N} \qquad s^2 = \frac{\Sigma X^2 - \frac{(\Sigma X)^2}{n}}{n-1}$$

 For grouped data:

$$\sigma^2 = \frac{\Sigma fX^2 - \frac{(\Sigma fX)^2}{N}}{N} \qquad s^2 = \frac{\Sigma fX^2 - \frac{(\Sigma fX)^2}{n}}{n-1}$$

 Standard deviation: take square root of variance.

II. **Relative Measure of Dispersion**
 Objective. To compare the dispersion in distributions having means that are either far apart, or that are measured in different units.
 A. **Coefficient of variation**
  1. Defined as the ratio of the standard deviation to the mean.
  2. Computed by

$$CV = \frac{s}{\overline{X}} (100)$$

III. Measure of Skewness
A. Coefficient of skewness
1. Objective. To describe departure from symmetry.
2. Computed by

$$sk = \frac{3 \ (\text{mean} - \text{median})}{\text{standard deviation}}$$

Its value usually ranges from -3 to $+3$. A value of 0 would indicate no skewness. That is, such a distribution is symmetrical.

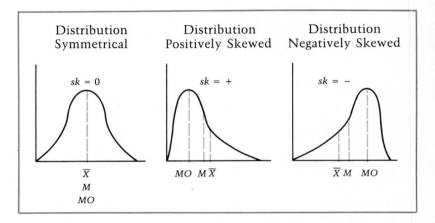

Chapter Exercises

15. The ages for a sample of ten women who gave birth to their first child are 18, 25, 31, 19, 22, 21, 19, 25, 16, and 27. Compute each of the following descriptive measures:
a. the range
b. the average deviation
c. the variance
d. the standard deviation
e. the coefficient of skewness

16. The number of yards gained rushing by the Rams football team during the first seven games of the season are 210, 203, 162, 134, 390, 184, and 211. These data constitute a population. Compute the following descriptive measures:
a. the range
b. the average deviation
c. the variance
d. the standard deviation
e. the coefficient of skewness

17. The percent of minorities enrolled in 12 New Orleans high schools are 32, 62, 58, 26, 61, 34, 46, 27, 56, 61, 64, and 24. Consider the data to be a population, and find the following descriptive measures:
a. the range
b. the average deviation
c. the variance
d. the standard deviation
e. the coefficient of skewness

18. The distribution of the number of children under 18 years of age in a sample of 40 families selected at random is shown.

Number of Siblings	Frequency
0	5
1	11
2	9
3	5
4	5
5	0
6	4
7	1

Find the following descriptive measures:
a. the range
b. the average deviation
c. the variance
d. the standard deviation
e. the quartile deviation
f. the coefficient of skewness

19. The United States was divided into 25 statistical regions, and the percent of households in which a female is the head of the household was determined for each region. These data were organized into the following frequency distribution:

Percent of Households	Frequency
5 - 9	5
10 - 14	7
15 - 19	9
20 - 24	3
25 - 29	1

Consider these data a population, and determine the following descriptive measures:
a. the range

b. the average deviation
c. the variance
d. the standard deviation
e. the quartile deviation
f. the coefficient of skewness

20. Two hundred physically disabled veterans are enrolled in a physical fitness program. As an initial test of strength, everyone has been asked to lift a weight. The maximum weight each person lifted has been recorded and organized into a frequency distribution.

Weight (in pounds)	Number of Persons
0 - 4	10
5 - 9	37
10 - 14	50
15 - 19	68
20 - 24	30
25 - 29	5

a. Determine the range.
b. Find the interquartile range and the quartile deviation.
c. Find the standard deviation and the variance.

The veterans in this group have also assembled an electronic panel "by the numbers," to test their mechanical reaction time. The mean length of time required to do the assembly was 6.0 minutes, the standard deviation 0.8 minutes. The median time was computed to be 5.6 minutes.

d. Compare the relative dispersion in the distribution of weights with that of the assembly time distribution.
e. Analyze the skewness in the two distributions.

21. According to the *U.S. News & World Report* cited in Chapter 3, the pay for 1,052 top executives in 396 of the largest U.S. corporations is as follows:

12 earned less than $100,000
213 earned $100,000 to $200,000
300 earned $200,000 to $300,000
241 earned $300,000 to $400,000
135 earned $400,000 to $500,000
76 earned $500,000 to $600,000
34 earned $600,000 to $700,000
41 earned $700,000 to $1,600,000

a. What is the range?

b. What is the quartile deviation?

c. What characteristic of the income distribution makes it difficult to compute the variance, the standard deviation, and the coefficient of skewness?

22. The number of players in the American League and the National League whose batting averages are between 150 and 374 are listed in the following table. Consider the data to be a population.

Average	American League	National League
350 - 374	2	0
325 - 349	10	5
300 - 324	17	9
275 - 299	28	25
250 - 274	36	27
225 - 249	22	17
200 - 224	12	10
175 - 199	1	4
150 - 174	4	0

a. Calculate the range, quartile deviation, mean deviation, variance and standard deviation for each league.

b. Compare the amount of dispersion in the two distributions.

Do all the problems. Then check your answers against those given in the Answer section of the book.

Chapter Achievement Test

I. True-False Questions. Indicate by circling the correct letter whether the statement is true or false. If false, provide the correct answer (3 points each).

1. T F The interquartile range is defined as half the distance between the third and first quartiles.

2. T F The mean deviation is based on the absolute difference between each observation and its mean.

3. T F The standard deviation is the square root of the variance.

4. T F If the frequency distribution is open-ended, the variance cannot be computed.

5. T F The range for a set of raw data is determined by finding the difference between the highest value and the lowest value.

6. T F If the mean of a frequency distribution is smaller than the median or mode, the distribution is positively skewed.

7. T F The coefficient of variation is expressed in terms of a percent.

8. T F The standard deviation of a frequency distribution is $10, $\overline{X}$ = $250, median = $250, mode = $250. This indicates that the distribution has a slightly positive skewness.

9. T F Distribution A has a mean of $60 and a standard deviation of $3. Distribution B has a mean of $150 and a standard deviation of $4.50. There is more relative dispersion in Distribution B.

10. T F Generally speaking, if the means of two distributions are about the same, the distribution with the smaller standard deviation does not have as great a spread as the other distribution.

Use the following statistics to answer questions 11 - 16. Statistics on the weights (in pounds) of three groups are

Statistic	Group 1	Group 2	Group 3
Mean	159	151	152
Median	156	154	152
Mode	150	160	152
Standard deviation	11	11	10
Mean deviation	8	9	8
Quartile deviation	7	8	7
Range	64	71	60

11. T F The distribution of the weights for Group 2 is negatively skewed.

12. T F The distance between the two extreme weights is somewhat greater for Group 2 than for Group 1.

13. T F For Group 1, half of the weights are below 156 pounds.

14. T F For Group 2, the weight that appears most frequently is 160 pounds.

15. T F For Group 3, the distribution is symmetrical.

16. T F For Group 3, the first quartile is 145 and the third quartile is 159.

II. Computation Problems. Use the following raw data for questions 17 - 20 (3 points each).

Several joggers ran these distances during the day:

Jogger	Distance (in miles)
Peter	2
Jan	6
Betty	3
Doug	4
Bill	5

17. What is the range?
18. What is the mean deviation?
19. What is the variance?
20. What is the standard deviation?

Use the following distribution for questions 21 - 25 (8 points each).

The appraised values for a sample of single-family dwellings in the Jefferson tax district are

Appraised Value (in thousands)	Number of Dwellings
$20 - $29	8
30 - 39	15
40 - 49	20
50 - 59	50
60 - 69	18
70 - 79	9

21. Determine the range.
22. Determine the quartile deviation.
23. Determine the variance and the standard deviation.

The distribution of the appraised values of the single-family dwellings in the Sanford tax district reveal: mean $87,000; median $84,000; mode $78,000; standard deviation $9,000.

24. Compare the relative dispersion in the two distributions.
25. Compare the skewness of the two distributions.

HIGHLIGHTS
From Chapters 1, 2, 3, and 4

In these first four chapters you have learned the basic vocabulary of statistics and the fundamentals of descriptive statistics. Major concepts covered are listed below, and are also defined in the Glossary at the back of the book.

Key Concepts

1. **Statistics** is the collection, organization, presentation, analysis, and interpretation of data for the purpose of making better decisions

2. Statistics may be divided into two areas, **descriptive** and **inferential.**

3. Data may be classified into four levels of measurement— **nominal, ordinal, interval,** or **ratio.**

4. A **population** is a collection or set of individuals, objects, or measurements. A **sample** is a part of the population. Calculations made from populations are **parameters,** and those made from samples are **statistics.**

5. A **frequency distribution** is a comprehensive summary of a set of observations. It separates the data into classes and shows the number of occurrences in each class.

6. **Histograms** and **frequency polygons** are graphic displays of frequency distributions.

7. A **less-than cumulative** frequency distribution shows the number of observations that are less than or equal to the upper limit of each class interval.

8. An **average** is a single representative value that is typical of the data considered as a whole. Four averages were discussed: **mean, weighted mean, median,** and **mode.**

9. When a data set contains extremely large or extremely small values, the resulting distribution may be **skewed.** A skewed distribution is one that is not **symmetrical.**

Key Terms

This list of terms is included in order for you to verify your recall of the material covered in Chapters 1 - 4. As you read each term, provide its definition in your own words. Then check your answers against the definitions given both in chapter and in the Glossary at the back of the book.

Descriptive statistics
Inferential statistics
Population
Sample
Mutually exclusive
Exhaustive
Nominal scale
Ordinal scale
Interval scale
Ratio scale
Raw data
Classes
Class frequency
True limits
Stated limits
Class midpoint
Frequency distribution

Histogram
Frequency polygon
Cumulative frequency polygon
Mean
Median
Mode
Weighted mean
Parameter
Statistic
Range
Quartile deviation
Interquartile range
Mean deviation
Variance
Standard deviation
Coefficient of variation
Coefficient of skewness

Key Symbols

$\overline{X}$ The mean of a sample.

n The number of observations in a sample.

Σ The symbol that indicates a group of values are to be added.

μ The mean of a population.

N The number of observations in a population.

$\bar{X}_w$ The weighted mean.

Q_3 The value of the third quartile.

Q_1 The value of the first quartile.

QD The quartile deviation.

MD The mean deviation or, as it is sometimes called, the average deviation.

σ^2 The variance of a population.

σ The standard deviation of a population.

s^2 The variance of a sample.

s The standard deviation of a sample.

CV The coefficient of variation.

sk The coefficient of skewness.

Review Problems

1. The following number of students failed freshman English in each section of the course offered at a midwestern university during the last fall quarter: 5, 1, 1, 4, 6, 7, 9, 7, 10, and 4.
   a. Is this example a sample or a population?
   b. Compute the mean.
   c. Compute the median.
   d. What is the mode?
   e. Compute the range.
   f. Compute the standard deviation.

2. The following are the audience ratings for the first 12 weeks of a new TV prime-time soap opera called "Houston": 12.6, 13.9, 13.3, 15.7, 12.0, 15.1, 13.7, 16.5, 17.5, 12.5, 14.7, and 15.0.
   a. Is this an example of a sample or a population?
   b. Compute the mean rating.
   c. Compute the median.
   d. What is the mode?
   e. Compute the range.
   f. Compute the standard deviation.

3. A study is being conducted to determine the number of TV sets per household. The following sample data were obtained for eight homes: 1, 3, 3, 4, 1, 0, 4, and 3.
   a. Compute the mean.
   b. Compute the median.
   c. What is the mode?
   d. Compute the standard deviation.

4. Lander Technical College is surveying students to find out how many hours per week they work. A sample of 50 students yielded the following information:

Hours Worked	Number of Students
0 - 9	5
10 - 19	9
20 - 29	15
30 - 39	10
40 - 49	9
50 - 59	2
	50

   a. Compute the mean.

   b. Compute the median.

   c. Compute the quartile deviation.

   d. Compute the variance.

5. A study made by Canard Caribbean Cruises entails recording the number of pieces of luggage checked by passengers. A sample of 24 passengers reveals the following breakdown: 1, 9, 9, 1, 10, 1, 6, 8, 9, 4, 11, 9, 4, 5, 5, 1, 8, 4, 6, 12, 4, 7, 5, and 8. Organize the data into a frequency distribution with 0 as the lower limit of the first class.

   a. Compute the mean.

   b. Compute the median.

   c. Compute the standard deviation.

   d. What is the range, using both methods?

   e. Compute the quartile deviation.

6. The Parry-Mutual Insurance Co. is studying the number of group insurance claims submitted each week during the last year. Company records have disclosed the following information (consider these data a population):

Number of Claims per Week	Weeks
0 - 2	6
3 - 5	15
6 - 8	19
9 - 11	9
12 - 14	3
	52

   a. Compute the mean.

   b. Compute the median.

   c. Compute the standard deviation.

   d. Compute the quartile deviation.

7. At the Silver Net Seafood Restaurant, which overlooks the scenic Conewango River, the mean dinner check for two people is $58.50, the median is $54.00, and the standard deviation $5.25. Compute the coefficient of variation and the coefficient of skewness.

8. The following are the weights of a sample of five Hereford Steer (in pounds) after two months in a feed lot (consider these as sample data): 968; 1,014; 1,247; 959; and 642.
   a. Compute the mean.
   b. Compute the standard deviation.
   c. What is the variance?

9. During the 1980 fall term there were 472 full-time faculty at the University of Toledo. The following table shows the number of full-time faculty in each of the six colleges. Construct a pie chart to depict these data:

College	Number of Faculty
Arts & Sciences	245
Business	45
Education	97
Engineering	45
Law	27
Pharmacy	13
Total	472

10. A student newspaper reported the attendance at the five home football games for last season was 2,520; 2,140; 29,600; 2,750; and 2,280.
   a. Calculate the mean.
   b. Calculate the median
   c. Assume that the third value was a misprint and should have been 2,960. Recalculate the mean and the median.
   d. Compare the effect of the misprint on the mean and the median.

Case Analysis

The McCoy's Market Case

The research director for McCoy's Market, a national chain of grocery stores, recently launched a broad study of their customers. Among other things, the study was to determine how much customers spend on food, when they prefer to shop (midweek or weekend), their sex, and the number of items purchased each shopping trip.

A sample of 65 customers revealed the data that follows. The second column shows the amount spent. The third column shows the code 1 for a midweek shopper and the code 2 for a weekend shopper. Next, the sex of the shopper (1 = female, 0 = male) is noted. The final column records the number of items purchased.

COLUMN	AMTSPENT	CODE	SEX	NO.ITEMS
COUNT	65	65	65	65
ROW				
1	1.940	1.	1.	9.
2	2.460	1.	1.	11.
3	2.580	2.	0.	10.
4	4.380	1.	1.	18.
5	5.540	1.	0.	9.
6	7.140	1.	0.	5.
7	7.940	1.	1.	13.
8	7.980	1.	1.	3.
9	8.700	1.	1.	17.
10	9.300	1.	1.	13.
11	11.380	1.	1.	14.
12	11.980	1.	1.	21.
13	12.760	2.	0.	13.
14	13.000	1.	1.	9.
15	14.380	1.	1.	5.
16	14.980	1.	0.	20.
17	16.240	1.	1.	16.
18	16.760	2.	0.	14.
19	17.760	2.	0.	14.
20	17.760	2.	1.	14.
21	17.800	1.	0.	16.
22	18.000	2.	1.	16.
23	19.700	1.	1.	22.
24	19.940	1.	1.	15.
25	20.960	1.	0.	7.
26	21.000	2.	1.	14.
27	23.940	1.	1.	21.
28	23.980	1.	1.	18.
29	24.980	1.	1.	15.
30	25.900	1.	1.	20.
31	25.960	2.	0.	20.
32	27.380	1.	1.	11.
33	27.760	1.	0.	31.
34	27.960	2.	0.	10.
35	29.760	2.	1.	20.
36	31.000	2.	1.	11.
37	35.960	2.	1.	32.
38	37.250	1.	1.	35.
39	39.880	2.	1.	23.
40	39.920	2.	1.	30.

COLUMN	AMTSPENT	CODE	SEX	NO.ITEMS
COUNT	65	65	65	65
ROW				
41	39.940	1.	0.	20.
42	42.500	1.	0.	21.
43	44.960	1.	0.	22.
44	45.900	1.	0.	16.
45	47.960	2.	1.	29.
46	47.960	2.	1.	29.
47	49.980	1.	0.	30.
48	53.560	2.	0.	28.
49	59.960	2.	0.	34.
50	65.200	2.	1.	44.
51	69.900	1.	0.	42.
52	71.900	1.	1.	47.
53	79.960	2.	1.	45.
54	83.380	1.	1.	32.
55	86.000	2.	1.	24.
56	95.400	2.	1.	49.
57	99.960	2.	0.	46.
58	99.960	2.	1.	51.
59	119.880	2.	1.	54.
60	123.960	2.	0.	67.
61	129.760	1.	1.	46.
62	134.580	1.	1.	42.
63	137.980	2.	1.	55.
64	161.420	2.	0.	62.
65	149.920	2.	0.	58.

In order to describe the data more concisely:

1. Construct two frequency distributions—one for the amount spent by midweek shoppers and one for the weekend shoppers.

2. Determine the means and standard deviations for both groups.

3. Find the coefficients of variation.

4. Experience indicates that shoppers who spend more than $90 tie up the checkout lanes. Draw a less-than cumulative frequency polygon for each group. Estimate the percent of customers spending more than $90.

5. Summarize your findings and make any suggestions that seem appropriate. (This case will appear again in subsequent Highlights.)

The St. Mary's Emergency Room Case

The Vice-President for Planning of St. Mary's Medical Center is concerned about the number of staff members and cost of services rendered in the hospital's emergency room and outpatient clinic. In order to gain a better understanding of the problem, she selects 50 records from the emergency-room files. She collects the following information: (a) the number of staff members who had dealings with a particular patient; (b) the age of the patient; (c) whether the patient was admitted to the hospital (code = 1), or treated and released (code = 0); (d) the shift during which the service occurred; and (e) the amount the patient was billed.

COLUMN	#OFSTAFF	PATI AGE	ADMITTED	SHIFT	COST
COUNT	50	50	50	50	50
ROW					
1	5.	60.	1.	3.	560.50
2	1.	84.	1.	1.	806.00
3	1.	19.	0.	3.	198.50
4	7.	68.	0.	3.	834.00
5	5.	82.	0.	3.	492.00
6	1.	56.	0.	2.	484.50
7	4.	85.	1.	3.	565.50
8	6.	58.	1.	2.	1017.50
9	1.	79.	1.	2.	597.00
10	2.	69.	1.	2.	806.50
11	5.	54.	1.	3.	984.00
12	2.	77.	0.	3.	144.00
13	6.	1.	0.	2.	789.50
14	3.	14.	0.	2.	557.00
15	2.	62.	0.	1.	287.00
16	7.	22.	1.	3.	1095.00
17	1.	17.	1.	2.	712.50
18	4.	64.	0.	3.	249.50
19	6.	66.	0.	3.	663.00
20	4.	29.	0.	3.	607.00
21	5.	61.	0.	2.	371.00
22	5.	71.	0.	1.	679.00
23	7.	13.	0.	2.	889.00
24	2.	84.	1.	1.	432.50
25	7.	85.	1.	2.	1056.50
26	3.	25.	1.	3.	526.50
27	3.	84.	0.	1.	381.00
28	6.	62.	0.	2.	465.00
29	6.	58.	1.	3.	709.50
30	6.	77.	0.	2.	327.50
31	4.	14.	0.	1.	365.00

COLUMN	#OFSTAFF	PATI AGE	ADMITTED	SHIFT	COST
COUNT	50	50	50	50	50
ROW					
32	2.	78.	1.	2.	575.50
33	1.	6.	1.	3.	470.50
34	1.	70.	0.	2.	143.50
35	1.	66.	1.	2.	564.00
36	5.	10.	0.	1.	420.50
37	2.	61.	0.	3.	573.00
38	5.	68.	1.	3.	824.50
39	2.	61.	0.	3.	584.00
40	2.	15.	0.	3.	177.00
41	7.	60.	0.	1.	663.50
42	2.	73.	1.	2.	614.00
43	7.	68.	1.	2.	770.50
44	6.	15.	0.	1.	531.00
45	7.	74.	0.	1.	449.00
46	6.	62.	1.	2.	753.50
47	2.	69.	1.	1.	619.50
48	6.	88.	1.	1.	1094.50
49	4.	13.	0.	3.	211.00
50	6.	79.	1.	2.	764.50

As one of her employees, you have been asked to do the following:

1. Construct a frequency distribution of the number of staff members required to service each patient. Compute the mean and standard deviation of this distribution.

2. Arrange the patients' ages into a frequency polygon and a cumulative frequency polygon. Compute the median age and the variance in the ages. Describe in words the shape of the distribution.

3. Prepare three charts showing the cost for *each* of the three shifts. Make a histogram of each. Find the mean and median for each distribution. Compute the range and variance for each distribution. Describe how the three distributions differ.

5
An Introduction to Probability

OBJECTIVES

When you have completed this chapter, you will be able to
- Define probability.
- Describe the functions of classical, relative frequency, and subjective concepts of probability.
- Calculate probabilities using the rules of addition and multiplication.
- Count the number of possible permutations and combinations.

Introduction

Chapters 2, 3, and 4 dealt with descriptive statistics. The emphasis in Chapter 2 was on organizing raw data into a frequency distribution. Recall, for example, the newlywed food budget which, when organized into a frequency distribution, revealed what the lowest and the highest amounts spent were, and in which class the largest concentration of food budgets could be found. In Chapters 3 and 4 measures of central tendency and dispersion were introduced. It was determined, for example, that the median newlywed food budget—that is, the amount spent by the 50th among the 100 newlyweds—was $56.92. The main focus in all four chapters was to describe a set of data that had already been observed.

The purpose of this chapter and of subsequent ones is to examine ways of calculating the probability that some event will occur. For instance, you may be interested in determining the probability that the incumbent governor in your state will be reelected in November, or in the probability that fewer than 80% of first-time speeding offenders will be fined a second time for this offense. Whatever your purpose, the ability to calculate probability will be a valuable tool in your everyday decision making.

Why Study Probability?

One of the major purposes of the science of statistics is to make it possible to infer, from the results of studies of just one part of a group, general statements about the entire group. As defined in Chapter 3, one portion of the entire group is called a **sample**, while the entire group is called the **population**. The process of reasoning from a set of sample observations to a general conclusion about a population is **statistical inference**. Probability is the "yardstick" used to measure the reasonableness that a particular sample result could also be valid for a certain population. At some point you may decide that the observed sample result is so improbable that you reject the claim that it applies to a specific population. Hence, in the science of statistics statements are "proved" or "disproved" by calculating how probable or improbable they are.

Probability will help us make decisions with respect to statistical inference. Throughout, we will use case studies like the two that follow to illustrate the potential use of probability in decision making.

Case 1. Senator Simeon claims that 75% of all taxpaying Americans favor tax reform. To investigate his contention, 500 taxpaying Americans were selected at random and interviewed. Of the 500 sampled, 70%, or 350, said they were in favor of tax reform. Can we now challenge the Senator's claim as incorrect? Or, could his contention still be fundamentally valid because the difference between his stated percent (75%) and that of the sample results (70%) may be attributable to chance and therefore would not be significant? In one of the forthcoming chapters we will show how calculating probabilities can help us make this kind of decision.

Case 2. Forty percent of the surnames on a very long list of eligible jury members are of Hispanic origin. Each name on the list has an equal chance of being selected for jury duty. How many out of 12 jury members would you expect to have Hispanic surnames? Would you be surprised if no persons with a Hispanic surname were selected for jury duty? You should be! The probability of selecting no Hispanic surname is $(1 - 0.4)^{12}$, or 0.0022. That is, the chance of selecting no Hispanic surname is only about 2/10 of 1%! In this case, probability is used to show that, with a fair method of selection, it would be quite unreasonable to expect no Hispanic-surname jury members to be selected out of 12.

Concepts in Probability

All of us have some idea of what a **probability** is—although it may be rather difficult to arrive at a precision definition. The weather forecaster notes that "there is a 30% *chance* of rain." Before the start of the baseball season, Jimmy the Greek and other Las Vegas odds makers might state that the *odds* of the Chicago White Sox representing the American League in the World Series are 250 to 1. And, the *likelihood* that humankind will land on Jupiter in the next four years is rather remote. Terms such as chance, odds, and likelihood are used interchangeably for the word probability.

What is a probability?

> **Probability**   A number that measures the likelihood that a particular event will occur.
>
> Chance

Three key words are used in the study of probability: **experiment, outcome,** and **event.** While they commonly appear in our everyday language, in statistics they have specific meanings.

> **Experiment**   The observation of some activity, or the act of taking some type of measurement.

This definition is more general than the one used in the physical sciences, where we picture researchers manipulating test tubes and microscopes. In statistics, an experiment has two or more different possible results, and it is uncertain which of them will occur.

> **Outcome**   A particular result of an experiment.

For example, the tossing of a coin is an experiment. You may observe the coin tossing, but you will be unsure whether it will come up "heads" or "tails." Similarly, asking 500 voters whether or not they intend to vote for a $3.1 million school bond is also an experiment. If a coin is tossed, one particular outcome might be a "head." Or, the outcome might be a "tail." As for the school-bond experiment, one outcome might be that 342 people favor its issuance. Another outcome might be that only 67 approve of it.

Still another outcome is that 203 favor the bond's issue. When one of the experiment's outcomes is observed, we call it an **event**.

> ✗ **Event**   A collection of one or more outcomes of an experiment.

Consider, for example, the single throw of a die. There are exactly six possible outcomes for this experiment. However, many events may be associated with it—the fact that the number of spots coming face up on the toss is odd, for example, or that the number that comes up is smaller than three. With the school-bond experiment, one event might be that a majority favor the $3.1 million issue. This event would occur whether the number of people in favor is 251, 252, 253, or any number up to and including 500. Notice that an event is not always simply an outcome.

Probabilities may be expressed as fractions (1/4, 5/9, 7/8), decimals (0.250, 0.556, 0.875), or percentages (25%, 56%, 88%). A probability is always between zero and one, inclusive. Zero describes the probability of something which cannot happen. If a corn seed is planted, the probability of having an elephant sprout from the seed is zero. At the other extreme, a probability of one represents a sure thing. In Portland, Maine, the mean high temperature is sure to be higher in July than in January.

Self-Review 5-1

A reminder: cover the answers in the margin.

A politician has just completed a major speech. He receives six pieces of mail commenting on it. He is interested in the number of writers who agree with him.

a. What is the experiment?
b. What are the possible outcomes?
c. Describe one possible event that might occur.

a. The count of favorable responses.
b. Any number between 0 and 6.
c. Several are
   1. The majority of comments (4, 5, or 6) are favorable.
   2. Nobody agrees (0).
   3. Everybody liked it (6).

Chapter Exercises

1. One card is drawn from a well-shuffled, standard 52-card deck of cards. If only the four suits are of interest (spades, hearts, diamonds, and clubs), what are all the possible outcomes?

2. Three persons are campaigning for mayor. Schwartz is a Democrat, White is a Republican, and Rossi is an Independent. Schwartz and Rossi are men. White is a woman.
   a. What is the experiment?
   b. Describe possible outcomes with respect to political party.
   c. Describe possible outcomes with respect to the sex of the candidates.

Types of Probability

There are three approaches to establishing the probability of events. The approaches are: classical, relative frequency, and subjective. The **classical** concept of probability is based on the assumption that several outcomes are *equally likely*.

> **Classical Concept of Probability**   The number of favorable outcomes divided by the total number of possible outcomes.

For example, what is the probability of getting a head on one toss of a coin? There are two possible outcomes to the experiment, a head and a tail. The two outcomes are equally likely, that is, the likelihood of a head is the same as the likelihood of a tail.

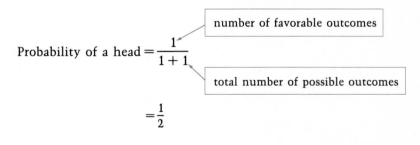

$$\text{Probability of a head} = \frac{1}{1+1}$$

number of favorable outcomes

total number of possible outcomes

$$= \frac{1}{2}$$

Self-Review 5-2

a. 1/52 or 0.0192
b. 4/52 or 0.0769
c. The classical concept.

One card is selected at random from a standard deck of 52 cards.

a. What is the probability that the card will be the ace of spades?
b. What is the probability that the card will be an ace?
c. What concept of probability does this problem illustrate?

The classical approach is useful when dealing with games of chance such as dice and card games, and other situations where the outcomes are equally likely or the outcomes are randomly selected, that is, every outcome has the same chance of occurring. Serious problems, however, develop when the outcomes are not equally likely. If you are an excellent student, for example, there is a good chance you will earn an A in this course. In your case, the outcome A does not have the same likelihood as the outcome F.

A second approach to probability is based on **relative frequencies.** This method uses the frequency of *past* occurrences to develop probabilities for the future.

> **Relative Frequency Concept of Probability**
> The number of times a particular event occurred in the past divided by the total number of observations.

First, the number of times a particular event happened in the past is computed. This value is then used to determine the likelihood that it will happen again. To illustrate, a mortality table revealed that out of 100,000 men aged twenty-five, 138 die within a year. Based on this experience, a life insurance company would estimate the probability of death for this age as

$$P(A) = \frac{138}{100,000} = 0.00138$$

This type of probability is used in so-called "actuarial tables" to help insurance companies establish the premiums to be charged for various types of life insurance—term, ordinary life, and so on.

In the mortality-table example, note that the probability of an event is denoted by a capital P. A code for the event is then written in parentheses. Commonly, capital letters or numbers are used to denote an event in the most concise manner possible. In this case, $P(A)$ stands for the probability a 25-year-old male will die during the year.

Self-Review 5-3

A political scientist randomly selected 200 eligible voters and determined the number of times they had voted in the last five general elections.

a. $P(1) = 30/200 = 0.15$
b. $P(3) = 40/200 = 0.20$
c. The relative frequency concept.

Number of Times Voted	Frequency of Occurrence
1	30
2	41
3	40
4	62
5	27
	200

a. What is the probability that a particular voter cast his or her ballot only once in the last five elections?
b. What is the probability that a particular voter cast his or her ballot exactly three times in the last five elections?
c. What concept of probability does this illustrate?

There are situations where there is little or no historical information from which to determine a probability. In such cases, only **subjective** probabilities can be employed. Using this approach, experts given the same information often differ in their estimates of the probability. They differ with respect to the likelihood of a major earthquake occurring in California during this decade, just as they differ in their selection of the most likely team to win the National League pennant this coming season. Subjective probability can be thought of as the *probability assigned by an individual or group based on whatever evidence is available.*

> **Subjective Concept of Probability**   The likelihood of an event assigned on the basis of whatever information is available.

Self-Review 5-4

a. Our estimate is 0.10; your estimate will no doubt be different.
b. The subjective concept.

a. What probability would you assign to a deep recession happening within a year?
b. What concept of probability does this illustrate?

In summary, there are three concepts of probability. The classical viewpoint assumes the outcomes are equally likely. If the outcomes are not equally likely, the relative frequency viewpoint is used. If no past experience is available, subjective judgment can be used to assign the probability an event will occur. Regardless of the viewpoint, the same laws of probability discussed in the following sections will be applied.

What concept of probability is used to assign the likelihood of the various events in the following experiments?

a. The probability of obtaining a 1 when rolling a single die.
b. The probability is 0.75 that the mean high temperature in Tampa, Florida is lower in June than in September.
c. The probability is 0.005 that a nuclear accident will occur at the Davis-Besse Plant within the next month.
d. Jones, Smith, Jackson, Keller, and Archer submit welfare claims. If only one is selected, the probability of selecting Jackson's claim is 0.20.

Self-Review 5-5

a. Classical.
c. Relative frequency.
c. Subjective.
d. Classical.

Elementary Probability Rules

In applications of probability often there is a need to combine the probabilities of events that are related in some meaningful way. In this section, two of the fundamental methods of combining probabilities—by addition and by multiplication—are discussed.

Rules of Addition

Special Rule of Addition.   We apply the **special rule of addition** if we wish to calculate the probability that at least one of several mutually exclusive events will occur. Recall that mutually exclusive means that when one of the events occurs, none of the other events can occur at the same time. If the result of a single roll of a die is a 4, it cannot be a 6 at the same time. Thus, the outcomes of a 4 and a 6 are mutually exclusive. Similarly, if the respondents to a questionnaire are classified as either male or female, then the events (male respondent, female respondent) are mutually exclusive. The special rule of addition for two events is

$$P(A \text{ or } B) = P(A) + P(B)$$

Recall that capital letters, such as A and B, refer to events.

Problem

What is the probability that an even number will result from one roll of a single die?

Solution

The event (an even number) is composed of three outcomes, namely:

The outcome of a 2  is event *A*.

The outcome of a 4 is event *B*.

The outcome of a 6 is event *C*.

The six possible outcomes are

The probability of each of the outcomes (2, 4, 6) is 1/6. The probability of the event "the outcome is an even number" is found by adding the three probabilities. That is, 1/6 + 1/6 + 1/6 = 3/6 or 0.50. In symbols, if *A* stands for the outcome of a 2, *B* represents the outcome of a 4, and *C* a 6, the probability of an even number appearing is computed by

$$P(\text{even}) \qquad = P(A) + P(B) + P(C)$$

$$P(A \text{ or } B \text{ or } C) = P(A) + P(B) + P(C)$$

$$= \frac{1}{6} + \frac{1}{6} + \frac{1}{6}$$

$$= \frac{3}{6} = 0.50$$

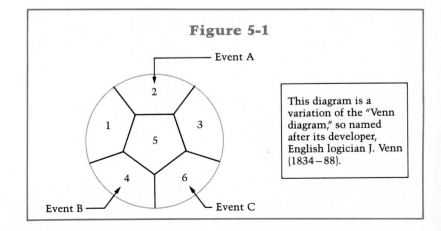

Figure 5-1

This diagram is a variation of the "Venn diagram," so named after its developer, English logician J. Venn (1834−88).

Recall that $P(\ )$ denotes the probability of the event described inside the parentheses.

Addition of the probabilities of the events of a 2, a 4, and a 6 in the previous die-rolling problem is allowed because the events are mutually exclusive. This notion of mutual exclusivity can be shown in diagram form (see Figure 5-1). The six regions represent all possible outcomes of an experiment.

Two hundred randomly selected prisoners in cell block M are surveyed and classified by type of crime committed.

Type of Crime	Number
Murder	48
Armed robbery	42
Rape	101
Kidnapping	7
Other	2

a. What is the probability that a particular prisoner selected in the sample is a convicted murderer?
b. What is the probability that a particular prisoner selected is a convicted kidnapper?
c. What is the probability that a particular prisoner selected is either a convicted murderer or a convicted kidnapper? What rule of probability was employed?

Self-Review 5-6

a. $48/200 = 0.24$
b. $7/200 = 0.035$
c. $55/200 = 0.275$, using the special rule of addition.

Chapter Exercises

3. A study was made of 138 children brought to metropolitan hospitals who exhibited evidence of abuse by an adult. The following table shows the abused child's position in the family.

Position	Frequency
Only child	34
Oldest	24
Youngest	50
Other	30

a. What is the probability of randomly selecting a child from the group just described and finding that the child was either the youngest or the oldest in a family?

b. What is the probability of selecting a child who is not an only child?

4. Shown in this table are the reported annual deaths of males aged 75 or over from one of the five leading types of cancer:

Site of Cancer	Number of Deaths
Lung	12,226
Prostate	10,835
Colon or rectum	8,426
Stomach	3,037
Pancreas	3,031
	37,555

For a randomly selected deceased male cancer victim
a. What is the probability that he died of one of the two primary causes?
b. What is the probability that he died of either cancer of the stomach or of the pancreas?
c. What is the probability that he did not die of lung cancer?

The General Rule of Addition.   We apply the **general rule of addition** to calculate the probability that at least one of several events will occur. They need not be mutually exclusive. For example, the current social security law has both a disability and a retirement provision. A welfare worker is studying the residents of the Sunshine Retirement Community. She finds that 20% of the residents are receiving disability payments and that 85% are receiving retirement income. If a retiree is randomly chosen for study, what is the probability that the person selected is receiving either disability payments or retirement income (or possibly both)?

The percents are converted to probabilities.

	Probability
Probability of receiving disability income	0.20
Probability of receiving retirement income	0.85
Total	1.05

Is the probability of 1.05 possible? It is not! A probability of more than 1.0 was ruled out in an earlier section. A probability is always 0 to 1, inclusive. What happened? Some of the people who receive payments have been "double-counted." That is, they receive both disability and retirement incomes.

To overcome this difficulty, the welfare worker must determine the percent of people who were counted twice and deduct this percent from the total. Suppose the investigator were to find that 15% of the people receive *both* disability and retirement incomes. Subtracting the corresponding probability of 0.15 from the total leaves 0.90. This is the probability of a person receiving at least one of these two types of income. The table summarizes these calculations.

	Probability
Probability of receiving a disability benefit	0.20
Probability of receiving a retirement benefit	0.85
Probability of receiving *both* a disability and a retirement benefit	−0.15
Probability of receiving one or the other benefit (or possibly both)	0.90

Symbolically, the general rule of addition is written

$$P(A \text{ or } B) = P(A) + P(B) - P(A \text{ and } B)$$

where A and B are two events. The word "or" takes into account the possibility that both A *and* B may occur and may need to be separated.

Applying this formula to the welfare problem:

Let A stand for the event "the retiree receives disability benefits."

Let B stand for the event "the retiree receives retirement benefits."

Let A and B stand for the event "the retiree receives both a retirement and a disability benefit."

If we want to find the probability of receiving a disability benefit *or* a retirement benefit then

$$
\begin{aligned}
P(A \text{ or } B) &= P(A) + P(B) - P(A \text{ and } B) \\
&= 0.20 + 0.85 - 0.15 \\
&= 0.90
\end{aligned}
$$

The general rule of addition applied to the welfare problem can be illustrated by a Venn diagram (see Figure 5-2). Note the overlapping of events.

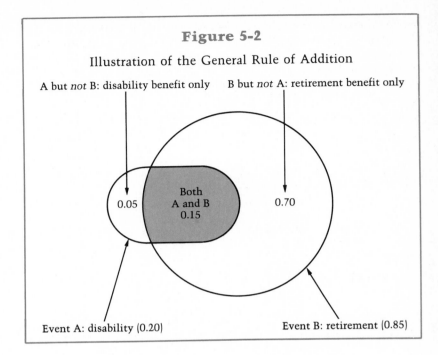

Figure 5-2

Illustration of the General Rule of Addition

A but *not* B: disability benefit only   B but *not* A: retirement benefit only

Both
A and B
0.15

0.05   0.70

Event A: disability (0.20)   Event B: retirement (0.85)

Self-Review 5-7

"Over 3.00" is event A, "employed" is event *B*.

$P(A$ or $B)$
$= P(A) + P(B) - P(A$ and $B)$
$= 0.45 + 0.25 - 0.10$
$= 0.60$

An analysis of the student records at Solid State University revealed that 45% of the students have a grade point average above 3.00. Twenty-five percent of the students are employed. Ten percent of the students are employed *and* have a grade point average above 3.00. What is the probability that a student selected at random will have a grade point average above 3.00 *or* be employed?

Chapter Exercises

5. A study of patient records at a public-health clinic showed that 15% of them had a dental examination. Forty-five percent had a general physical examination. Five percent had both. If a patient's record is randomly selected, what is the probability that he or she received either a dental examination or a physical examination?

6. Twenty percent of the members serving on boards of directors of large corporations are women. Five percent are persons connected with a university. Two percent of large corporations have female board members with university ties. If a corporation is randomly selected, what is the probability that its board of directors will have a member who is either a woman or someone from a university?

Rules of Multiplication

General Rule of Multiplication. What if we wanted to find the probability of both A and B occurring? For example, we might wish to find the probability of a student having a grade point average above 3.00 and also being employed. Such situations are different from those for which the addition rule can be used. Instead of computing the probability of one of the two outcomes occurring, we wish to find the probability that they *both* happen. This is termed a **joint probability.**

> A **joint probability** measures the likelihood that two or more events will happen concurrently.

Another probability concept is called **conditional probability.** Recall that probability measures uncertainty. But the degree of uncertainty may change as new data become available. Conditional probability is the tool that describes the new probability corresponding to some event B after it is known that some other event A has occurred. Symbolically, it is written $P(B|A)$. The vertical ($|$) slash does not mean division. Often, it is read "given," as in "the probability of B given A."

> A **conditional probability** is the likelihood an event will occur, assuming that another event has already occurred.

In the social security illustration, the probability that a person will receive a retirement benefit $P(B)$ is 0.85. However, if first we learn that the person is receiving a disability benefit, our estimate of the probability that he or she is also receiving a retirement benefit changes. Only 20% of the total receive disability benefits [$P(A) = 0.20$]. Out of that 20%, 15% also receive a retirement benefit. Hence, the conditional probability is 15/20 or 0.75 [$P(B|A)$]. The knowledge that a person received a disability benefit reduced the likelihood from 0.85 to 0.75 that he or she would receive a retirement benefit. In symbols, this calculation is written as follows:

$$P(B|A) = \frac{P(A \text{ and } B)}{P(A)}$$

This same relation gives us a way to compute $P(A \text{ and } B)$ when $P(A)$ and $P(B|A)$ are known. We simply multiply $P(A) \cdot P(B|A)$. This is the **general rule of multiplication.** It is written

$$P(A \text{ and } B) = P(A) \cdot P(B|A)$$

Problem

A study is conducted to determine the possible correlation between the level of education attained and a person's position on abortion. A random sample of one hundred individuals reveals the results shown in Table 5-1.

Table 5-1
Level of Education and Position on Abortion

Position on Abortion	Less than Four Years High School	High-School Graduate	Some College	College Graduate	Total
Favor	5	15	15	25	60
Oppose	10	10	10	10	40
Total	15	25	25	35	100

The question to be explored is: What is the probability of selecting an individual with less than four years of high-school education who favors abortion? The events occurring at the same time are "less than four years of high school" and "favors abortion."

Solution

Applying the general rule of multiplication:

$$P(B \text{ and } A) = P(A) \cdot P(B|A)$$

where

A stands for the event "a person favors abortion."

B stands for a person with less than four years of high-school education.

P(*B*|*A*) represents the probability of selecting a person who has less than a high-school education when it is known (or given) that the person favors abortion. The vertical line is read as "given that." The problem reads "the probability that a person selected has less than four years of high-school education and favors abortion is found by multiplying the probability that the person favors abortion by the probability that the person has less than a high-school education, given that the individual favors abortion." Again referring back to Table 5-1 for the probabilities:

$$P(B \text{ and } A) = P(A) \cdot P(B|A)$$
$$= 60/100 \cdot 5/60$$
$$= 0.60 \cdot 0.0833$$
$$= 0.05$$

This probability could also have been computed by determining the probability of selecting an individual with less than a high-school education [$P(A) = 15/100 = 0.15$] and then the probability that the person favors abortion, given that he or she has less than a high-school education [$P(B|A) = 5/15 = 0.33$].

$$P(A \text{ and } B) = P(A) \cdot P(B|A)$$
$$= (0.15) \cdot (0.33)$$
$$= 0.05 \text{ (the same as observed and calculated earlier)}$$

If more than two simple events are involved, the general rule of multiplication can be extended. For example, $P(A \text{ and } B \text{ and } C) = P(A) \cdot P(B|A) \cdot P(C|A \text{ and } B)$.

In the previous example about people's attitudes toward abortion, the joint probability could have been read directly from the table. It was pointed out that the general rule of multiplication gives the same answer. Sometimes, however, the information we have is expressed in percents instead of counts as in the previous illustration. The general rule of multiplication must be used to solve such problems.

Problem

The police in a small municipality know that 25% of the homeowners leave their doors unlocked. Crime records show that 4% of the homes whose doors are left unlocked are burglarized. What is the probability that a home is both left with its doors unlocked *and* burglarized?

Solution

The general rule of multiplication applies.

Let A represent the event "left with doors unlocked."
Let B represent the event "burglarized."

We know:

$$P(A) = 0.25$$
$$P(B|A) = 0.04 \leftarrow \boxed{\text{4\% of the homes } \textit{with unlocked doors} \text{ are burglarized}}$$

Solving:

$$
\begin{aligned}
P(A \text{ and } B) &= P(A) \cdot P(B|A) \\
&= (0.25)(0.04) \\
&= 0.01
\end{aligned}
$$

This indicates that 1% of *all* the homes in this small city *both* are left with their doors unlocked and are burglarized.

Self-Review 5-8

a. $P(A \text{ and } M) = P(A) \cdot P(M|A)$
b. 0.25, found by $23/60 \times 15/23$.

A sociologist conducted a study of a sample of 60 students to determine how many of each sex smokes marijuana. The results were

	Male	Female	Total
Smoke marijuana	15	8	23
Do not smoke marijuana	20	17	37
Total	35	25	60

Using the general rule of multiplication, determine the probability of selecting a male who smokes marijuana.

a. Letting A be the event "the student smokes marijuana" and M the event "a male is selected," supply the appropriate formula.
b. Determine the joint probability.

An interesting technique used to show probabilities, joint probabilities and conditional probabilities, is to plot them on a so-called **tree diagram.** Table 5-1 is repeated here to illustrate the tree diagram's application.

Position on Abortion	Less than Four Years High School	High-School Graduate	Some College	College Graduate	Total
Favor	5	15	15	25	60
Oppose	10	10	10	10	40
Total	15	25	25	35	100

In the tree diagram in Figure 5-3, notice that there are two main branches going out from the trunk on the left. The upper branch is labeled "Favor," the lower branch "Oppose." Note that the probability written on the "favor abortion" branch is 60/100 and on the "oppose abortion" branch is 40/100.

Problem
Using the tree diagram in Figure 5-3, locate the probability that a person favors abortion and has less than a high-school education.

Solution
1. Find the upper branch representing the event "favor abortion." As noted before, its probability (60/100) is written on the tree branch.
2. Continuing along the same path, find the branch "Less than 4 yrs. high school." Written on this branch is the conditional probability of 5/60.

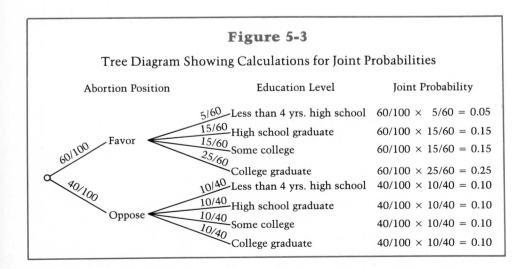

Figure 5-3
Tree Diagram Showing Calculations for Joint Probabilities

Abortion Position	Education Level	Joint Probability
	Less than 4 yrs. high school	60/100 × 5/60 = 0.05
	High school graduate	60/100 × 15/60 = 0.15
Favor	Some college	60/100 × 15/60 = 0.15
	College graduate	60/100 × 25/60 = 0.25
	Less than 4 yrs. high school	40/100 × 10/40 = 0.10
	High school graduate	40/100 × 10/40 = 0.10
Oppose	Some college	40/100 × 10/40 = 0.10
	College graduate	40/100 × 10/40 = 0.10

3. Multiplying those two probabilities gives 0.05, shown at the end of the path. This is the joint probability of selecting a person who both favors abortion and has less than four years of high-school education.

Self-Review 5-9

a. 25/60
b. 8/25
c.

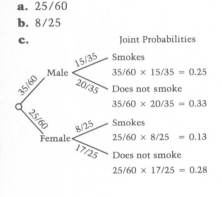

The study from Self-Review 8 is repeated here:

	Male	Female	Total
Smoke marijuana	15	8	23
Do not smoke marijuana	20	17	37
Total	35	25	60

a. What is the probability of selecting a female?
b. Assuming the person selected is a female, what is the probability that she smokes marijuana?
c. Draw a tree diagram showing all possible joint probabilities.

Chapter Exercises

7. A study was undertaken to correlate students' mathematical ability with their interest in statistics. The results were

Math Ability	Low	Interest in Statistics Average	High	Total
Low	40	9	11	60
Average	15	16	19	50
High	6	10	25	41
	61	35	55	151

a. What is the probability of selecting a student with both low math ability and a low interest in statistics?
b. What is the probability of selecting a student with both average math ability and a high interest in statistics?
c. Draw a tree diagram showing all possible joint probabilities.

8. Freshmen were classified according to two traits: whether they had won a high-school letter for athletics, and college grade point average.

High School	Low	College GPA Average	High	Total
Letter winner	50	30	50	130
Nonletter winner	20	30	20	70
	70	60	70	200

A student is selected.
a. What is the probability of selecting a student with a low college GPA who won a letter in high school?
b. What is the probability of selecting a student with a high college GPA who won a letter in high school?
c. Draw a tree diagram showing all possible joint probabilities.

Special Rule of Multiplication.   If there are two *independent* events A and B (meaning that the occurrence of event A does not affect the occurrence of event B), the **special rule of multiplication** is used to find the probability of A and B happening. This can be written

$$P(A \text{ and } B) = P(A) \cdot P(B)$$

Problem
There are six faces to a die. Suppose we have two of these dice. One is colored red, the other one white. They are rolled on the floor. What is the probability of obtaining a pair of 2's, that is, a 2 on the red die and a 2 on the white die?

Solution
The outcome of the white die is not dependent on the outcome of the red die; the two events are independent. Thus, the special rule of multiplication can be applied.

One of the six sides of a die is a 2 spot. The probability of a 2 spot, therefore, is 1/6. The probability of a 2 spot coming face up on the red die is 1/6, and the probability of a 2 spot on the white die is also 1/6. The probability of both coming face up is 1/36, found by:

$$P(A \text{ and } B) = P(A) \cdot P(B)$$

$$= \frac{1}{6} \times \frac{1}{6}$$

$$= 1/36$$
$$= 0.0278$$

Figure 5-4 illustrates this concept. Note:

1. There are 36 points (dots).
2. A is the event "a 2 spot appears on the red die."
3. B is the event "a 2 spot appears on the white die."
4. The shaded intersection of A and B is the point where a 2 spot appears on both the red and the white dice.

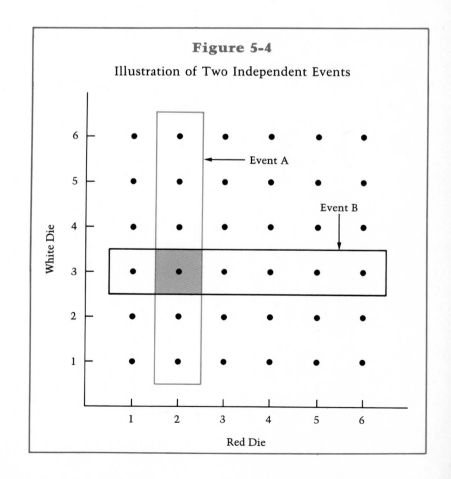

Figure 5-4

Illustration of Two Independent Events

Note that while there are 36 points (dots) in Figure 5-4, the shaded intersection includes only 1 point. Hence, 1 out of 36 points meets the stipulations, and the probability is 0.0278, found by 1/36. This agrees with the result found by the special rule of multiplication.

The probability of a boy being born is 1/2, the probability of a girl is 1/2. Let A be the event "the first child born to Jane and Doug Schmitz is a boy," and B is the event "the second child is also a boy."

a. Supply the correct formula and determine the probability that the first two children born are boys.

b. What is the probability that the first three children are boys?

a. $P(A$ and $B) = P(A) \cdot P(B) = $ $1/2 \times 1/2 = 1/4 = 0.25$

b. $1/2 \times 1/2 \times 1/2 = 1/8 = 0.1250$

Chapter Exercises

9. Out of every 100 cars that start in the Texas Grand Prix race, only 60 finish. Two cars are entered by the Penske team in this year's race.
   a. What is the probability that both will finish?
   b. Are the two events independent?
   c. What rule of probability does this illustrate?
   d. What concept of probability does this problem illustrate?
   e. What is the probability that neither of the 2 cars will finish the Grand Prix?

10. The Penn Bank has two computers. The probability that the newer one will break down in any particular month is 0.05. The probability the older one will break down in any particular month is 0.10.
   a. Are these events independent?
   b. What is the probability that both will break down during the month of October?
   c. What is the probability that neither will break down during October?

Some Principles of Counting

When the number of possible outcomes is small, as with the previous experiments, it is not difficult to list all the possibilities. For example, to list the possible outcomes for two children in a family is quite simple. The possibilities are two boys, one boy, and one girl, and two girls. Suppose, however, you had to list and count the possible outcomes for a family of 15 children! Obviously, a more efficient method of counting the outcomes would be needed. Three principles of counting that facilitate calculations in such cases will be explored next: (1) the multiplication formula, (2) the permutation formula, and (3) the combination formula.

Multiplication Formula

The **multiplication formula** states:

Multiplication Formula If a choice consists of two steps, the first of which can be made in m ways and the second in n ways, then there are $m \cdot n$ possible choices.

Problem

The Sears art department is designing an advertisement for the new catalog. The artist must show all possible variations on a new outfit consisting of two blouses and three skirts. The blouse options are long sleeve or short sleeve. The skirt options are solid color, Scotch plaid, or long evening skirt. How many different interchangeable outfits are there?

Solution

The letter m represents the number of blouse options. The letter n represents the number of skirt options. Using the multiplication formula:

$$\begin{aligned} \text{Number of options} &= m \times n \\ &= 2 \times 3 \\ &= 6 \end{aligned}$$

These six options are shown in Figure 5-5 on the following page. This counting method can be extended to more than two events.

Problem

A new-car buyer has a choice of five body styles, two engines, and eight different colors. How many different car choices does the buyer have?

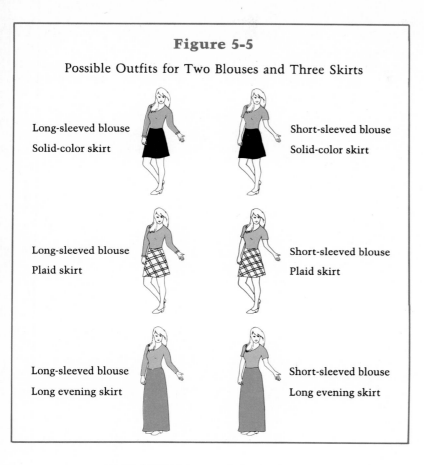

Figure 5-5

Possible Outfits for Two Blouses and Three Skirts

Long-sleeved blouse
Solid-color skirt

Short-sleeved blouse
Solid-color skirt

Long-sleeved blouse
Plaid skirt

Short-sleeved blouse
Plaid skirt

Long-sleeved blouse
Long evening skirt

Short-sleeved blouse
Long evening skirt

Solution
There are $5 \times 2 \times 8 = 80$ different choices among the cars that could be ordered.

An operating team generally consists of a surgeon, an anesthetist, and a nurse. A team must be scheduled for a delicate operation at Harbor Hospital. There are two surgeons capable of performing the operation. Three anesthetists are available, and five nurses are on call. How many different operating teams could be assembled?

Self-Review 5-11

$2 \times 3 \times 5 = 30$

Chapter Exercises

11. A restaurant offers five sandwiches, four cold drinks, and six desserts as part of its luncheon special. How many different specials are there?

12. One of Canada's western provinces is considering using only numbers on their automobile license plates. If only four digits are to appear on the plates (such as 8313), how many different plates are possible? Hint: the first digit could be any number between 0 and 9.

Permutation Formula

In the previous section, one item was chosen from each of several groups (one blouse from the blouses and one skirt from among the skirts, for example). In contrast, if more than one item is to be selected from the *same* group and if the *order of the selection is important,* the resulting arrangement is called a **permutation** of the items.

> **Permutation**   An ordered arrangement of a group of objects.

The permutation formula is

$$_nP_r = \frac{n!}{(n - r)!}$$

where n represents the number of objects, and r represents the number of objects being considered. For example, if there are five persons and three are to fill the offices of president, vice-president and secretary, $n = 5$ and $r = 3$.

The notation $n!$ is called "n factorial." It is the product of $n \cdot (n - 1) \cdot (n - 2), \ldots , (1)$. So 4! (four factorial) is 24, found by $(4) \cdot (3) \cdot (2) \cdot (1)$. 5! = 120. Zero factorial, written 0!, is treated as a special case and is defined to equal 1.

Problem

Five avid football fans organize the Yellow Jacket Booster Club. The five fans are Smith, Topol, Jackson, Lopez, and McNeil. Three officers are to be selected: a president, a secretary, and a treasurer. One group of officers might consist of President Lopez, Secretary Topol, and Treasurer McNeil. How many distinct slates (permutations) are possible?

Solution

   n is the number of avid boosters. (There are 5.)

   r is the number of offices to be filled. (There are 3.)

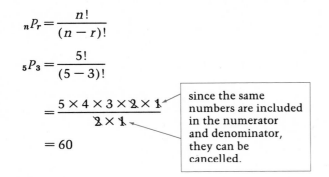

$$_nP_r = \frac{n!}{(n-r)!}$$

$$_5P_3 = \frac{5!}{(5-3)!}$$

$$= \frac{5 \times 4 \times 3 \times 2 \times 1}{2 \times 1}$$

since the same numbers are included in the numerator and denominator, they can be cancelled.

$$= 60$$

Problem

How many three-letter "words" can be made from the letters A, B, C, D, E, and F? No letter may be repeated in a word (such as DBD).

Solution

$$_nP_r = \frac{n!}{(n-r)!}$$

$$_6P_3 = \frac{6!}{(6-3)!}$$

$$= \frac{6 \cdot 5 \cdot 4 \cdot 3 \cdot 2 \cdot 1}{3 \cdot 2 \cdot 1}$$

$$= 120$$

Computer passwords that are three symbols long are to be made from four characters *, +, /, and ¢ without repeating any of the symbols. For example, the three symbols *¢+ could make up one possible password. How many distinct passwords of three-unit length are possible from the four characters?

Self-Review 5-12

$$n = 4$$

$$r = 3$$

$$_4P_3 = \frac{4!}{(4-3)!}$$

$$= 4 \times 3 \times 2 = 24$$

Chapter Exercises

13. "The Triple" at the local racetrack consists of correctly picking the order of finish of the first three horses in the eighth race. Ten horses have been entered in today's race. How many "Triple" outcomes are possible?

14. A real estate developer has eight basic house designs. Zoning regulations in his community do not permit look-alike homes on the same street. The developer has five lots on Mallard Road. In how many different ways can the new homes be arranged?

Combination Formula

In the previous section dealing with permutations, the *order* of the outcomes was important. For example, one slate of Yellow Jacket officers might be President Jackson, Secretary Topol, and Treasurer McNeil. Another slate might have President McNeil, Secretary Jackson, and Treasurer Topol. Every time there is a change in the order, the count of possibilities is increased by one.

In another problem the order may not be important. If for example, Gonzales, Clay, and Higbee were selected to serve as social committee, it would be the same committee if the order were Clay, Higbee, and Gonzales. In other words, the committee of Higbee, Clay, and Gonzales is counted just *once,* regardless of the order of names. Technically, such a group is considered a **combination.**

> **Combination** One particular arrangement of a group of objects or persons selected from a larger group without regard to order.

If order is not important, then, one can use the *combination formula* to count the possible number of combinations.

$$_nC_r = \frac{n!}{(n-r)!\,r!}$$

Problem

The Yellow Jacket Booster Club wants to select two persons to serve as a membership committee. How many different committees could be selected? One possible committee might consist of Lopez and McNeil. Of course, that committee would be the same as McNeil and Lopez.

Solution

There are ten possible membership committees:

Lopez, McNeil   McNeil, Smith   Topol, Jackson   Jackson, Smith
Lopez, Smith   McNeil, Topol   Topol, Smith
Lopez, Topol   McNeil, Jackson
Lopez, Jackson

Using the combination formula to determine the total number of possible membership committees of two ($r = 2$), from the five-member booster club ($n = 5$):

$$_nC_r = \frac{n!}{(n-r)!\,r!}$$

$$_5C_2 = \frac{\overset{2}{5 \cdot \cancel{4} \cdot \cancel{3} \cdot \cancel{2} \cdot \cancel{1}}}{(\cancel{3} \cdot \cancel{2} \cdot \cancel{1})(\cancel{2} \cdot \cancel{1})\underset{1}{}}$$

$$= 10 \text{ (same as listed by name)}$$

A group of seven mountain climbers wish to form a mountain-climbing team of five. How many different teams could be formed?

Self-Review 5-13

$$_7C_5 = \frac{7!}{(7-5)!\,5!}$$

$$= \frac{7 \cdot 6}{2 \cdot 1} = 21$$

Chapter Exercises

15. How many different poker hands are possible from a deck of 52 cards? (Note: you are dealt 5 cards.)

16. The director of welfare has just received 20 new welfare cases for investigation. A caseworker will be assigned eight new cases. How many different groups of eight cases could caseworker Klein be assigned?

Contrasting a Permutation with a Combination

To reemphasize the difference between a permutation and a combination, recall that a **permutation** is an arrangement where order is important. That is, a, b, c is one permutation; c, a, b a second permutation; and b, a, c a third. The permutation formula counts these different orders as different permutations. There are three permutations of the three letters a, b, and c taken three at a time.

If the order of selection is unimportant, the number of orders is called a **combination.** That is, if it is immaterial whether the three letters are written, a, b, c; c, a, b; or b, a, c, the combination formula will count these as a single combination.

Self-Review 5-14

a. Permutation.

b. $_nP_r = \dfrac{4!}{(4-2)!} = 12$

c. Combination.

d. $_4C_2 = \dfrac{4!}{(4-2)!2!} = 6$

The tuna fleet has four flags of different colors. Two flags are arranged on a mast as a signal to the ships in the fleet.

a. If you wished to count the number of different signals that can be constructed using the four flags hoisted two at a time, would you use a permutation or a combination? (Each new order has a different meaning.)

b. A blue flag on top and a yellow one below means "fish sighted." However, a yellow flag on top and a blue flag below means "stormy weather ahead." How many different signals can be constructed using the four flags two at a time?

c. Suppose it had been decided that any two flags, such as a green flag on top and a red one below, *or* a red flag on top and a green one below, have the same meaning. If you wished to count the number of signals that could be used, would you apply the permutation formula or the combination formula?

d. Refer to **c.** How many different signals could be constructed using the four flags two at a time?

Summary

Probability measures the likelihood that a particular event will occur. A probability may range between 0 and 1, with 0 representing the likelihood that an event cannot happen, and with 1 representing something that is absolutely certain. A probability may be in the form of a fraction, such as 1/4; a decimal, 0.25; or a percent, 25%.

Three concepts of probability were examined. They were the classical viewpoint, the relative frequency viewpoint, and the subjective viewpoint. The classical approach is based on the assumption that the outcomes are equally likely. On the roll of a die, the outcomes ⚀ ⚁ ⚂ ⚃ ⚄ ⚅ equally likely. The probability of any of these outcomes is found by dividing the number of favorable outcomes by the total number of outcomes.

The relative frequency approach to probability is based on past experience. If, for example, past experience reveals that 38 out of every 1,249 prisoners in honor farms walk off the farm and escape, the probability of an escape is $38/1,249 = 0.0304$.

If no past experience is available to determine a probability, subjective judgment is used. This probability is based on whatever evidence is available. It is not surprising, therefore, that ten football "experts" disagree with respect to the probability that the Dallas Cowboys, the New York Giants, or the Denver Broncos will go to the Super Bowl next year.

Four rules of probability were examined. The special rule of addition requires that the outcomes of an experiment be mutually exclusive. For two events: $P(A \text{ or } B) = P(A) + P(B)$. The general rule of addition is appropriate when the events are not mutually exclusive. For two events: $P(A \text{ or } B) = P(A) + P(B) - P(A \text{ and } B)$.

The general rule of multiplication allows us to determine the probability that several events may occur at the same time. For example, the rule would be used to find the probability that a person is a Republican and earns over $100,000 a year. The rule for two events is $P(A \text{ and } B) = P(A) \cdot P(B|A)$. If one event is not dependent on another, that is, if it is independent, which means $P(B) = P(B|A)$, then the special rule of multiplication is applied to find the probability that both event A and event B will occur. The rule for two independent events is $P(A \text{ and } B) = P(A) \cdot P(B)$.

Three counting formulas were introduced. The multiplication formula states that if there are m distinct results from doing one thing and n distinct results from doing a second thing, there is a total of $m \times n$ possible results. The permutation formula is used to count if the order of an arrangement is important. The formula is

$$_nP_r = \frac{n!}{(n-r)!}$$

If the order of an arrangement is not important, the combination formula is used to count the number of possible combinations. It is

$$_nC_r = \frac{n!}{(n-r)!\,r!}$$

An Introduction to Probability

Chapter Outline

I. Probability

A. Objective. To measure the likelihood of outcomes as a basis for further reasoning.

B. Definition. A number that expresses the likelihood that a particular event will occur.

C. Rules of Probability

1. The *special rule of addition* requires that the events be mutually exclusive. Written for three events:

$$P(A \text{ or } B \text{ or } C) = P(A) + P(B) + P(C)$$

Illustration: What is the probability that an ace of hearts, a king of diamonds, or a jack of spades will be drawn from a standard deck of cards containing 52 cards? Answer:

$$P(\text{ace or king or jack}) = 1/52 + 1/52 + 1/52$$
$$= 3/52 = 0.0577$$

2. The *general rule of addition* is used when the events are not mutually exclusive. Written for two events:

$$P(A \text{ or } B) = P(A) + P(B) - P(A \text{ and } B)$$

Illustration: What is the probability that a heart or a jack will be selected from a standard deck of 52 cards? Answer:

$$P(A \text{ or } B) = 13/52 + 4/52 - 1/52 = 16/52 = 0.3077$$

The probability of the jack of hearts, 1/52, was subtracted to prevent double-counting it.

3. The *general rule of multiplication* for two events states that

$$P(A \text{ and } B) = P(A) \times P(B|A)$$

where $P(B|A)$ stands for the probability B will occur, given that A occurred. Illustration: a barrel contains ten small boxes. Only two of the boxes contain prizes. What is the probability that two prizes will be drawn on the first two draws from the barrel (without replacement)?

$$P(A \text{ and } B) = 2/10 \times 1/9 = 2/90 = 0.022$$

4. The *special rule of multiplication* for two independent events is $P(A \text{ and } B) = P(A) \cdot P(B)$. Illustration: past experience indicates that for an automobile less than one year old, the probability of a tire being replaced is 0.04. The probability of a muffler being replaced is 0.02. The probability that both a tire and a muffler will be replaced is $P(A \text{ and } B) = P(A) \cdot P(B) = 0.04 \times 0.02 = 0.0008$.

II. Three Principles of Counting

A. *Multiplication* is a principle used when there are m distinct results from doing one thing and n different results from doing another thing. The total possible number of results is found by $m \times n$.

B. *Permutation* is an arrangement where order is important. For three objects, a, b, c is one permutation; b, c, a is another permutation; and c, a, b another. To count the number of permutations of n objects taken r at a time:

$$_nP_r = \frac{n!}{(n-r)!}$$

Illustration: six colors have been selected. Three out of the six colors are to be used to decorate rooms in a motel. How many different rooms can be decorated, assuming one color goes on the floor, another on the walls, and the third color on the ceiling? n = number of colors, r = number of colors used.

$$_nP_r = \frac{6!}{(6-3)!} = \frac{6 \times 5 \times 4 \times 3 \times 2 \times 1}{3 \times 2 \times 1} = 120$$

C. *Combination* is an arrangement where the order of selection is immaterial. For example, a, b, c and b, a, c are considered as one combination. Illustration: there are nine basketball players of about the same size and talent. Only five can start the game. How many different starting "fives" can the coach have? n = 9 players, r = 5 starters.

$$_nC_r = \frac{n!}{(n-r)!r!} = \frac{9!}{(9-5)!5!} = 126$$

Chapter Exercises

17. A dealer receives a shipment of four TV sets. One of the sets is known to be defective. If one of the sets is sold to a customer, what are the possible outcomes?

18. A car dealer has ten new cars in stock. Four are subcompact, four are compact, and two are luxury models. A car is sold. What is the experiment? What are the possible events regarding the type of car? If the luxury models are four-door sedans and all other cars are two-door, describe the possible outcomes with respect to the number of doors.

19. What concept of probability is used in each of the following events to assign the likelihood of the outcomes?

a. AMC decided to market a new subcompact car. The company has never offered a car in this competitive field before. Therefore no comparable data are available. The probability AMC will sell 2,000,000 cars next year is 0.70.

b. Dave Winfield, an outfielder for the New York Yankees, has a season's batting average of .285. What is the probability he will get a base hit the next time he comes to bat?

c. A student is randomly selected from your class. The probability that his or her birthday is in July is $1/12 = 0.083$.

20. Mr. Blume and Ms. White are planning a date on Saturday night. The probability they will see a basketball game, a movie, or a horse race are 0.25, 0.20, and 0.10 respectively. They will not do more than one of these activities. Of course, they may also decide to do something else entirely. What is the probability they will do one of these things? What is the probability they will choose one of the two sporting events?

21. A review of the records at Lander Community College revealed the following ethnic breakdown of the student population:

Caucasian	1,200
Black	640
Hispanic	280
Oriental	80
Native (Indian)	60
Other	20

What is the probability a randomly selected student is either caucasian or black? What is the probability the selected person is not a member of either of the two largest categories?

22. A study by the Florida Tourist Commission revealed that 70% of tourists entering the state visit DisneyWorld, that 50% visit Busch Gardens, and that 40% visit both. What is the probability that a particular tourist visits at least one of the attractions?

23. A certain large city has a morning newspaper, *The Mirror*, and an afternoon paper, *The Observer*. A study shows that 30% of households subscribe to the morning paper and 40% to the afternoon paper. A total of 20% of households subscribe to both. What percent of households subscribe to at least one of the papers?

24. The owner of a ski resort has been informed by a private weather service that the probability of an abundant snow base is 0.80. If there is an abundant snow, the probability that the owner will make more than a normal profit is 0.85. What is the probability that there will be an abundant snow base and that the owner will make more than a normal profit?

25. A survey of students at Pemberville Tech revealed the following employment breakdown:

	Work More than 20 hrs. per Week	Don't Work	
Male	75	125	200
Female	25	275	300
	100	400	500

What is the probability of selecting a female student who works more than 20 hours? What is the probability of selecting a male student who works?

26. Mr. Benzy just retired at age 65. His wife is 62 years old. If the probability is 0.50 that a man aged 65 will live another ten years, and the probability is 0.70 that a woman aged 62 will live another ten years, what is the probability that both Mr. and Mrs. Benzy will live another ten years? (Assume that the life expectancy of the husband is independent of the wife's life expectancy.)

27. Titusville Oil Company is currently drilling at two new sites. Based on the available geological evidence, the probability of striking oil at Site I (in Oklahoma) is 0.40, and at Site II (in Pennsylvania) is 0.50. What is the probability that both drilling operations will be successful? What is the probability neither will be successful?

28. Consider the generating of telephone numbers. Within a given exchange, say 874, how many different numbers are possible? (Assume that the numbers 874-0000 and 874-9999 are acceptable.)

29. Three scholarships are available to students at Salem College. The values are $500, $600, and $700. If 15 qualified students applied, in how many different ways could these scholarships be awarded? (No student may receive more than one scholarship.)

30. A mail-order company sells eight different cheeses. As part of a special Christmas holiday package, customers may select three different cheeses for their package. How many different gift packages are possible?

31. A hospital has two independent energy sources. Historical records show there is a 0.98 chance that the primary source will operate during severe weather conditions. The probability that the backup power source will operate under severe weather conditions is 0.95. What is the probability both will fail? What is the probability that the primary source will fail and the backup source will operate properly?

32. Tests employed in the detection of lung cancer are 90% effective—that is, they fail to detect the disease correctly 10% of the time. If three persons, all known to have cancer, are tested, what is the probability that the disease will be detected? What is the probability that two cases will be detected and one case not detected?

33. A zoologist has five male and three female guinea pigs. She randomly selects two for an experiment.
   a. What is the probability that both are males?
   b. What is the probability that both are females?
   c. What is the probability that the two are not of the same sex?
   d. Are these events independent?

Chapter Achievement Test

Do all the problems. Then check your answers against those given in the Answer section of the book.

I. True-False Questions. Classify each of the following statements as either true or false (4 points each).

   1. A probability is a number that can assume any value from 0 to 1, inclusive.
   2. According to the relative-frequency definition, the probability of an event is the ratio of the number of favorable outcomes to the total number of outcomes.
   3. The general rule of multiplication is $P(A \text{ or } B) = P(A) \cdot P(B|A)$.
   4. The probability that two randomly selected people were both born on a Sunday is 1/49.
   5. The number of permutations of five things taken three at a time is less than the number of combinations of five things taken three at a time.
   6. If $P(A \text{ and } B) = P(A) \cdot P(B)$, then A and B are mutually exclusive.
   7. In the general rule of addition, $P(A) + P(B) - P(A \text{ and } B)$, the $P(A \text{ and } B)$ represents the joint probability of the conditions of both A and B being met.
   8. The terms "mutually exclusive" and "independent" mean the same thing.

9. If 50 women are competing for the Miss America Title, the winner and the four runners-up can be selected in 50!/45! ways.

10. Because the subjective estimates of probability contain personal judgments, the general rule of addition does not apply.

11. The combination formula would be used to calculate the number of different ways that eight executives could park their cars in a lot having exactly eight spaces.

II. Computation Problems (11 points each).

The United States Department of Labor conducted a study of 29,355,000 families with children under 18 years old. Two traits were noted for each family, namely, whether the family was headed by a female and whether the mother worked. These are the results:

Status of Mother	Male Family Head	Female Family Head
In labor force	10,907,000	2,438,000
Not in labor force	14,368,000	1,642,000

If a family were selected at random from this population

12. What is the probability the mother is in the labor force?

13. What is the probability the family is headed by a female?

14. A recent newspaper article stated that the probability of downing an attacking airplane at each of four independent missile stations was 0.25. If an attacking plane had to pass over all four stations and the four stations were independent, what is the probability the plane would reach the target?

15. A combination lock will open when the correct three-digit combination is selected. Each digit can be any number from 0 to 9. How many different combinations are possible? (Assume a combination of 0 - 0 - 0 or 9 - 9 - 9 is acceptable.)

16. A student is taking two courses, History and Math. The probability the student will pass the math course is 0.60, and the probability of passing the history course is 0.70. The probability of passing both is 0.50. What is the probability of passing at least one of the two courses?

6
Probability Distributions

OBJECTIVES

When you have completed this chapter, you will be able to
- Define a random variable.
- Define a probability distribution.
- Distinguish between a discrete and a continuous probability distribution.
- List the characteristics of the binomial probability distribution.
- Compute probabilities for the binomial probability distribution.

Introduction

With Chapter 5 we started our investigation of statistical inference. The main objective in statistical inference is to be able to make decisions about an entire population based on a study of just part of that population—the sample. We pointed out and illustrated that calculating the probability of an event is a very useful technique in this type of decision making. Recall that probability is the likelihood that a particular outcome will happen.

In the previous chapter we also calculated the probability of specific events. For example, we determined the probability of selecting no persons with a Hispanic surname for a jury of twelve. Now we will study the entire range of events that may result from an experiment. In order to describe the likelihood of each outcome of this range of events, a probability distribution is used.

What Is a Probability Distribution?

In order to draw accurate conclusions about the population from which a sample is taken, we need to know the probability of the various outcomes when the sample is selected.

A **probability distribution** describes all the possible outcomes of some random event and their corresponding likelihoods. It is very much like a relative frequency distribution. However, instead of being merely descriptive of what *has* occurred, it is used to project into the future and describe what *will* probably occur.

Probability Distribution   A listing of the outcomes that may occur and of their corresponding probabilities.

For example, if a coin is tossed twice, the possible outcomes are

First Toss	Second Toss
H	H
H	T
T	H
T	T

where H represents a ''head'' and T represents a ''tail.''

If only the *number* of tails that will occur is of interest, and not the order in which they occur, the following table can describe that situation.

The probability distribution for the number of tails that may occur on two tosses of a coin is shown graphically by the histogram in Figure 6-1. The number of tails (X) is shown on the X-axis and the probability of X on the Y-axis.

As stated, then, a probability distribution lists the values that may occur (0, 1, or 2 on the X-axis) and their corresponding probabilities (0.25 or 1/4, 0.50 or 1/2, and 0.25 or 1/4 on the Y-axis). As we shall see in later chapters, probability distributions provide the

---

Table 6-1

Probability Distribution for the Number of Tails Resulting from Two Tosses of a Coin

Number of Tails X	Probability of Two Tails $P(X)$
0	1/4 = 0.25
1	1/2 = 0.50
2	1/4 = 0.25
Total	4/4 = 1.00

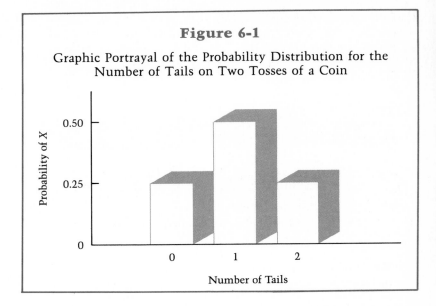

Figure 6-1

Graphic Portrayal of the Probability Distribution for the Number of Tails on Two Tosses of a Coin

theory upon which is built statistical inference, that is, reasoning from a sampled portion of a population to characteristics of the entire population. Two important features of a probability distribution are

1. The probability of a particular outcome X must always be between 0 and 1, inclusive.

2. The sum of the probabilities of all possible mutually exclusive outcomes is 1. In the coin-tossing experiment, each outcome had a probability between 0 and 1, inclusive, and the sum of the probabilities of 1/4, 1/2, and 1/4 is 4/4, or 1.00.

Self-Review 6-1

A reminder: cover the answers in the margin.

As an experiment, a coin is to be tossed three times. Some of the possible outcomes are

	Toss	
1	**2**	**3**
H	H	H
H	H	T
H	T	H

a. In any order, the complete list is

H	H	H
H	H	T
H	T	H
H	T	T
T	H	H
T	T	H
T	H	T
T	T	T

a. Complete the list of possible outcomes.

b. Develop a probability distribution showing the number of tails possible.

c. Portray the probability distribution in a graph.

Discrete and Continuous Random Variables

The number of tails that may occur when a single coin is tossed twice is also an example of a **random variable.** The random variable, in that instance, was designated by the letter X, and could assume any one of the three numbers (0, 1, or 2) assigned to it.

> **Random Variable** A quantity that assumes one and only one numerical value as a result of the outcome of an experiment.

A random variable may take two forms: discrete or continuous. A **discrete random variable** can assume only certain distinct values. Usually, it is a result of counting. The number of male children in a family with five children can only be 0, 1, 2, 3, 4, or 5; likewise, the number of highway deaths in Kansas during the Thanksgiving weekend can only be counted in increments of 1. Both are examples of discrete random variables. Note that there can be 0, 1, 2, 3, . . . , but there cannot be $2\frac{1}{4}$ children or deaths. This does not rule out the possibility that a discrete random variable would assume fractional values. However, the fractional values must have distance between them. For example, mortgage interest rates and stock prices are discrete random variables that are commonly expressed in dollar fractions.

Continuous random variables will be discussed in detail in Chapter 7. One example of them would be the weight of young men in your class. A weight might be 176 pounds, or 176.1 pounds, or 176.13 pounds, or 176.134 pounds, and so on depending on the accuracy of the scale.

A **discrete probability distribution** is based on a discrete random variable. The difference between a random variable and a probability distribution is that the random variable lists only the possible outcomes, whereas *the probability distribution includes both the list of possible outcomes and the probability of each outcome.*

b.

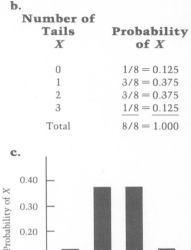

Number of Tails X	Probability of X
0	1/8 = 0.125
1	3/8 = 0.375
2	3/8 = 0.375
3	1/8 = 0.125
Total	8/8 = 1.000

c.

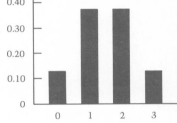

Self-Review 6-2

State which of the following random variables are discrete and which are continuous:

Experiment	Random Variable
a. A baseball player bats five times during a game.	He can have 0, 1, 2, 3, 4, or 5 base hits.
b. Counting the number of patients entering the emergency room of a hospital for treatment during a day.	There can be 0, 1, 2, 3, . . . , patients.
c. Fly via commercial airline from Detroit to Sarasota, Florida.	Time it took flight 102 could be 4 hours 26 minutes, or 4 hours 26 minutes $3\frac{2}{5}$ seconds, and so on depending on the accuracy of the timing device.

a. Discrete.
b. Discrete.
c. Continuous.

Why Study Probability Distributions?

As remarked earlier, we need to calculate the likelihood of a particular sample outcome so that conclusions about the entire population can be made. Thus, the probability of each possible outcome of some experiment is organized into a distribution. The following example gives a hint of the manner in which a probability distribution can be used to make an inference about, or "test" for, racial

Table 6-2
Number of Black Families in a Sample of Ten

Number X	P(X)	
0	0.028	
1	0.121	
2	0.233	
3	0.267	most probable
4	0.200	
5	0.103	
6	0.037	
7	0.009	
8	0.001	most improbable
9	0.000	
10	0.000	

bias. Consider a community composed of 70% white families and 30% black families. Suppose ten families had been randomly selected to be interviewed on the advisability of rezoning a school district. The probability distribution for the number of black families among the ten selected is shown in Table 6-2 on the previous page.

Observe that the most probable event is either 2, 3, or 4 black families. In other words, most of the time the selection process will result in 2, 3, or 4 black families being included. The event that 7, 8, 9, or 10 black families out of the ten will be selected is a result that is quite improbable. Its occurrence would cause you to suspect racial bias in the selection of the sample.

The Binomial Probability Distribution

One widely used discrete probability distribution is the **binomial distribution.** The binomial distribution accurately describes, to a reasonable degree, many experiments that require the probability of the number of successes or failures in a sample or a set of repeated trials. The following characteristics identify the binomial probability distribution:

1. Each outcome is classified into one of two mutually exclusive categories. An outcome is either a "success" or a "failure." For example, a lottery ticket is either a winning number (a success), or not a winning number (a failure). The outcomes are mutually exclusive, meaning that a ticket cannot be both a success and a failure at the same time. We do not record the amount of winnings, only the fact of winning.

2. The binomial distribution is a count of the number of successes. It is a discrete distribution. X can only assume certain integer values, namely, 0, 1, 2, 3, . . . , successes. There cannot be, for example, $2\frac{1}{2}$ successes.

3. Each "trial" is *independent*. This means that the outcome of one trial does not affect the outcome of any other trial. It is common to use the word "trial" when discussing the binomial probability distribution. Usually, the number of trials is the size of the sample (n) that was selected.

4. The probability of success on each trial is the same from trial to trial. For a lottery ticket, the probability of having a winning number on one ticket is the same as the probability of having a winning number on another ticket.

Self-Review 6-3

1. Mutually exclusive: a baby cannot be a girl and a boy at the same time.
2. Only two possible outcomes: boy or girl.
3. Trials are independent. If boy is born, this does not affect sex of next child.
4. Probabilities remain the same for each birth, 0.50 and 0.50.

The probability that expectant parents will have a girl is 1/2, or 0.50. Likewise, the probability they will have a boy is 0.50. The probability of having 0, 1, 2, 3, 4, and 5 girls in a family of five children is

Number of Girls X	Probability P(X)
0	0.0312
1	0.1562
2	0.3125
3	0.3125
4	0.1562
5	0.0312

Explain why this distribution qualifies as a binomial probability distribution.

Constructing a Binomial Probability Distribution

To construct a binomial probability distribution it is necessary to know (1) the total number of trials, and (2) the probability of success on each trial.

Problem

A senator from Connecticut estimates that 80% of the voters in her state favor tax reform. A random sample of six voters is selected. Construct a binomial distribution for the probability that exactly 0, 1, 2, 3, 4, 5, and 6 out of the six voters will favor tax reform. In this problem the number of trials is six and the probability of success (will vote for tax reform) is 0.80.

Solution

For the sake of illustration, let us first compute the probability that exactly five of the six voters selected in the sample will favor tax reform. If F represents the fact that a person favors tax reform and O represents someone who opposes it, one outcome is that the first voter contacted opposes tax reform and the rest favor it. In symbols, this outcome could be written O, F, F, F, F, F. The voter preferences are presumed independent, so the probability of this particular joint occurrence is the product of the individual probabilities. Hence, the likelihood of finding a voter who is opposed to tax reform followed by five who favor tax reform is

$$(0.2)\,(0.8)\,(0.8)\,(0.8)\,(0.8)\,(0.8) = 0.0655$$

However, the problem does not require that the first voter be the one who is opposed to tax reform. The voter opposing tax reform could be any one of the six people sampled. Any of the six outcomes listed result in exactly five people favoring tax reform and one opposing it.

Order of Occurrence	Probability of Occurrence
O, F, F, F, F, F	$(0.2)\,(0.8)\,(0.8)\,(0.8)\,(0.8)\,(0.8) = 0.0655$
F, O, F, F, F, F	$(0.8)\,(0.2)\,(0.8)\,(0.8)\,(0.8)\,(0.8) = 0.0655$
F, F, O, F, F, F	$(0.8)\,(0.8)\,(0.2)\,(0.8)\,(0.8)\,(0.8) = 0.0655$
F, F, F, O, F, F	$(0.8)\,(0.8)\,(0.8)\,(0.2)\,(0.8)\,(0.8) = 0.0655$
F, F, F, F, O, F	$(0.8)\,(0.8)\,(0.8)\,(0.8)\,(0.2)\,(0.8) = 0.0655$
F, F, F, F, F, O	$(0.8)\,(0.8)\,(0.8)\,(0.8)\,(0.8)\,(0.2) = \underline{0.0655}$
	0.3930

Any one of these six possibilities has exactly five voters who favor tax reform. Hence, the probability that exactly five of the six voters sampled will favor tax reform is equal to the sum of the six possibilities or 0.3930.

The probabilities of the other outcomes, such as 0, 1, 2, and so on, of the six sampled voters favoring tax reform could be computed in a similar fashion. A more direct method, however, is to use the **binomial formula.**

$$P(X) = {}_nC_X\, p^X (1-p)^{n-X} \quad \text{or} \quad P(X) = \frac{n!}{X!(n-X)!}\, p^X (1-p)^{n-X}$$

where

    n is the total number of trials. It is 6 in this problem.

    p is the probability of success on each trial. It is 0.80.

    X is the number of observed successes. Here it is the number favoring reform.

$_nC_X$ is the formula for the number of successes in n trials. This is the combination formula.

Substituting these values in the formula to find the probability that exactly five will favor tax reform:

$$P(X) = \frac{n!}{X!(n-X)!}\, p^X (1-p)^{n-x}$$

$$P(5) = \frac{6!}{5!(6-5)!}\, (0.80)^5 (1-0.80)^{6-5}$$

Recall from Chapter 5 that 6! (six factorial) means $6 \times 5 \times 4 \times 3 \times 2 \times 1$. When the same number is included in both the numerator and the denominator, it can be cancelled:

$$= \frac{6 \cdot \cancel{5} \cdot \cancel{4} \cdot \cancel{3} \cdot \cancel{2} \cdot \cancel{1}}{\cancel{5} \cdot \cancel{4} \cdot \cancel{3} \cdot \cancel{2} \cdot \cancel{1} \cdot 1} (0.80)^5 (0.20)^1$$

$$= (6)\,(0.80)^5 (0.20)^1$$

$$= 0.3930$$

This is the same probability obtained earlier. The remaining probabilities for 0, 1, 2, . . . , were computed in similar fashion, and are shown in Table 6-3.

Table 6-3

Binomial Probability Distribution for an n of 6 and a p of 0.80

Number in Favor of Tax Reform X	Probability of Occurrence $P(X)$
0	0.000
1	0.002
2	0.015
3	0.082
4	0.246
5	0.393
6	0.262
Total	1.000

A fire chief estimates that the probability an arsonist will be arrested is 0.3. What is the probability that exactly three arsonists will be arrested in the next five deliberately set fires?

Chapter Exercises

1. An insurance representative has appointments with four prospective clients tomorrow. From past experience she knows that the probability of making a sale on any appointment is 0.20. What is the probability that she will sell a policy to three of the four prospective clients?

2. James Jewellers permits the return of diamond rings, provided the return occurs within ten days of the purchase date. In the store's experience, 10% of the diamond rings are returned. Five rings are bought by five different customers. What is the probability that none will be returned?

A Computer Output

The computer can be used as a tool to calculate binomial probabilities. The following output from a program written in the BASIC programming language gives the same probabilities as those shown in Table 6-3. The complete listing for the BASIC program follows on pages 180 and 181.

The Use of Binomial Tables

Another way to determine the probabilities needed to construct a binomial distribution is to use tables that give the probabilities for various values of n and p. Such a table is given in Appendix A; a small portion of it, for the case where $n = 6$, is duplicated as Table 6-4 on page 180.

In the previous problem involving a sample of six voters, a p of 0.80 was used to illustrate the use of a binomial table. Refer to Table 6-4. The probability of exactly five out of six voters being in favor of tax reform can be found by the following procedure.

1. Find the section where the sample size or the number of trials is given under the first heading N. In this case it is 6.

2. Within that block, locate the row headed by the number of successes under the column headed X. In this case it is 5.

Self-Review 6-4

$n = 5 \quad X = 3 \quad P = .3$

$$P(3) = \frac{5!}{3!2!}(0.3)^3(0.7)^2$$
$$= (10)(0.027)(0.49)$$
$$= 0.132$$

$$P(x) = \frac{n!}{x!(x-1)!}(P)^x(1-p)^{n-x}$$

$$P(3) = \frac{5!}{3! \times 2!} \times .3^3 \times .7^2$$

$$P(3) = \frac{120}{6 \times 2} \times .027 \times .49$$

$$P(3) = .1323$$

$P = .2 \quad n = 4 \quad x = 3$

$$P(x) = \frac{n!}{x!(x-1)!}(P)^x(1-p)^{n-x}$$

$$P(3) = \frac{4!}{3! \times 2!} \times .2^3 \times .8^1$$

$$P(3) = \frac{24}{6 \times 2} \times .008 \times .8$$

$$P(3) = .0128$$

Program Output

```
RUN
BIN2      03:52 PM           22-Mar-82

INPUT THE NUMBER OF TRIALS AND THE PROBABILITY OF A SUCCESS,
SEPARATED BY A COMMA
? 6,.8

        SUMMARY INFORMATION

THE NUMBER OF TRIALS IS 6
THE PROBABILITY OF A SUCCESS IS .8
THE MEAN IS 4.8   THE STANDARD DEVIATION IS .979796

   BINOMIAL DISTRIBUTION

   SUCCESSES          PROBABILITY

       0                .000
       1                .002
       2                .015
       3                .082
       4                .246
       5                .393
       6                .262
```

Table 6-4

Binomial Probability Distribution, $n = 6$

X	.05	.1	.2	.3	.4	.5	.6	.7	.8	.9	.95
0	.735	.531	.262	.118	.047	.016	.004	.001	.000	.000	.000
1	.232	.354	.393	.303	.187	.094	.037	.010	.002	.000	.000
2	.031	.098	.246	.324	.311	.234	.138	.060	.015	.001	.000
3	.002	.015	.082	.185	.276	.313	.276	.185	.082	.015	.002
4	.000	.001	.015	.060	.138	.234	.311	.324	.246	.098	.031
5	.000	.000	.002	.010	.037	.094	.187	.303	.393	.354	.232
6	.000	.000	.000	.001	.004	.016	.047	.118	.262	.531	.735

N= 6 PROBABILITY

Program Listing

```
LIST
BIN2      10:39 AM          07-Apr-82
5 REM THIS PROGRAM COMPUTES THE COMMMPLETE BINOMIAL DISTRIBUTION
6 REM FOR A GIVEN NUMBER OF TRIALS AND A PROBABILITY OF SUCCESS
30 PRINT
31 PRINT
35 PRINT"INPUT THE NUMBER OF TRIALS AND THE PROBABILITY OF A SUCCESS,
36 PRINT"SEPARATED BY A COMMA"
40 INPUT N,P
41 PRINT
43 PRINT"            SUMMARY INFORMATION"
44 PRINT
45 PRINT  THE NUMBER OF TRIALS IS";N
46 PRINT "THE PROBABILITY OF A SUCCESS IS";P
47 PRINT "THE MEAN IS";N*P;" THE STANDARD DEVIATION IS";SQR(N*P*(1-P))
48 PRINT
50 PRINT " BINOMIAL DISTRIBUTION"
55 PRINT
85 PRINT "SUCCESSES";TAB(15);"PROBABILITY"
90 A$="   ###             .###"
91 PRINT
100 FOR I=0 TO N
105    F,G=0
110    FOR A=I+1 TO N
120    F=LOG(A)+F
130    NEXT A
140    FOR A=2 TO N-I
150    G=LOG(A)+G
160    NEXT A
165    L1=F-G
170    Z1=L1+(I*LOG(P))+((N-I)*LOG(1-P))
180    IF Z1<-13.8155 THEN Z=0
190    Z=EXP(Z1)
200 PRINT USING A$,I,Z
220 NEXT I
999 END
```

3. Move across the row until you reach the column headed by the desired probability, or value of p. It is 0.80 in this case (marked .8 in Table 6-4).

4. The desired probability is where the row and column of interest intersect. In Table 6-4, for $n = 6$, the value 0.393 is located at the intersection of row 5 and column .8. This result agrees with the earlier calculation.

Self-Review 6-5

$n = 5$, $p = 0.3$

X	P(X)
0	0.168
1	0.360
2	0.309
3	0.132
4	0.028
5	0.002

In the previous self-review the fire chief estimated the probability of an arsonist's being arrested at 0.3. The question asked was: What is the probability that, for the next five arson-related fires, three arsonists will be arrested? The probability was computed to be 0.132. Using Appendix A, complete the binomial probability distribution for this problem.

Problem

Referring back to Table 6-4, what is the probability that five or more voters favor tax reform? What is the probability that fewer than six favor tax reform?

Solution

The first question can be answered by adding the probabilities for five and six. $P(5) + P(6) = 0.393 + 0.262 = 0.655$. The probability of fewer than six can be determined either by $P(0) + P(1) + P(2) + P(3) + P(4) + P(5) = 0.000 + 0.002 + 0.015 + 0.082 + 0.246 + 0.393 = 0.738$, or by $1 - P(6) = 0.738$. The latter approach—that is, finding the probability of an event not happening and subtracting it from 1, is called the **complement rule**. It is written: $P(\text{not } A) = 1 - P(A)$.

Self-Review 6-6

a. 0.162, found by $P(3) + P(4) + P(5) = 0.132 + 0.028 + 0.002$.

b. 0.832, found by either $1 - P(0) = 1 - 0.168$, or $P(1) + P(2) + P(3) + P(4) + P(5)$ (slight discrepancy due to rounding).

The binomial probability distribution developed for arsonists in the previous self-review is

X	P(X)
0	0.168
1	0.360
2	0.309
3	0.132
4	0.028
5	0.002

a. What is the probability that three or more arsonists will be arrested in the next five fires?

b. What is the probability that one or more arsonists will be arrested in the next five fires?

Chapter Exercises

3. The instructor in Political Science 101 gives a weekly ten-question multiple-choice quiz. For each question there are five choices, but only one of them is the correct answer. A student did not attend class or read the text assignments (a common occurrence). He did, however, take the weekly quiz.

   a. Using the formula for computing the binomial probabilities, compute the probability that the student will guess exactly three of the ten correct answers.

   b. Using the binomial table in Appendix A, develop a binomial probability distribution.

   c. The instructor had announced that students who score six or more correct out of ten pass the quiz. What is the probability that a student who neither attended class nor read the assignments will pass the test?

4. As noted earlier, chances are 50 - 50 that a newborn baby will be a girl. For families with five children:

   a. What is the probability of having three girls and two boys?

   b. What is the probability that all the children are girls?

   c. What is the probability of having at least one girl?

5. Thirty percent of all automobiles sold in the United States are foreign-made. Four new automobiles are randomly selected.

   a. Referring to Appendix A, what is the probability that none of the four are foreign-made?

   b. Construct a binomial probability distribution showing the probabilities of 0, 1, 2, 3, and 4 out of four being foreign-made.

   c. Portray the binomial probability distribution in the form of a graph.

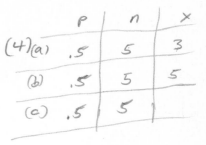

A probability distribution is a listing of the outcomes of an experiment and of their corresponding probabilities. For example, the following probability distribution gives the likelihoods of having 0, 1, and 2 girls for a family planning to have two children:

Number of Girls	Likelihood (probability)
0	1/4 = 0.25
1	2/4 = 0.50
2	1/4 = 0.25
	4/4 1.00

A *discrete* probability distribution has a countable number of outcomes. For example, the probability distribution just cited can have only three outcomes: 0, 1, or 2 girls. A *continuous* probability distribution is based on a continuous random variable. That is, the variable may take on an infinite number of values (within a given range), depending on the accuracy of the measuring device. Temperatures in Maine, for example, take on values ranging from 50°F to 95°F during the summer months. A temperature reading may be recorded as 58°, 58.2°, 58.23°, or 58.237°F, depending on the accuracy of the recording device.

The binomial probability distribution is discrete. Its characteristics are: (1) there are only two possible outcomes to a trial—success or failure, (2) the outcomes of a trial are mutually exclusive, (3) each trial is independent, and (4) the probability of a success remains the same from trial to trial. The probability of having a girl, for example, is the same for each child born.

In order to construct a binomial probability distribution, the total number of trials and the probability of success on each trial are needed. For example, in developing a probability distribution for the number of girls likely to be born to a family planning to have two children, the total number of trials is 2, and the probability of having a girl each trial is 1/2. The list of probabilities can be determined by the binomial formula:

$$P(X) = {}_nC_X \, p^X(1-p)^{n-X} \quad \text{or} \quad \frac{n!}{X!(n-X)!} \, p^X(1-p)^{n-X}$$

As an alternative to calculating all the probabilities needed to construct a binomial probability distribution, a table containing the binomial probabilities may be used (see Appendix A). Many computer programs are also available to do the routine calculations.

Chapter Outline

Probability Distributions

I. Probability Distributions

    A. Objective. To describe the entire range of outcomes that may result from an experiment. This is the basis for further reasoning about the population from which a sample is selected.

    B. Definition. A listing of all values of a random variable and the corresponding probability of occurrence.

    C. *Discrete probability distribution:* based on a discrete random variable. Discrete random variables can assume a countable number of numerical values. Example: for a sample of ten persons, the number of persons favoring an increase in the state's income tax can only be 0, 1, 2, 3, . . . , 10.

D. *Continuous probability distribution:* based on a continuous random variable. Continuous random variables can assume an infinitely large number of values within a given range. The number depends on the accuracy of the measuring device. Example: the outside diameter of a tree might be 6.7 meters, 6.72 meters, 6.727 meters, and so on, depending on the degree of accuracy of the tape measure being used.

II. Binomial Probability Distributions

A. Objective. To describe experiments that deal with the probability of the number of successes or failures in a set of repeated trials.

B. Characteristics

1. Each outcome can be classified into one of two mutually exclusive categories. Example: right or wrong.
2. It is a discrete distribution resulting from a count of the "successes."
3. Each trial is independent—that is, the outcome of one trial does not affect the outcome of any other trial.
4. The probability of success is the same from trial to trial.

C. The probability of an outcome can either be determined from Appendix A or it can be computed as follows:

$$P(X) = \frac{n!}{X!(n-X)!} p^X (1-p)^{n-X} \quad \text{or} \quad {}_nC_x \, p^X (1-p)^{n-X}$$

Illustration: What is the probability that exactly eight answers out of a ten-question, true-false examination will be guessed correctly?

$$P(8) = \frac{10!}{8!(10-8)!} (0.50)^8 (0.50)^{10-8} = 0.0439$$

Chapter Exercises

Test question

6. If a baseball player with a batting average of .300 comes to bat five times in a game, what is the probability he will get
   a. exactly two hits? .309
   b. fewer than two hits? .528
   c. at least one hit? .832

(a) $P = .3$ $n = 5$ $X = 2$

(b) $P = .3$ $n = 5$ $X = 0 \& 1$

(c) $P = .3$ $n = 5$ $X = 1, 2, 3, 4, 5$

7. Four persons are stopped at random and asked the date and year of their birth. From this information, the day of the week on which they were born is determined. What is the probability that two of the four were born on a Monday?

8. A new pilot TV show has been created. The network president thinks 60% of the population will like the new show. If the pilot is shown to 15 people, and the president is correct, what is the probability that ten or more will indicate they liked it?

9. A student accuses an instructor of being chauvinistic and of harassing female students. As evidence, the accuser points to the fact that on only four of the last ten occasions has he called on a male student. Thirty percent of the large class are women. What is the probability that the result cited might have occurred by chance alone?

10. A poplar tree, if less than three feet high and when transplanted in the spring, has a 40% chance of survival. If six such trees are transplanted, what is the probability that exactly five will survive? What is the probability that 2, 3, 4, or 5 will survive?

11. A particular type of birth-control device is effective 90% of the time when used as directed. If a sample of 15 people use the device for one year, what is the probability that none of the devices will have failed?

Chapter Achievement Test

Do all the problems. Then check your answers against those in the Answer section of the book.

I. Multiple Choice (5 points each). Select the correct response.
1. A listing of all values of a random variable and the corresponding probability of occurrence is called
   a. an experiment
   b. a probability distribution
   c. an event
   d. a probability
   e. none of the above
2. For a discrete probability distribution, which of the following are correct statements?
      I. The sum of all possible probabilities is 1.0.
      II. Decimal probabilities such as 0.87 are not possible.
      III. Negative probabilities are not possible.
   a. all are correct
   b. I and II are correct
   c. I and III are correct
   d. I is the only correct answer
   e. none are correct
3. Which of the following variables is *not* an example of a continuous random variable?
   a. flight time between New York and Miami
   b. average age of the senior class

    c. average weight of defensive linemen in the National Football League

    d. number of outpatient surgical patients on a particular day at St. Luke's Hospital

    e. all are correct

4. One feature of a discrete probability distribution is that

    a. the sum of the probabilities of all possible mutually exclusive outcomes is 1.0

    b. the probability of any particular outcome must be between 0 and 1.0

    c. the probability of any event cannot be a negative number

    d. all of the above are correct

5. Which one of the following is *not* a characteristic of the binomial distribution?

    a. each outcome can be classified into one of two mutually exclusive categories

    b. the probability of a success is the same from trial to trial

    c. each trial is independent

    d. it is a discrete distribution

    e. it may take on an infinitely large number of values within a given range

6. The neighborhood paperboy knows that the probability is 2/3 that a patron will be home when he collects. If he stops at five homes, what is the probability that he will find no patrons at home?

    a. 0.0041

    b. 0.0412

    c. 0.0453

    d. 0.1317

    e. 0.8683

    f. none of the above

7. What is the probability that four or more will be at home?

    a. 0.4609

    b. 0.3292

    c. 0.0041

    d. 0.8683

    e. none of the above

Three situations are described in Questions 8 through 10. If the binomial is appropriate, record a T. If the binomial is not appropriate, record NT and state why it isn't appropriate.

8. A store receives a shipment of 25 stereo speakers of which ten have been damaged in shipment. A customer buys six of these speakers without examining them first. What is the probability that three are defective? Is the binomial appropriate?

9. Health records indicate that 75% of all first-graders are fully immunized by the time they reach age six. A class of 15 first-grade students is being studied. What is the probability that more than ten are fully immunized? Is the binomial appropriate?

10. A bank processes ten personal loan applications per day. The probability that a particular application will be denied is 0.30. What is the probability that more than two will be denied on the same day? Is the binomial appropriate?

II. Solve the following problems (25 points each).

11. A certain medicine is 70% effective; that is, out of every 100 patients who take it, 70 are cured. A group of 12 patients are given the medicine. Find the probability that
    a. 8 are cured
    b. fewer than 5 are cured
    c. 10 or more are cured

12. It is known that 30% of all high-school graduates go to a four-year college. If the graduating class of a very small school of 20 is considered, find the probability that
    a. more than 8 go to a four-year college
    b. fewer than 6 go to a four-year college
    c. more than 9 go to a four-year college

7
The Normal Probability Distribution

OBJECTIVES

When you have completed this chapter, you will be able to
- List the important characteristics of the normal distribution.
- Compute a z value and compare a particular observation with an entire group of observations.
- Calculate the probability that an observation will occur between two points.
- Determine a point beyond which a stated percent of the observations will occur, or a point beyond which there is a stated probability of occurrence.

Introduction

In Chapter 2 raw data were organized into a frequency distribution. Next, frequency polygons and other graphs were constructed to describe the location and shape of the frequency distribution. In Chapters 3 and 4 we discussed the part of statistics known as "descriptive statistics" and introduced various measures that describe the central tendency and dispersion in the data. With Chapter 5 we began our exploration of the idea of probability, which measures the likelihood that some uncertain event will occur. The concept of a probability distribution was developed in Chapter 6. A probability distribution extends the notion of a frequency distribution to include a description of some experiment and of all of its possible outcomes. Chapter 6 described a very important *discrete* distribution—the binomial.

In this chapter we will examine a very important *continuous* probability distribution known as the **normal distribution.** Its importance is due to the fact that, in practice, experimental results very often seem to follow this mounded and "bell-shaped" pattern. The normal probability distribution is also important because most of the sampling methods developed later in this book are based on it. Incidentally, this distribution came to be known as the "normal" probability distribution because around 1800 absolutely every

Table 7-1

Distribution of Ages of
Surgical Patients

Age of Patient	Percent of Total
0 up to 10	9
10 up to 20	10
20 up to 30	19
30 up to 40	20
40 up to 50	17
50 up to 60	14
60 up to 70	11

Source: St. Luke's Hospital, Maumee,
Ohio.

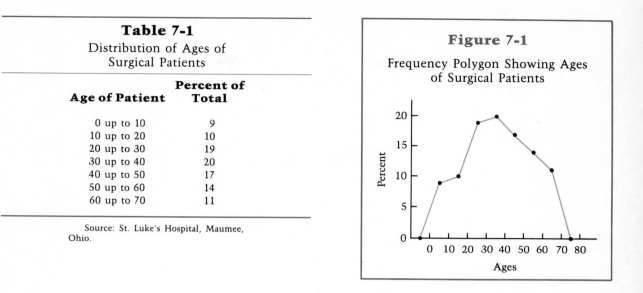

Figure 7-1

Frequency Polygon Showing Ages
of Surgical Patients

set of data had to follow it, or else statisticians would think something was wrong—not "normal"—with the data. The normal distribution requires that the data be of interval scale. The following
are examples of the type of interval-level measurements that frequently array themselves into a bell-shaped pattern when large samples are selected:

- weights of seven-year-old girls;
- heights of adult males;
- IQ scores;
- blood-pressure readings.

Table 7-1 and Figure 7-1 illustrate this bell-shaped or "normal" tendency for a real set of data: the ages of surgical patients at a hospital
in northwestern Ohio.

Characteristics of a Normal Probability Distribution

The normal distribution has the following characteristics:

1. The normal curve has a single peak at the precise center of the
distribution. The mean, median, and mode—which in a normal
distribution are equal—are all located at the peak. Therefore,

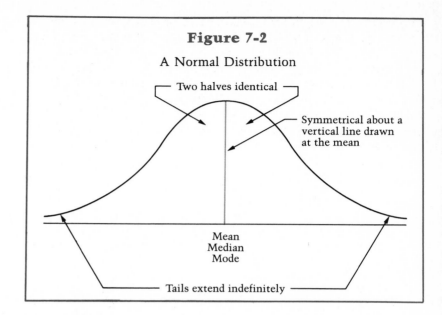

Figure 7-2

A Normal Distribution

exactly one-half, or 50%, of the area created by the bell-shaped curve is below the center of the distribution, and exactly one-half of the area is above it.

2. A normal probability distribution is *symmetrical* about its mean. If you were to "fold" the probability distribution along its central value, the two halves would be identical.

3. The normal curve falls off smoothly in a "bell shape" and the two tails of the probability distribution extend indefinitely in either direction. In theory, the curve never actually touches the *X*-axis (see Figure 7-2).

The "Family" of Normal Probability Distributions

Strictly speaking, there is not just one normal probability distribution. Rather, there is an entire "family" of related normal probability distributions. If you are studying the annual incomes of a group of male employees, that probability distribution might follow the normal distribution with a mean of $22,394 and a standard deviation of $842. The annual incomes of a group of female employees could, on the other hand, follow the normal distribution with a mean of $17,652 and a standard deviation of $797. In general, for each value of a mean and a standard deviation there is a particular normal probability distribution. When either the mean or the standard deviation changes, a new normal probability distribution is formed.

Figure 7-3 illustrates three normal probability distributions. Each of them has a mean of 30 but different standard deviations. Figure 7-4 shows a set of normal probability distributions that have different means but the same standard deviation of 6. Finally, Figure 7-5 portrays two normal probability distributions for which both the means and the standard deviations are different. Although each of

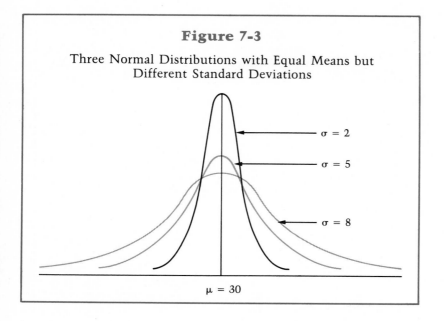

Figure 7-3

Three Normal Distributions with Equal Means but Different Standard Deviations

$\sigma = 2$

$\sigma = 5$

$\sigma = 8$

$\mu = 30$

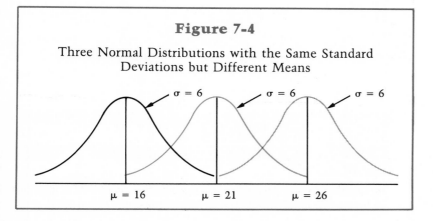

Figure 7-4

Three Normal Distributions with the Same Standard Deviations but Different Means

$\sigma = 6$ $\sigma = 6$ $\sigma = 6$

$\mu = 16$ $\mu = 21$ $\mu = 26$

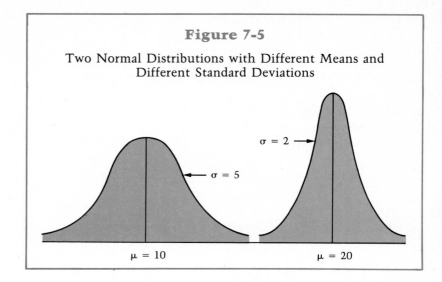

Figure 7-5

Two Normal Distributions with Different Means and
Different Standard Deviations

the normal probability distributions displayed in Figures 7-3, 7-4,
and 7-5 differs somewhat in appearance, each is still a member of
the "family" of normal probability distributions.

Areas Under the Normal Curve

Now we wish to calculate the probability that a normally distributed
quantity will fall within a specified interval. For example, we might
need to know the probability that Mr. Brown will stay in the labor
force until he reaches the age of 70. This information could be
used to plan contributions to a retirement fund.

Probabilities for continuous variables correspond to areas under
a smooth curve. Consequently, computing probabilities in such instances is merely a question of finding a desired area under the
normal curve. In dealing with normal probability distributions, three
relationships are used extensively:

1. About 68% of the area under the normal curve is within one
   standard deviation of the mean.
2. The area within two standard deviations of the mean is about
   95% of the total.
3. Virtually all (99.73%) of the area under the normal curve is
   within three standard deviations of the mean.

For example, if a normal probability distribution has a mean of 20 and a standard deviation of 4, then

1. About 68% of the area is between 16 and 24, found by 20 ± 1(4).
2. About 95% of the area is between 12 and 28, found by 20 ± 2(4).
3. Virtually all the area is between 8 and 32, found by 20 ± 3(4).

Problem

The mean daily temperature for a city in Hawaii is 21°C. The standard deviation is 2.5°C. What percent of the temperatures are within 1, 2, and 3 standard deviations of the mean?

Solution

1. About 68% of the mean daily temperatures are between 18.5° and 23.5°C, found by $\mu \pm 1(\sigma) = 21 \pm 1(2.5)$.
2. About 95% of the mean daily temperatures are between 16° and 26°C, found by $\mu \pm 2(\sigma) = 21 \pm 2(2.5)$.
3. Virtually all of the mean daily temperatures are between 13.5° and 25.5°C, found by $\mu \pm 3(\sigma) = 21 \pm 3(2.5)$.

A reminder: cover the answers in the margin.

A test is given to males between the ages of 17 and 19 to measure anxiety. The mean score is computed to be 50, the standard deviation 6. About 95% of the males have an anxiety score within what range?

Self-Review 7-1

38 and 62: $\mu \pm 2(\sigma)$
$= 50 \pm 2(6)$
$= 50 \pm 12$

The Standard Normal Probability Distribution

As noted, there are many normal probability distributions—one for each pair of values for a mean and a standard deviation. While this makes the normal probability distribution very versatile in describing many different real-world situations, it would be very awkward to provide tables of areas for each such normal probability distribution. An efficient method for overcoming this difficulty is available. This method calls for *standardizing* the distribution. To find the area between a value of interest (X) and the mean (μ), first we compute the distance between the value (X) and the mean (μ), then we express that difference in units of standard deviation. In other words, we compute the value

$$z = \frac{X - \mu}{\sigma}$$

where

X is any observation of interest.

μ is the mean of the normal distribution.

σ is the standard deviation of the normal distribution.

Finally, the desired area under the curve, or probability, is found by referring to a table whose entry corresponds to the calculated value of z.

The value of z just computed is alternately called the **standardized value,** the **standard normal deviate,** or **z score.** The value of z actually follows a normal probability distribution with a mean of zero and a standard deviation of one unit. This probability distribution is known as the **standard normal probability distribution.**

Problem

The ages of patients admitted to the coronary-care unit of a hospital are normally distributed with a mean of 60 years and a standard deviation of 12 years. What is the computed z value (standardized value) for a patient (a) age 78? (b) age 45?

Solution

(a) $X = 78$ Computing z:

$\mu = 60$

$\sigma = 12$

$$z = \frac{X - \mu}{\sigma}$$

$$= \frac{78 - 60}{12}$$

$$= 1.5$$

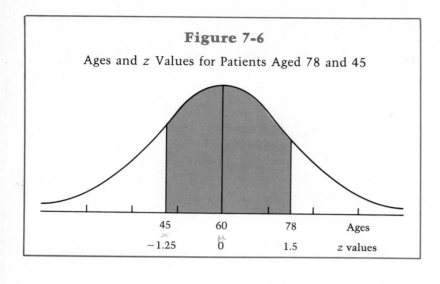

Figure 7-6

Ages and z Values for Patients Aged 78 and 45

| 45 | 60 | 78 | Ages |
| -1.25 | 0 | 1.5 | z values |

(b) $X = 45$
$\mu = 60$
$\sigma = 12$

Computing z:

$$z = \frac{X - \mu}{\sigma}$$

$$= \frac{45 - 60}{12}$$

$$= -1.25$$

Note that a z value merely transforms a selected value to a deviation from the mean expressed in standard deviation units. The location of the two ages (78 and 45) is shown in Figure 7-6.

Referring to the ages of patients, what is the z value ($\mu = 60$ years, $\sigma = 12$ years) for a patient:

a. aged 68?
b. aged 33?

Self-Review 7-2

a. 0.67

$$\frac{68 - 60}{12} = 0.67$$

b. $\frac{33 - 60}{12} = \frac{-27}{12} = -2.25$

Chapter Exercises

1. The monthly food expenditures of families of five on welfare were studied. The mean amount spent was $125 and the standard deviation $20. Assuming the monthly expenditures are normally distributed:

a. Standardize the expenditure of $105. That is, convert the expenditure of $105 to a z value.
b. Standardize the expenditure of $145.
c. Remembering that the distribution of expenditures is normal, what percent of the welfare families will spend between $105 and $145 a month?

2. A mathematics instructor studied the lengths of time required for students to complete the final examination. She found that the mean time was 90 minutes and the standard deviation 10 minutes. Assuming the lengths of time are normally distributed:
a. 95% of the lengths of time will fall between what two times?
b. 99.7% of the students will complete the final examination between what two time intervals?

Comparing Scores on Different Scales

Scores or values of different scales, or in different units, can be compared by converting them to z values. This can best be explained by an illustration.

Problem
A mentally disabled person takes a special anxiety test and scores 84 on it. The scores for this test are normally distributed, with a mean of 80 and a standard deviation of 8. He also takes a mechanical aptitude test designed especially for the disabled and scores 28 on it. The scores of this test are normally distributed about a mean of 20. The standard deviation is 6. Transform the scores to z values to compare the performance of this handicapped person on the two tests.

Solution
The z value for the anxiety score is 0.5, found by

$$z = \frac{X - \mu}{\sigma}$$

$$= \frac{84 - 80}{8}$$

$$= 0.5$$

The z value for the mechanical aptitude test score is 1.33, found by:

see p. 194

(a) 90 + 20 min
90 − 20 min

70 min to 110 min

(b) 90 + 30
90 − 30

60 min to 120 min

A) $\frac{.4250}{2).9500} = 1.96$

90 ± (1.96)(10)
90 ± 19.6
90 ± 19.6
109.6 − 70.4

b) 90 ± (2.97)(10)
90 ± 29.7
119.7 − 60.3

$\frac{.4985}{2).9970}$

$$z = \frac{X - \mu}{\sigma}$$

$$= \frac{28 - 20}{6}$$

$$= 1.33$$

Interpretation: the performance of the handicapped person on the anxiety test is slightly above average. This means that relative to other handicapped persons who took the test, this person's score is 0.5 standard deviations above the mean (above average). The performance on the mechanical aptitude test is well above average. That is, relative to other handicapped persons taking the test this person's score is 1.33 standard deviations above average.

It should be noted that a negative z value would indicate a below-average performance.

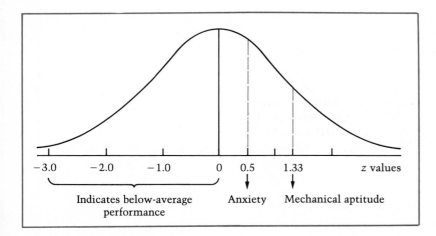

The mean age of prisoners in a state prison is 40 years and the standard deviation 10 years. The ages are normally distributed. The scores on a test measuring the social consciousness of the prisoners are also normally distributed, with a mean of 500 and a standard deviation of 100.

A prisoner age 38 scored 750 on the test. Compare his relative position within the two distributions. $\frac{750 - 500}{100} = 2.5$

$\frac{38 - 40}{10} = -0.2$

$\frac{X - M}{Sd}$

Self-Review 7-3

Age. z value is −0.2, slightly below average age.

Social consciousness: z value +2.50, indicating well above average.

Chapter Exercises

3. Martha and Frank Clark are newlyweds. Both are under 25 years of age and both are college graduates. Their combined income is $30,000 a year. The combined average income of all newlyweds under 25 and who are college graduates is $28,000. The standard deviation of that distribution is $5,000 and it is normal. Comment on the relative position of the Clarks' income.

4. The hourly wages of two people working in the trades are compared. Neil Holzmann, a carpenter, earns $12.00 per hour. Joe Bevilacqua, a plumber, earns $14.00 per hour. A survey of both trades in the same city reveals the following information (assume both distributions are normal):

	Plumber	Carpenter
Average	$15.00	$10.00
Standard deviation	1.75	1.25

Compare the relative position of the two tradesmen within their respective trades.

z values are also useful in determining what percent of a group of observations will be located *between* two values, or the probability that an observation will occur between two values. Again, we will use a problem to illustrate.

Problem

A research study revealed that the amount spent by persons seeking a seat on the city council in medium-sized cities is normally distributed, with a mean of $6,000 and a standard deviation of $1,000. The question to be explored is: What percent of the candidates seeking office spend between $6,000 and $7,250?

Solution

The z value corresponding to $7,250 is 1.25, found by

$$z = \frac{X - \mu}{\sigma}$$

$$= \frac{\$7,250 - \$6,000}{\$1,000}$$

$$= 1.25$$

The actual amounts spent, and their corresponding z values, are depicted by the graph that follows. The upper scale shows the actual amounts spent and the lower scale the z values. Notice that the mean converts to 0 on the z scale.

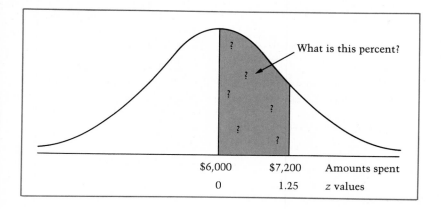

With that information, we can now determine what percent of candidates spend between $6,000 and $7,250. Percents derived from z scores have already been computed and organized into standard tables (in this text, see Appendix C, "Normal Probability Distribution"). The values found in such tables are actually *areas under the standard normal curve, or probabilities.*

To find the percent under the normal curve corresponding to a z value of 1.25, first go down the left column of Appendix C to a z of 1.2. Then move horizontally to the column headed 5 (1.2 or 1.20 + .05 = 1.25, the value you are seeking) and read the decimal.

Table 7-2

A Portion of a z Table (The Normal Probability Distribution, Appendix C)

	SECOND DECIMAL PLACE OF z									
z	0	1	2	3	4	5	6	7	8	9
⋮	⋮	⋮	⋮	⋮	⋮	⋮	⋮	⋮	⋮	⋮
1.0	.3413	.3438	.3461	.3485	.3508	.3531	.3554	.3577	.3599	.3621
1.1	.3643	.3665	.3686	.3708	.3729	.3749	.3770	.3790	.3810	.3830
1.2	.3849	.3869	.3888	.3907	.3925	.3944	.3962	.3980	.3997	.4015
1.3	.4032	.4049	.4066	.4082	.4009	.4115	.4131	.4147	.4162	.4177
1.4	.4192	.4207	.4222	.4236	.4251	.4265	.4279	.4292	.4306	.4319
⋮	⋮	⋮	⋮	⋮	⋮	⋮	⋮	⋮	⋮	⋮

It is .3944. To convert to percents, the number in the table is multiplied by 100, or more simply the decimal point is moved two places to the right. Converted to a percent, then, .3944 is 39.44%. Interpreting this outcome, 39.44% of the candidates for city council in medium-sized cities spend between $6,000 and $7,250 on their campaigns. Table 7-2 on page 201 summarizes this procedure.

---

* HW 2/26/85

Self-Review 7-4

The weights of boxes of Crunchy breakfast cereal are normally distributed with a mean of 450 grams and a standard deviation of 2 grams. What percent of the boxes weigh between

a. $z = \dfrac{X - \mu}{\sigma}$

a. 450 and 454 grams?
b. 447.3 and 450 grams?

$= \dfrac{454 - 450}{2}$

$= 2.00$

47.72% from Appendix C.

b. $\dfrac{447.3 - 450.0}{2} = -1.35$

41.15%.

Recall that one of the characteristics of a normal distribution is its symmetry—that is, the left half of the curve is identical to the right half. Thus, the percent of total observations between the z values of 0 and +1.00 (34.13%), is the same as the percent between 0 and −1.00 (also 34.13%). Shown diagrammatically:

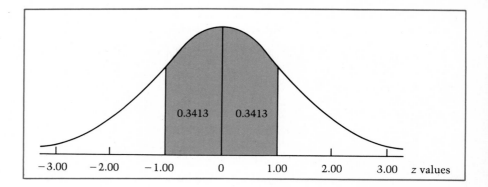

What percent of the z values are between −1.00 and +1.00? The answer is obtained by $0.3413 + 0.3413 = 0.6826 = 68.26\%$.

Problem

To illustrate this concept further, suppose Cardy Halzey decides to run for city council. What is the likelihood (probability) that she will spend between $5,000 and $8,000?

Solution

Note that, because the question asked about a probability (not a percent), the answer must also be in the form of a probability—that is, a decimal. The problem can be divided into two parts.

1. What is the probability of a campaign expenditure between $5,000 and $6,000 ($6,000 is the mean)?

$$z = \frac{X - \mu}{\sigma}$$

$$= \frac{\$5,000 - \$6,000}{\$1,000}$$

$$= -1.00$$

The decimal for -1.00 (from Appendix C) is 0.3413.

2. What is the probability of a campaign expenditure between $6,000 (the mean) and $8,000?

$$z = \frac{X - \mu}{\sigma}$$

$$= \frac{\$8,000 - \$6,000}{\$1,000}$$

$$= 2.00$$

The decimal for 2.00 (from Appendix C) is 0.4772.

Adding 0.3413 and 0.4772 gives 0.8185. Thus, the probability that Halzey will spend between $5,000 and $8,000 is 0.8185. The various components of this problem are shown in the following diagram:

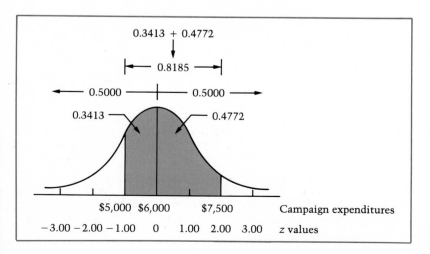

Probabilities for Unbounded Intervals

In the previous section we computed the percent of observations occurring between two values, or—stated a different way—the probability that a particular observation would occur between two points. In this section, this process is reversed. We will be interested in the percent of the observations above or below a given point.

Problem

What percent of the candidates spend $8,000 *or more* on their campaign for city council?

Solution

Note in the previous diagram that 50% (0.5000) of the candidates spend $6,000 or more. The decimal representing the amounts spent between $6,000 and $8,000 is 0.4772. Subtracting 0.4772 from 0.5000 gives 0.0228, or 2.28%. This is the percent of candidates who spend $8,000 or more on the election.

Self-Review 7-5

a. 7.93%

$$z = \frac{2.35 - 2.50}{0.75} = 0.2$$

Refer to appendix C for 0.0793.

b. 9.54%, found by

$$z = \frac{2.45 - 2.50}{0.75}$$

$$= 0.07$$

$$z = \frac{2.63 - 2.50}{0.75}$$

$$= 0.17$$

Decimals from Appendix C are 0.0279 + 0.0675 = 0.0954

c. 0.18%, found by

$$z = \frac{3.5 - 2.5}{0.75} = 1.33$$

Then 0.5000 − 0.4082 = 0.0018
= 0.18%

The grade point averages at La Siesta University are normally distributed with a mean of 2.5 and a standard deviation of 0.75.

a. What percent of the students have a grade point average between 2.35 and 2.5?
b. What percent of the students have a grade point average between 2.45 and 2.63?
c. What percent of the students have a grade point average of 3.5 or above?

Chapter Exercises

5. The nicotine content of a certain brand of king-sized cigarette is normally distributed with a mean of 2.0 mg and a standard deviation of 0.25 mg. What is the probability that a cigarette has a nicotine content
a. of 1.6 mg or less?
b. between 1.6 and 2.1 mg?
c. of 2.1 mg or more?

6. The mean yearly amount of sap collected for making maple syrup is ten gallons per tree. The distribution of the amounts collected per tree is normal with a standard deviation of 2.0 gallons.

a. What is the probability that a particular tree produces 13.0 gallons of syrup or more a year?
b. What is the probability that a particular tree produces between 9.0 and 12.5 gallons of syrup a year?

Finding Areas from a Known Probability

In our previous work on the normal curve we were asked to compute the probability of an observation occurring between two values. For example, we determined the probability that Cardy Halzey would spend between $6,000 and $8,000 on her campaign. Now the process is reversed. We are given a probability and asked to compute the range of X values.

Problem

Returning once more to the expenditures of the candidates for city council, where $\mu = \$6,000$ and $\sigma = \$1,000$, 10% of the candidates spent what amount or more on their campaigns? The components of this problem are shown graphically:

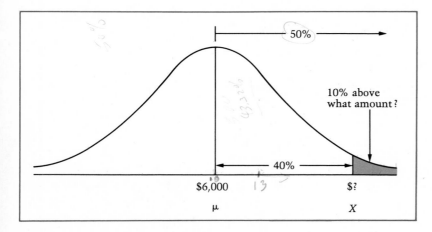

Solution

As noted, 50% of the expenses are $6,000 or more, and 10% are some unknown amount or more. The unknown amount is designated by X. Logically, 40% of the campaign expenditures are between $6,000, or μ, and X. The z value corresponding to 0.4000 is obtained by searching in Appendix C. The closest decimal is 0.3997. This corresponds to a z value of 1.28.

Handwritten notes in right margin:

6(a)

$z = \dfrac{X - \mu}{\sigma}$

$z = \dfrac{13 - 10}{2}$

$z = \dfrac{3}{2}$

$z = 1.5$ .5000

From Table → .4332

.0668

or 6.68%

(b) $z = \dfrac{X - \mu}{\sigma}$

$z = \dfrac{12.5 - 10}{2} = 1.25$

$z = \dfrac{9.0 - 10}{2} = .5$

Values from table

$z = 1.25$ .3944

$z = .5$ + .1915

.5859

58.59%

$M \pm z\sigma$

$6000 + (1.28)(1000)$

Inserting the z value of 1.28 and solving for X we find

$$z = \frac{X - \mu}{\sigma}$$

$$1.28 = \frac{X - \$6,000}{\$1,000}$$

Then

$$\$1,280 = X - \$6,000$$

$$X = \$7,280$$

Thus, about 10% of the candidates spend \$7,280 or more on their bids for election to the city council.

Self-Review 7-6

50% - 3%
= 47%

A grade point average of 3.91.

$$1.88 = \frac{X - 2.50}{0.75}$$

$$1.88(0.75) = X - 2.50$$
$$1.41 = X - 2.50$$
$$X = 3.91$$

From the previous self-review: the mean grade point average of students at La Siesta University was 2.5, the standard deviation 0.75. The top 3% among students are to be given special recognition. What grade point average (or above) does a student need in order to receive special recognition?

Chapter Exercises

7. The heights of adult males follow a normal distribution with a mean of 70 inches and a standard deviation of 2.6 inches. How high should a doorway be constructed so that 98% of men can pass through it without having to bend?   *-26.54*

HW

8. A certain brand of passenger-car tire has a mean tread life of 40,000 miles. The tread life is normally distributed with a standard deviation of 3,000 miles. Five percent of the tires will lose all their tread before they go what distance?

50% - 5% = 45% or .45 from table

(8) $\mu = 40,000$ mi
$\sigma = 3,000$ mi

$-1.64 = \frac{X - 40,000}{3,000}$

$-4920 = X - 40,000$

$35,080 = X$
mi

The Normal Approximation to the Binomial

The binomial distribution was discussed in the previous chapter. Recall that the binomial is a discrete probability distribution. The binomial is characterized by p, the probability of a success, and n, the number of trials. Figure 7-7 shows four binomial distributions,

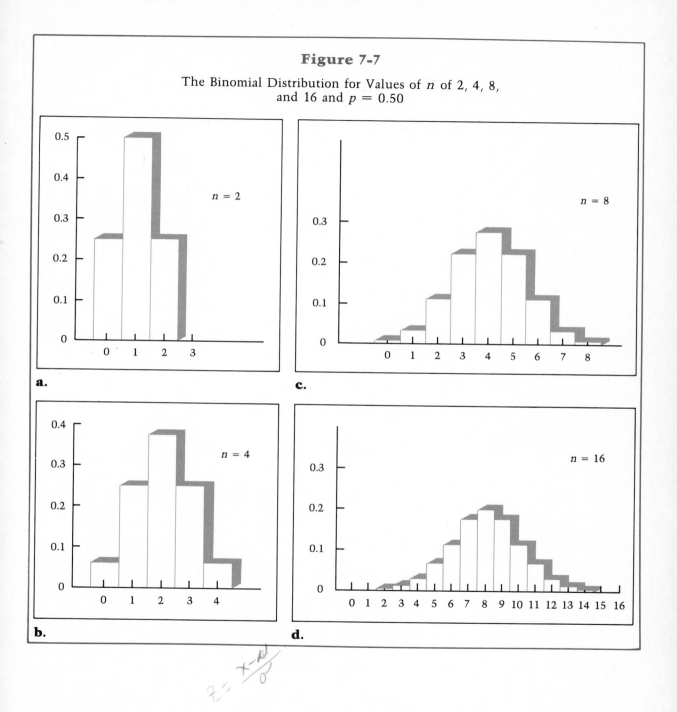

Figure 7-7

The Binomial Distribution for Values of n of 2, 4, 8, and 16 and $p = 0.50$

where p is 0.50 and n is 2, 4, 8, or 16. Notice that as the size of n increases, the distribution begins to approximate the normal probability distribution.

Thus, the normal distribution may be used to estimate binomial probabilities. As a general rule of thumb, when np and $n(1 - p)$ are both at least five, the normal probability distribution is a very good approximation for the binomial distribution.

Problem

The city's legal affairs director reports that, based on past experience, 60% of automobiles reported stolen are recovered and returned to their owners. In a month in which 100 automobiles are stolen, what is the probability that 65 *or more* will be recovered and returned to their owners?

Solution

Recall from Chapter 6 that there are four characteristics of a binomial distribution:

1. Each outcome can be classified into one of two mutually exclusive categories. The outcome (the recovery of the automobile) is either "a success" or "a failure."
2. The binomial distribution results from a count of the number of successes (the number recovered, in this case).
3. Each trial is independent, meaning that the outcome of one trial (successfully recovering the car or failing to recover it) does not affect the outcome of any other attempt.
4. The probability of success on each trial (0.60) is the same from trial to trial. This means that the probability of your stolen car being recovered is 0.60 and the probability of your friend's car being recovered is also 0.60.

For this binomial probability distribution the total number of trials is 100, and the probability of success on each trial is 0.60. The normal probability distribution may be used to approximate the binomial distribution because both np and $n(1 - p)$ exceed five.

$$np = 100(0.60) = 60$$
$$n(1 - p) = 100(1 - 0.60) = 40$$

The steps needed to determine the probability that 65 or more cars will be recovered and returned are:

Step 1 Compute the mean and the variance of the binomial distribution. They are computed as follows (μ is used as the designation of the mean, σ^2 as the designation for the variance):

$$\mu = np$$
$$\sigma^2 = np(1-p)$$

The mean is 60 and the variance 24, found by

$$\mu = 100(0.60) = 60$$
$$\sigma^2 = (100)(0.60)(0.40) = 24$$

Step 2 Determine the z value corresponding to 65 using the standard normal distribution.*

$$z = \frac{X - \mu}{\sigma}$$

where σ is the standard deviation of the distribution found by $\sqrt{\sigma^2}$; the square root of the variance of 24 is 4.90. Computing z:

$$z = \frac{X - \mu}{\sigma}$$

$$= \frac{65 - 60}{4.90}$$

$$= 1.02$$

Step 3 Find the probability of a z value of 1.02 or greater occurring. The area, or probability, between 0 and 1.02 standard normal deviates is determined by referring to Appendix C. Go down the left column to 1.0 and then read the probability headed by the column marked 2. The probability is 0.3461. Thus, the area beyond 1.02 standard normal deviates is 0.1539, found by 0.5000 − 0.3461. Shown graphically:

* Technically speaking, the z value should be based on an X of 64.5 instead of 65. If this were done, it would be referred to as a "continuity correction," because we are approximating a discrete distribution using a continuous distribution. However, when the sample size is large, this correction will have little or no effect.

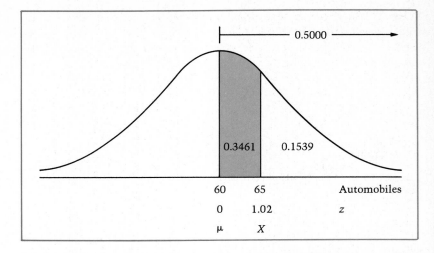

Interpreting, the probability is 0.1539 that in a month in which 100 automobiles are stolen, 65 *or more* will be recovered and returned. To put it another way, 65 or more automobiles will be recovered and returned 15.39% of the months in which 100 automobiles are stolen.

A Computer Output

The program that follows, written in the BASIC programming language, gives the complete distribution of the probabilities of recovery for 0, 1, 2, 3, . . . , 100 automobiles stolen during the month. (Recall that the probability of recovery is 0.60.) Notice that this binomial distribution is very close to a normal distribution.

Binomial Distribution

Successes	Probability	Successes	Probability
0	.000	46	.001
1	.000	47	.003
2	.000	48	.004
3	.000	49	.007
•	•	50	.010
•	•	51	.015
•	•	52	.021
41	.000	53	.029
42	.000	54	.038
43	.000	55	.048
44	.000	56	.058
45	.001	57	.067

Successes	Probability
58	.074
59	.079
60	.081
61	.080
62	.075
63	.068
64	.059
65	.049
66	.039
67	.030
68	.022
69	.015
70	.010
71	.006
72	.004
73	.002
74	.001
75	.001
76	.000
77	.000
•	•
•	•
•	•
98	.000
99	.000
100	.000

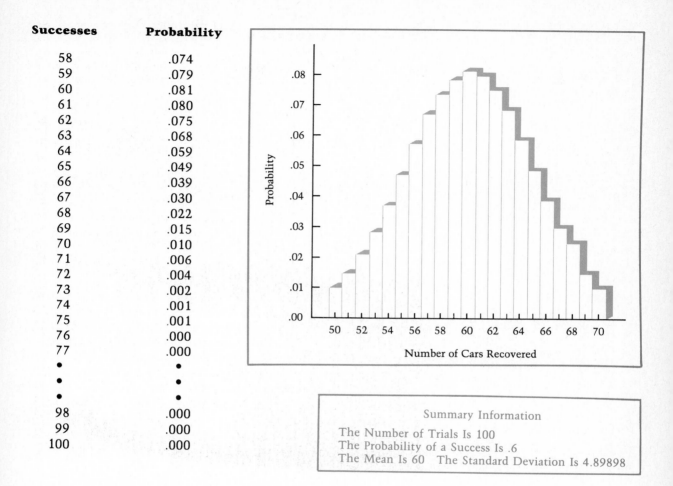

Summary Information

The Number of Trials Is 100
The Probability of a Success Is .6
The Mean Is 60   The Standard Deviation Is 4.89898

Self-Review 7-7

a. Yes. $n = 100$, $p = 1/5 = 0.20$
Both np and $n(1 - p)$ are greater than 5.
$np = 100(0.20) = 20$
$n(1 - p) = 100(1 - 0.20) = 80$

b. $\mu = np = (100)(0.20) = 20$
$\sigma^2 = np(1 - p)$
$= (100)(0.20)(80)$
$= 16$
$\sigma = \sqrt{16} = 4$

The instructor in Geology 115 gives only a final examination. It consists of 100 multiple-choice questions with five possible answers for each. The instructor announces that at least 30 correct answers will be required to pass the course.

Assume that you are enrolled in the course but have never attended class or read any of the assignments. You decide, however, to take the final examination.

a. Can the normal approximation to the binomial be used? Why?
b. Determine the mean and the standard deviation.
c. Calculate the probability that you will pass Geology 115.
d. Portray the probabilities and other parts of the problem graphically.

c. $z = \dfrac{X - \mu}{\sigma} = \dfrac{30 - 20}{4} = 2.5$

Appendix C, probability for z of 2.5 = 0.4938. Then 0.5000 − 0.4938 = 0.0062. Your chance of passing is only 6/10 of 1%.

d.

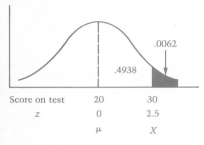

Score on test	20	30
z	0	2.5
	μ	X

Chapter Exercises

9. The Wayward Inn, a 300-room resort hotel, experiences an 85% occupancy rate, on the average, in January. Use the normal approximation to the binomial to find
   a. The probability that at least 260 rooms are occupied in January.
   b. The probability that fewer than 240 rooms are occupied in January.

10. An airline manager estimated that, on the average, 8% of the passengers flying across the Atlantic experience some airsickness. What is the probability that on a transatlantic flight of 150 passengers, at least five will experience some airsickness?

Summary

The normal distribution often describes an observed frequency distribution such as weights, heights, scores, and other human characteristics. A normal distribution is described by the mean μ and the standard deviation σ. There are many normal distributions—a different one for each combination of a mean and standard deviation.

A normal distribution has the following characteristics: it has a single peak, it is bell shaped, symmetrical, and the tails drop off indefinitely. The mean, median, and mode are equal.

Computations involving a normal distribution employ the standard normal distribution. An observation can be standardized by subtracting the mean from the observation and dividing the difference by the standard deviation. In symbols:

$$z = \frac{X - \mu}{\sigma}$$

where

   z is the standardized normal value.
   X is the selected observation.
   μ is the mean of the normal distribution.
   σ is the standard deviation of the normal distribution.

Three broad classes of applications were considered:

1. z values were used to compare a particular observation relative to the entire distribution of observations. Illustration:

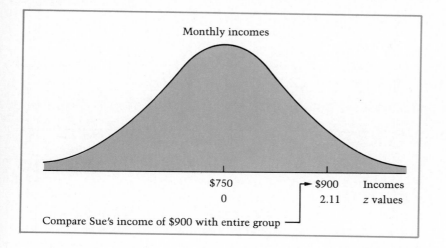

2. The percent of observations occurring between two values, or the probability of an observation occurring between two points, were computed. Illustration:

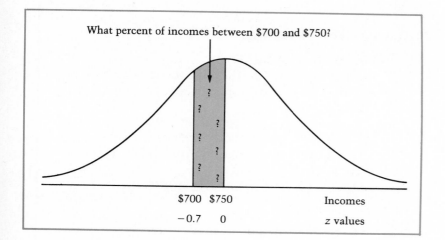

3. When the percent of observations under the normal curve were given, a corresponding X value, or values, were determined. Illustration:

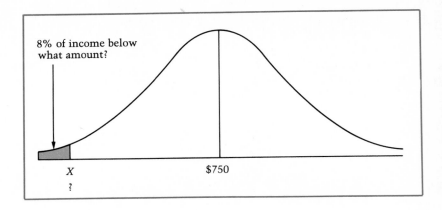

8% of income below
what amount?

X
?

$750

The normal probability distribution can be used instead of the binomial probability distribution when both np and $n(1 - p)$ are at least 5. To construct a normal approximation to the binomial, the mean μ and the variance σ^2 of the binomial distribution are needed. The mean μ is found by np, and σ^2 by $np(1 - p)$. Then the probability for each possible outcome, designated by X, is found by

1. Computing the z value using $z = \dfrac{X - \mu}{\sigma}$

2. Referring to the standard normal distribution (Appendix C) for the probability corresponding to the computed z value.

Chapter Outline

Normal Distribution

I. Normal Distribution

A. Objective. To describe the most widely occurring probability distribution and to apply it when appropriate.

B. Characteristics

1. It requires at least interval level data.

2. There is a "family" of normal distributions. There is one distribution for a mean of 2,000 and a standard deviation of 100 and another for a mean of 2,000 and a standard deviation of 250, and so on.

3. The mean and standard deviation are used for its construction.

4. It is symmetrical and bell shaped in appearance.

C. Widely used relationships

1. About 68% of all the observations will fall within one standard deviation of the mean.

2. About 95% occur within two standard deviations of the mean.
3. Virtually all the observations are within three standard deviations of the mean.

II. Applications of the Normal Distribution

A. Standardized score, called z value, found by

$$z = \frac{X - \mu}{\sigma}$$

B. Typical problems
① A customer waited 8 minutes at a grocery store checkout line. Compare this time with all customers.
② What percent of the customers will wait between 5 and 6 minutes?
③ Ten percent of the customers will check out in less than how many minutes?

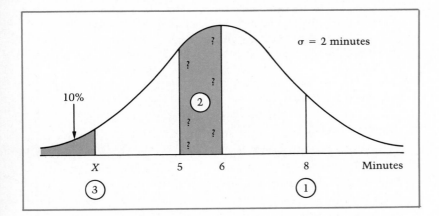

III. Normal Approximation to the Binomial Distribution

A. Used when both np and $n(1 - p)$ are at least 5.
B. Procedure. For a particular outcome designated as X:
1. Compute z by $\frac{(X - \mu)}{\sigma}$ where $\mu = np$ and $\sigma^2 = np(1 - p)$.
2. Finding the probability associated with the z value by referring to Appendix C.

Chapter Exercises

11. An applicant for a position with the Norton Corporation earned a score of 60 on the Sales Aptitude Test, 150 on the Personnel Human Factors Evaluation Test, and 90 on the Financial Management Test. The scores of each of these tests approximate a normal distribution. The means and standard deviations of the tests are

Test	Mean	Standard Deviation
Sales Aptitude Test	50	7
Personnel Human Factors Test	120	25
Financial Management Test	85	5

    a. Compute a z value for each of the applicant's test scores.
    b. On which of the tests did the applicant do the best relative to the entire group?
    c. Regarding the Sales Aptitude Test, what percent scored higher than the applicant?
    d. Based on the applicant's performance on the three tests, to what area would you assign her—sales, personnel, or finance? Why?

12. The number of hours a week a college student devotes to study is normally distributed, with a mean of 30 hours and a standard deviation of eight hours.
    a. What percent of the students will study less than 20 hours?
    b. What percent will study more than 35 hours?
    c. Out of a class of 200 students, how many will study between 25 and 35 hours?

13. Memorial Hall is used as a site for both student-sponsored concerts and intercollegiate basketball games. Attendance figures for both the concerts and the basketball games are normally distributed. The means and standard deviations are

	Concerts	Basketball
Mean attendance	8,600	7,200
Standard deviation	560	600

    a. What percent of the basketball games have an attendance of 8,000 or over?
    b. Tickets to concerts are priced so that, if 7,000 tickets are

sold, expenses are covered. If fewer than 7,000 are sold, the students lose money, and if more than 7,000 are sold they make a profit. What percent of the time will the students lose money?

14. A juice dispenser is set to fill cups with an average of 7 ounces of fruit juice. The standard deviation of the process is 0.3 ounces.
    a. If 8-ounce cups are used, what percent of them will overflow?
    b. What percent of the cups will have less than 6.5 ounces of juice in them?

15. Crash Airlines is studying its service from Detroit to Tampa. Historical data show that the average number of passengers per flight is 235.6, and the standard deviation of the normal distribution is 36.3 passengers.
    a. What is the probability a particular flight will carry more than 260 passengers?
    b. What is the probability a particular flight will have fewer than 180 passengers?
    c. What is the probability a particular flight will have between 240 and 250 passengers?
    d. What is the probability a particular flight will have fewer than 240 passengers?

16. The owner of a fast-food restaurant keeps records of the daily hamburger demand, which is normally distributed with a mean of 260 pounds and a standard deviation of 20 pounds.
    a. What percent of days will the owner need more than 310 pounds of hamburger?
    b. The owner does not want to run out of meat more than 1% of days. How many pounds should he order every day?

17. A researcher reports that the average heart rate of rats is 120 beats per minute and that 45% of all rats tested had heart rates in the range 120 to 140. Assume that these rates are normally distributed.
    a. What standard deviation is implied by these data? (Hint: use the formula for z to compute the standard deviation.)
    b. What percent of the animals have heart rates in the range 100 to 120?
    c. What percent of the rats have heart rates in excess of 140?
    d. What are the standard scores corresponding to 120 and 140?

18. If an elm tree is less than three feet high and is transplanted in the spring, it has a 40% chance of survival. If 50 such trees are transplanted, what is the probability that 25 or more will survive? What is the probability that between 18 and 20 will survive?

19. A particular type of birth-control device is effective 90% of the time when used correctly. If the device is employed 300 times, how many times would you expect the device to fail? What is the probability it would fail 35 or more times?

Chapter
Achievement Test

Do all the problems. Then check your answers against those given in the Answer section of the book.

I. Multiple-Choice Questions. Select the correct answer (5 points each).

1. Which of the following is a normal distribution?

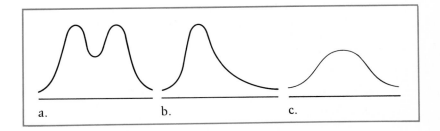

a.   b.   c.

   a. only picture a
   b. only picture b
   c. only picture c
   d. all of the above
   e. only b and c

2. A *z* value is
   a. the deviation from the mean divided by the standard deviation
   b. the mean of a distribution
   c. always positive
   d. all of the above are true
   e. none of the above are true

3. Which of the following is *not* a correct statement regarding a normal distribution?
   a. half of the observations are greater than the mean and half are less than the mean
   b. the mean and median are equal
   c. the same number of observations can be expected between the mean and plus one standard deviation as between plus one standard deviation and plus two standard deviations
   d. a normal distribution approaches the *X*-axis but never reaches it

4. In a normal distribution about 95% of the observations are contained within these limits:
   a. $\mu \pm 0.50\ \sigma$
   b. $\mu \pm 1.00\ \sigma$
   c. $\mu \pm 2.0\ \sigma$
   d. $\mu \pm 3.0\ \sigma$
5. The standard normal z value
   a. may be used to compare the relative position of an observation on two normal distributions
   b. is calculated by the formula $(X - \mu)/\sigma$.
   c. is used to calculate the probabilities associated with any normal distribution
   d. all of the above are correct
   e. none of the above are correct
6. The area under the normal curve between the mean and a z value of -1.34 is
   a. 0.0901
   b. 0.4032
   c. 0.4099
   d. 0.4236
   e. none of the above
7. Calculate the unknown z value for the following standard normal curve:

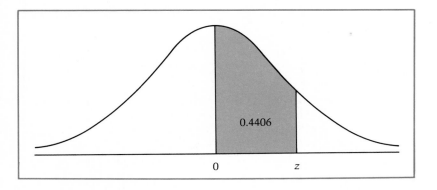

0.4406

0   z

   a. 2.00
   b. 1.96
   c. -1.56
   d. 1.56
   e. none of the above

A package of 75-watt standard light bulbs is labelled "Average life is 1,000 hours." Assume that the life of a light bulb is normally distributed with a standard deviation of 250 hours.

8. What is the probability that a bulb will last 1,300 hours or less?
   a. 0.9500
   b. 0.3849
   c. 0.1151
   d. 0.8849
   e. none of the above

9. What is the probability that a bulb will last 570 hours or less?
   a. 0.4573
   b. 0.9573
   c. 0.0436
   d. 0.0427
   e. none of the above

10. Eighty percent of the bulbs will last at least how many hours?
    a. 790
    b. 1,490
    c. 1,210
    d. 1,412.5
    e. none of the above

II. Computation Problems (25 points each).

11. In 1979, the number of daily admissions at Riverside Hospital followed a normal distribution with a mean of 39.52 admissions per day and a standard deviation of 6.29 admissions per day.
    a. What is the standardized z value corresponding to 50 admissions?
    b. What is the standardized z value corresponding to 25 admissions?
    c. What percent of days had more than 50 admissions?
    d. What percent of days had fewer than 25 admissions?
    e. On the busiest 10% of the days, what was the minimum number of admissions?

12. Suppose the U.S. Postal Service were to claim that 80% of letters mailed in New York City destined for Los Angeles are delivered within three working days. To verify this claim, you mail 200 letters to various destinations in the Los Angeles area. Compute the following probabilities:
    a. What is the probability that more than 150 of the letters will be delivered within three working days?
    b. Fewer than 148?
    c. Between 150 and 160?
    d. The probability is 10% that what number or more will be delivered within a minimum of three working days?

HIGHLIGHTS
From Chapters 5, 6, and 7

In the last three chapters you were introduced to the fundamental concepts of probability. The various methods for determining a probability, the rules for combining several probabilities, and two different kinds of probability distributions were described, discussed, and illustrated.

Key Concepts

1. A **probability** is a number that expresses the likelihood that a particular event will happen. There are three types, or definitions, of probability:
   a. **Classical probability** Each of the possible outcomes is equally likely. If there are n outcomes, the probability of a particular outcome is $1/n$.
   b. **Relative frequency** The total number of times the event has occurred in the past, divided by the total number of observations.
   c. **Subjective** The assignment of probability is based on whatever information is available—personal opinion, hunches, and so on.

2. The two fundamental rules of probability are the **rule of addition** and the **rule of multiplication.**
   a. Rule of addition. If two events A and B are mutually exclusive, the probability that one or the other of the events will occur is

$$P(A \text{ or } B) = P(A) + P(B)$$

   This is called the **special rule of addition.** If the events are *not* mutually exclusive, the probability that one or the other will occur is

$$P(A \text{ or } B) = P(A) + P(B) - P(A \text{ and } B)$$

Key Concepts (*continued*)

where $P(A \text{ and } B)$ is the probability of the joint occurrence of the two events. This is called the **general rule of addition.**

b. Rule of multiplication. If two events A and B are unrelated (independent), the probability of their joint occurrence is the product of the two probabilities.

$$P(A \text{ and } B) = P(A) \cdot P(B)$$

This is called the **special rule of multiplication.** If the two events are related (not independent), the probability of their joint occurrence is

$$P(A \text{ and } B) = P(A) \cdot P(B|A)$$

where $P(B|A)$ refers to the probability that the event B occurs, given that A has already happened. This is called the **general rule of multiplication.**

3. A **probability distribution** is a listing of the outcomes that may occur and the corresponding probability associated with each of the outcomes.

4. The two critical features of a probability distribution are
   a. The likelihood of a particular outcome must be between 0 and 1.0.
   b. The sum of all possible mutually exclusive outcomes must total 1.0.

5. There are two types of **random variable**—discrete and continuous. A **discrete** random variable can assume only certain distinct values and is usually the result of counting. A **continuous** random variable may assume an infinite number of values within a given range.

6. The **binomial distribution** is an example of a discrete random variable. It has the following characteristics:
   a. Each outcome is classified in one of two mutually exclusive categories.
   b. Each trial is independent.
   c. The probability of a success remains the same from trial to trial.
   d. It results from counting the number of successes in the total number of trials.

Key Terms

Probability
Experiment
Outcome
Event
Classical probability
Relative frequency
Subjective probability
Special rule of addition
General rule of addition
Mutually exclusive events
Independent events
Joint probability
Complement rule

Conditional probability
Special rule of multiplication
General rule of multiplication
Permutations
Combinations
Probability distribution
Random variable
Discrete probability distribution
Continuous probability distribution
z value
z score
Standard normal distribution

Key Symbols

$P(A)$ The probability of the event A happening.

$P(B|A)$ The conditional probability the event B will occur, given that A has already happened.

$_nP_r$ The number of permutations of n objects taken r at a time.

$_nC_r$ The number of combinations of n objects taken r at a time.

z The z score, or the value of the standard normal distribution.

Review Problems

1. A study is made of younger students' understanding of television commercials. A sample of 200 youngsters are first shown various television commercials, then they are questioned about each commercial, and it is determined whether or not they understood it. Results are as follows:

	Age 10 - 12	12 - 14	16 - 18	Total
Understood	30	40	50	120
Did not understand	40	30	10	80
Totals	70	70	60	200

   If a youngster is randomly selected
   a. What is the probability that the commercial was understood?
   b. What is the probability that a youngster is both 10 - 12 years old and understood the commercial?
   c. Given that the student did not understand the commercial, what is the probability that he or she was 10 - 12 years old?
   d. What is the probability that a youngster either did not understand the commercial or is 10 - 12 years old?

2. A social club at a large university has 300 members who are registered in one of three different colleges. Their colleges of registration and grade point averages, on a 4-point scale, are summarized in the following table:

College	Grade Point Average (GPA) Greater than 3.0	Between 2.0 and 3.0	Lower than 2.0	Total
Arts and Sciences	20	40	30	90
Business	60	50	10	120
Education	20	60	10	90
Total	100	150	50	300

   A student is selected at random from the list of club members.
   a. What is the probability that the student is registered in the Business College?
   b. What is the probability that the student has a GPA greater than 3.0 *and* is in the College of Education?
   c. What is the probability that the student has a GPA lower than 2.0 *or* is in the College of Education?

3. The small town of Sugar Grove has two ambulances. Records show that the first ambulance is in service 80% of the time and the second 70% of the time.
   a. What is the probability that both are in service when needed?
   b. What is the probability that at least one is in service when needed?

4. The security manager of a large building reports the probability is 0.05 that a fire alarm will not operate when needed. If there are three alarms in the building, what is the probability none of the alarms will operate during a particular fire? What is the probability at least one will operate during a particular fire?

5. An appliance dealer sponsors advertisements on both radio and TV. A study of 200 customers revealed that 80 had seen the advertisement on TV, 120 had heard the radio advertisement, and 40 had both seen the TV and heard the radio advertisements.
   a. What is the probability that a customer heard both the radio and TV advertisements?
   b. What is the probability that a customer heard the advertisement *either* on the radio *or* on TV?

6. The sheriff needs new tires for his road patrol cars. The probability that he will purchase Michelin, Goodyear, or Uniroyal brands of tires are, respectively, 0.20, 0.30, and 0.40. What is the probability that he will not purchase any of these brands?

7. According to police records, 70% of the reckless drivers are fined, 50% have their driver's licenses revoked, and 40% are both fined and have their licenses revoked. What is the probability that a particular reckless driver will either have her license revoked or be fined?

8. There are 20 men and 10 women in a statistics class. Three students are to be randomly selected to take a trial test. What is the probability that all three are men? What is the probability that at least one is a man?

9. The suicide-prevention unit in a particular city estimates that 20% of the callers are serious about taking their lives. If on a particular day the unit received ten calls, what is the probability that none of the callers were serious? What is the probability that at least two were serious?

10. The area covered by a painter with one gallon of paint is normally distributed with a mean of 400 square feet and a standard deviation of 60 square feet. If the manufacturer specifies that one gallon should cover between 375 and 450 square feet, what percent of the time will the painter exceed the upper limit of the manufacturer's specification? What percent of the time will he be within the manufacturer's limit?

11. It is estimated that 70% of the law-school graduates in a particular state pass the state bar examination on the first try. What is the probability that in a group of 12 students nobody passes? At least one student passes? More than half pass?

12. It is estimated that 80% of all household plants are overwatered by their owners. In a group of 50 plants, what is the probability that more than 35 were overwatered? That up to 45 were overwatered?

13. The mean life of socks used by the Army is 60 days with a standard

deviation of 12 days. Assume the life of the socks is normally distributed. If 1 million pairs are issued, how many would need replacement after 50 days? After 70 days?

14. Past records indicate that 30% of the students enrolling at a particular university graduate within five years of their entrance. In a group of 14 newly enrolled students, what is the probability that fewer than half will graduate within the five years? More than 8? .906 + .008

15. The number of sandwiches sold by the Deli Bar in the student union is normally distributed with a mean of 215 and a standard deviation of 20. On what percent of days does the Deli Bar sell more than 200 sandwiches? Less than how many sandwiches are sold 10% of the time?

Case Analysis

The McCoy's Market Case

(Data for these two cases can be found in the first Chapter Highlights, pp. 126 - 130.)

1. (a) What is the probability that a randomly selected customer is a female?
   (b) What is the probability that a randomly selected customer shops during the midweek period?

2. (a) What is the probability that a randomly selected customer is either female or a midweek shopper? (b) What is the probability the customer is both female and a midweek shopper? (c) Do these events appear to be independent?

3. Would the distribution of the customers' gender be most likely to follow a binomial or a normal probability distribution?

4. Would the distribution of the number of items selected most likely follow a discrete or a continuous probability distribution.

The St. Mary's Emergency Room Case

1. (a) What is the probability that a patient will be admitted? (b) What is the probability that the patient will arrive during the first shift?

2. (a) What is the probability that a patient will either be admitted or arrive during the first shift? (b) What is the probability that a patient will arrive during the first shift and be admitted? (c) Do these two events appear to be independent?

3. What distribution comes closest to fitting the number of staff required?

4. What distribution would describe the cost per patient?

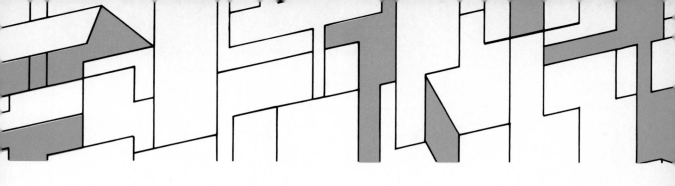

8

Sampling Distributions and the Central Limit Theorem

OBJECTIVES

When you have completed this chapter, you will be able to
- Describe the purposes and methods of sampling.
- Describe the central limit theorem.
- Develop and describe the distribution of sample means.
- Compute a confidence interval.

Introduction

What do the following four problems have in common? (1) The President of the United States wants to know how many voters in Texas will vote for him in November; (2) Revlon wants to know how many women will buy newly-developed Siren Red lipstick, if it is marketed; (3) the General Services Administration needs to know how many government-owned computers are not fully utilized; and (4) the National Organization for Women (NOW) wishes to know what percent of all banks in the United States have at least one female director.

These problems have one thing in common: the information would be very difficult to obtain. It would be almost impossible (and very expensive) for the President to contact every potential voter in Texas by November. Likewise, it would be impossible and, again, very expensive, for Revlon to contact all women in the world (or even just in the United States) and ask each one to try a complimentary stick of Siren Red. NOW would find it difficult to contact every bank in the United States.

The usual, less costly, way to obtain this kind of information is to take a **sample.** As mentioned earlier, a sample is a smaller group selected from the population of interest. The objective of studying the smaller group is to obtain information about the whole **population.** The President may hire a polling service, who in turn may sample 2,000 Texans on their political preference in the forthcoming election. NOW might select, say, 50 banks at random and determine what percent have at least one female director. If 15 out of the 50, or 30%, have at least one female director, NOW might reasonably conclude that 30 percent of *all* banks in the United States have at least one female director.

These four situations illustrate how information obtained from a sample can be used to say something about the population. As noted earlier, this is the process called statistical inference. Recall from Chapter 1 that statistical inference is the process of reasoning from specific instances or data to general conclusions about the entire group or population. In this chapter and those that follow, you will be introduced to statistical techniques that are based on **probability sampling.**

Probability Sampling Each member of the population of interest has a known likelihood of being included in the sample.

If probability sampling is not used, sample results may not be representative of the entire population. In such cases, it is said that the results are **biased.** To illustrate, Revlon might contact 400 women in New York City about Siren Red mainly because the group's location is convenient to Revlon's New York office. The results, however, might not be representative of all women in the United States; the color red may suggest "warmth" to women in New Mexico and "aggressiveness" to those in New York, or vice versa.

In this chapter we will first discuss several methods of selecting a sample and situations in which each method might be used. Then the sampling distribution of a widely used statistic, namely the sample mean, will be examined. Finally, the notion of a confidence interval will be developed.

Methods of Probability Sampling

There are several types of probability sampling. We will study simple random sampling, systematic random sampling, and stratified random sampling. Each of these types of sampling has a similar goal: *to allow chance to determine the items or persons that make up the sample.* While each sample outcome may not be predictable when taken alone, groups of samples are quite predictable. The population of interest can consist either of items (such as all cassette decks produced by Pioneer during the past month) or of persons (such as all the registered voters in Precinct 9). Other examples of a population might be all the banks in the United States, all the fish in a pond, or all the students now attending Yale University.

Random Sampling

The most widely used method of probability sampling is **simple random sampling** or, as it is often called, random sampling.

> **Random Sample**   A sample chosen so that each member of the population has the same chance of being selected for the sample.

A random sample does not just happen, nor should just any collection of objects carelessly be called a random sample. The selection of a random sample must be planned with care. One method we can use is similar to a lottery. First we write the name of each member of the population, or an identifying number—such as a social security number—on a slip of paper. Then the slips of paper are thoroughly mixed. Finally, the desired number of slips of paper are drawn. As an illustration, let us look at NOW's interest in the percent of all banks having at least one female director. The group could obtain a list of banks in the United States and assign a number to each of them. If number 121 were drawn first, that bank (say, the First National Bank of Arizona) would be contacted. This procedure would be repeated until the desired sample size had been selected.

It is very important that the population of interest be precisely defined. NOW, for example, would have to decide whether branch banks with their own board of directors should be included on the list; a member of the population should be listed only once.

An easier way to select a random sample employs a **table of random digits.** A portion of such a table generated by a computer and found in Appendix B is

71529	51996	99289	44268	42759	72434	54402
11776	17395	61317	63290	17067	18408	08992
82437	75248	23715	61194	62175	11149	44793
14997	08398	37662	90175	65331	02562	38020
55317	50018	64380	49047	57111	41641	25427
47422	53721	11419	38616	72171	21523	80967
09540	89442	52381	35035	15884	64273	96028

The table is generated in a random fashion, that is, each of the ten digits has the same chance of being included. Hence, blind chance determines the outcome of the selection process and bias does not enter the procedure.

The entire population from which the sample will be drawn is arranged in some systematic fashion (perhaps alphabetically); next, each item is assigned a number. Let us assume our research involves the response of psychiatric patients at Palm Hospital to a new drug. The population consists of 70 psychiatric patients who receive the drug. An identification number is assigned to each patient, starting with 00 and ending with 69. First, a starting point in the table is randomly selected. You could close your eyes and place a pencil down on the page. Suppose, for example, that number were 14 (see the table).

71529	51996	99289	44268	42759	72434	54402
11776	17395	61317	63290	17067	18408	08992
82437	75248	23715	61194	62175	11149	44793
14997	08398	37662	90175	65331	02562	38020
55317	50018	64380	49047	57111	41641	25427
47422	53721	11419	38616	72171	21523	80967
09540	89442	52381	35035	15884	64273	96028

starting point · second patient · third patient

The patient identified by number 14 becomes part of the sample. You can move in any direction you choose. Perhaps you could look at the second hand of a clock and move in the direction in which it is pointing. Moving horizontally to the right, the first two digits in the next column are 08. Therefore, patient 08 is also part of the sample. The next patient in the sample is number 37. The following random number is 90. Since no patient is assigned that number, it is omitted. This procedure is continued until the desired number of patients is obtained for the sample.

A reminder: cover the answers in the margin.

Following is a class roll for a beginning course in social science. Three students are to be randomly selected to investigate and report on a new community program for the mentally retarded. Suppose you had written numbers 1 through 38 on slips of paper and then had randomly selected numbers 29, 5, and 11. Which students would be included in the sample?

Self-Review 8-1

29 Vicki Rista
 5 Timothy Cowan
11 James House

WINTER QUARTER   PRELIMINARY CLASS ROSTER

SS 101   03   INTRO TO SOC SC

2:00 PM   3:40 PM   MW   UH 422   W MARCHAL

NAME	RANK		NAME	RANK
1. August, Nancy M.	FR		20. McFarlin, Ireatha	FR
2. Benner, Robert A.	JR		21. Meinke, Denise M.	JR
3. Brenner, Susan M.	SO		22. Morrison, David D.	JR
4. Clark, Richard C.	FR		23. Navarre, Garry G.	JR
5. Cowan, Timothy J.	JR		24. Oyer, David	SO
6. Cross, Jill M.	SO		25. Pastor, Virginia Marie	SO
7. Daschner, John H.	SO		26. Pickens, Mitch A.	SO
8. Figliomeni, Michael A.	JR		27. Price, Doug C.	SO
9. Grady, Walter P.	JR		28. Rawson, Jeryl L.	FR
10. Heinrichs, James M.	SO		29. Rista, Vicki A.	SO
11. House, James D.	JR		30. Schmidt, Randy F.	SO
12. James, Phyllis E.	JR		31. Sherman, Mike J.	SO
13. Kimmel, Kurt D.	SO		32. Shull, Karen A.	SO
14. Lach, Jerry William	JR		33. Snow, Sue A.	FR
15. Lehman, Tim J.	SO		34. Straub, Jeff J.	SO
16. Lenz, Matthew H.	SO		35. Turco, Greg W.	SO
17. Martin, Diane M.	SO		36. Von Hertsenberg, Kevin	JR
18. Mason, Craig D.	SO		37. Wagner, Holly S.	SO
19. McCullough, Randy N.	SO		38. Yamada, Jay A.	JR

Systematic Random Sampling

If the population is large, say the 15,599 students enrolled at the University of Utah, a list of all students could be obtained and a random starting point selected. Suppose the 28th student were chosen. To save the effort of selecting additional random numbers, a constant number could be added to the starting number. If we decided to add 100 to the starting number, the students identified by numbers 28, 128, 228, 328, and so on would become members of the sample. This systematic procedure is aptly called **systematic random sampling.**

The results of a systematic sample will be just as representative as those from a simple random sample. It should not be used, however, if there is any possibility of bias in the ordered list. For example, if you were doing a study on absenteeism, it would be unwise to take a systematic sample of every seventh day. The results would be unduly affected by the starting day. If a Monday had been selected as the starting day, then all the other days selected would also be Mondays, and it is well known that absences are higher on Mondays.

> **Systematic Random Sample**   The members of the
> population are arranged in some fashion. (They may be num-
> bered 1, 2, 3,   . . . , listed alphabetically, or ordered by
> some other method.) A random starting point is selected.
> Then every kth element is chosen for the sample.

Refer back to the previous self-review. Suppose that this sample is
to consist of every ninth student enrolled in the class after a starting
student has been randomly selected from students numbered 1
through 9. Suppose this starting point is the fourth student. Which
students will be in the sample?

Self-Review 8-2

4	Richard Clark
13	Kurt Kimmel
22	David Morrison
31	Mike Sherman

Stratified Random Sampling

In planning some types of surveys, it is desirable that the sample
be representative not just of the population as a whole, but of certain
subdivisions or groups within it. For example, Revlon might want
to ensure that women in each of ten regions of the United States
be included in their research project to determine the market potential of Siren Red lipstick. To accomplish this goal, we would divide
the country into ten geographical regions and randomly select a
sample of women within each region. Each woman selected would
be asked to try Siren Red and to report her reaction. Dividing the
country into regions is called **stratifying the population.**
Other traits commonly used to form strata are age, income level,
and political party affiliation. The manner in which the sample is
gathered may be either nonproportional or proportional to the total
number of members in each stratum.

> **Stratified Random Sample**   After the population of
> interest is divided into logical strata, a sample is drawn from
> each stratum or subgroup.

A **stratified random sample,** therefore, is formed by identifying the natural subgroupings in the population and by selecting
an independent random sample from each stratum or subgroup. Obviously, it is important that we define the strata carefully to ensure
that each member belongs to only one subgroup.

The advantage of stratified sampling is that one member of a subgroup is usually quite similar to other members of that subgroup while being quite different from the members of other subgroups. If, for example, the research project involved surveying executives on the role of government in business, the population could be stratified into bankers, executives of large firms, executives of small firms, and so on. Bankers tend to think alike about the role of government in business, but their opinions might differ drastically from those of small business executives. Unless the subgroups differ significantly from each other, nothing is gained by stratification. For example, if our research were concerned with the attitude of college students about compulsory military training, a stratification of the population (all college students) into those from the Far West, East, and so on, would not seem justified unless the researcher believes that geographic location affects attitude toward military service.

Self-Review 8-3

Refer back to Self-Review 8-1. Separate the population into three strata: freshmen (FR), sophomores (SO), and juniors (JR). Suppose it had been decided that one freshman, five sophomores, and four juniors would constitute the sample. Randomly select the required number from each stratum.

There are a large number of possibilities. One is:

FR	5	Snow
SO	9	Mason
	3	Daschner
	11	Oyer
	12	Pastor
	8	Martin
JR	7	Lach
	10	Navarre
	4	Grady
	3	Figliomeni

No doubt the composition of your sample will be different.

Cluster Sampling

A popular specialized kind of stratified sampling is known as **cluster sampling.** Cluster sampling sometimes becomes necessary in studies that cover large geographic regions, simply in order to keep the cost of sampling down to a reasonable amount. This method saves travel time from one location to another by forcing the sample to be grouped within a particular geographic region. As an illustration, assume our research objective were to record the opinions of Democrats, Republicans, and Independents in a large city with respect to a number of political issues. As a first step, we would subdivide the city into smaller units (precincts). Such smaller units are often called **primary units.** Next, several primary units could be selected at random—let us say, primary units 2, 5, 8, and 11. Finally, persons in each of these primary units would be selected at random and interviewed. By concentrating our efforts in four primary units, we would have saved considerable travel time and expense.

The previous discussion of scientific sampling techniques is greatly oversimplified. Should you become involved in a research project requiring sampling, we urge you to consult books specializing in sampling methods.

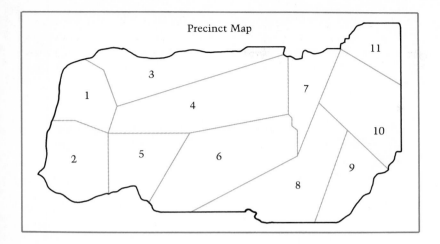

Precinct Map

Chapter Exercises

1. A study of hotel accommodations in a metropolitan area reveals the existence of thirty such facilities. The city's Convention and Visitors Bureau is surveying charges per day for single-occupancy rooms. The daily rates for this statistical population are $25, $35, $22, $25, $30, $24, $25, $20, $25, $24, $28, $24, $28, $25, $27, $25, $35, $25, $30, $25, $17, $25, $21, $18, $16, $24, $21, $21, $13, $19.

   a. Using the random numbers in Appendix B, draw a simple random sample of size six from this population.

   b. Select a systematic random sample by randomly choosing a starting point among the first five hotels and then including every fifth observation.

   c. If the last ten hotels on the list are all "cut-rate" hotels, describe how you could select a sample of four regular hotels and two "cut-rate" hotels.

Sampling Error

The previous discussion of various scientific sampling methods emphasized the importance of trying to choose a sample so that every member of the population has a known chance of being selected. In other words, the sample should be representative of the population. It would be unreasonable, however, to expect the sample characteristics to match the population *exactly*. The mean of the sample might be different from the population mean by *chance alone*. The standard deviation of the sample will probably be different from the population standard deviation. We can, therefore, expect some divergence between the **sample statistics** (such as the mean and standard deviation) and the corresponding population values, known as **parameters.** This difference is known as the **sampling error.**

> **Sampling Error** The difference between the value of the population parameter and its corresponding sample statistic.

The idea of sampling error can be illustrated with a very simple example. Suppose your five grades (the population) to date in this course were 69, 86, 82, 70, and 98. A sample of two grades is selected at random from this population of grades to estimate your mean grade. They are, let us say, 70 and 86. The mean of this sample is 78. The mean of another sample of two grades (69 and 98) is 83.5. The mean of all five grades (the population) is 81. Notice that sampling errors of −3 and 2.5 are made in estimating the population mean.

Given this potential sampling error, how can political polls, for example, make accurate predictions about the behavior of the voting population based only on sample results? How can a quality-control inspector in a manufacturing plant make a decision about the quality of a product after inspecting only a sample of ten parts? The answers to these questions are explained by developing a sampling distribution for the sample mean.

The Sampling Distribution of the Mean

As noted in the previous section, the means of samples of a specified size selected from a population vary somewhat from sample to sample. When all the sample means that might occur are organized into a probability distribution, the resulting distribution is called the **sampling distribution of the mean.**

> **Sampling Distribution of the Mean** A probability distribution of all possible sample means of a given size selected from a population.

Table 8-1

Lengths of Service of Six Social Workers

Social Worker	Length of Service	
Don	4	the population mean μ is 3.16667, found by
Gary	3	
Sue	3	
Bob	2	$\mu = \dfrac{4+3+3+2+4+3}{6}$
Kirk	4	
Sandra	3	$= 3.16667$

The problem that follows illustrates how such a distribution is constructed.

Problem

The six social workers in our district are to be studied with respect to length of service, income, education, and so on. The respective lengths of their service are listed in Table 8-1, above. (Note that these data constitute a population.)

Suppose that, as an experiment, we were to select a sample of four social workers and calculate the mean length of their service. What would be the sampling distribution of the mean? Stated differently, what values could result and how likely are they?

Solution

All possible samples of size four and their means are

Sample Names				Length of Service	Sample Mean $\overline{X}$
Don	Gary	Sue	Bob	4, 3, 3, 2	12/4 = 3.00
Don	Gary	Sue	Kirk	4, 3, 3, 4	14/4 = 3.50
Don	Gary	Sue	Sandra	4, 3, 3, 3	13/4 = 3.25
Don	Gary	Bob	Kirk	4, 3, 2, 4	13/4 = 3.25
Don	Gary	Bob	Sandra	4, 3, 2, 3	12/4 = 3.00
Don	Gary	Kirk	Sandra	4, 3, 4, 3	14/4 = 3.50
Don	Sue	Bob	Kirk	4, 3, 2, 4	13/4 = 3.25
Don	Sue	Bob	Sandra	4, 3, 2, 3	12/4 = 3.00
Don	Sue	Kirk	Sandra	4, 3, 4, 3	14/4 = 3.50
Don	Bob	Kirk	Sandra	4, 2, 4, 3	13/4 = 3.25
Gary	Sue	Bob	Kirk	3, 3, 2, 4	12/4 = 3.00
Gary	Sue	Bob	Sandra	3, 3, 2, 3	11/4 = 2.75
Gary	Sue	Kirk	Sandra	3, 3, 4, 3	13/4 = 3.25
Gary	Bob	Kirk	Sandra	3, 2, 4, 3	12/4 = 3.00
Sue	Bob	Kirk	Sandra	3, 2, 4, 3	12/4 = 3.00

The mean of the sample means is 3.16667, found by

$$\frac{3.00 + 3.50 + 3.25 + \cdots + 3.00}{15}$$

$$= \frac{47.50}{15}$$

Note that the mean of the sample means is the same as the population mean computed in Table 8-1.

The sample means are presented in the form of a probability distribution. (See Table 8-2.) Logically, it is called the **sampling distribution of the mean.**

The sampling distribution of the means in Table 8-2 is merely the probability of the sample means for all possible samples of size four taken from the population of six social workers. The distribution is also plotted in Figure 8-1. Immediately following Figure 8-1 is the distribution of the original population values.

Note these predictable patterns:

1. The mean of the sampling distribution and the mean of the population are equal (3.16667 in this case).

2. The spread in the distribution of the sample means is smaller than the spread in the population values; the sample means range from 2.75 to 3.50, while the population ranges from a low of 2 to a high of 4.

3. The shape of the sampling distribution of the means and the shape of the population values are different. The shape of the distribution of sample means tends to be "bell shaped" and approximate the normal probability distribution.

In summary, we took random samples from a population and for each sample calculated a sample statistic (the mean length of service). Since each possible sample has a known chance of selection, the probability of a mean length of service being 2.75 years, 3.0 years, and so on can be determined. The distribution of these mean lengths of service is aptly called the sampling distribution of the mean.

Table 8-2

Sampling Distribution of the Mean

Sample Mean	Frequency	Probability
2.75	1	1/15 = 0.0667
3.00	6	6/15 = 0.4000
3.25	5	5/15 = 0.3333
3.50	3	3/15 = 0.2000
	15	1.0000

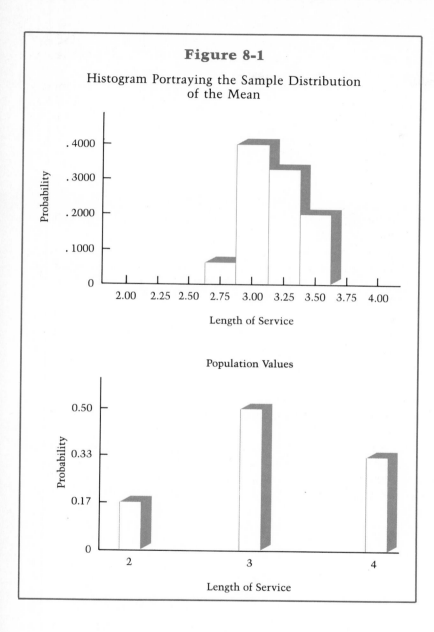

Figure 8-1

Histogram Portraying the Sample Distribution
of the Mean

Even though in practice we only see one particular random sample, in theory, any sample could arise. Consequently, we view the sampling process as repeated sampling of the statistic from its sampling distribution. This sampling distribution is then employed to measure how reasonable or likely a particular outcome might be.

Self-Review 8-4

A population consists of five prisoners in cell block M; the length of time each has spent in prison is shown below:

Name	Years
Dow	5
Smith	3
Artz	6
Kim	2
Batt	4

a. Compute the mean length of imprisonment for the population.
b. Select all possible samples of size two from the population. Compute the mean of each sample.
c. Does the mean of the sample means equal the population mean?
d. Give the sampling distribution of the means.
e. Plot the sampling distribution of the means and the population.
f. Is the sampling distribution tending to be bell shaped and beginning to approximate a normal distribution?
g. Cite evidence to show there is less spread in the sampling distribution compared with the population values.
h. Is the population normally distributed or nonnormal?

a. $20/5 = 4.0$

b.

	$\overline{X}$
Dow, Smith	4.0
Dow, Artz	5.5
Dow, Kim	3.5
Dow, Batt	4.5
Smith, Artz	4.5
Smith, Kim	2.5
Smith, Batt	3.5
Artz, Kim	4.0
Artz, Batt	5.0
Kim, Batt	3.0

c. Yes $40/10 = 4.0$, same as $\mu = 20/5 = 4.0$.

d.

X	f	Probability
2.5	1	$1/10 = 0.1000$
3.0	1	$1/10 = 0.1000$
0.5	2	$2/10 = 0.2000$
4.0	2	$2/10 = 0.2000$
4.5	2	$2/10 = 0.2000$
5.0	1	$1/10 = 0.1000$
5.5	1	$1/10 = 0.1000$
	10	1.0000

e.

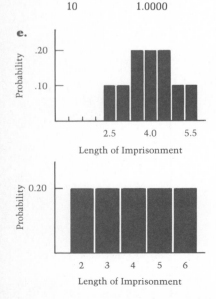

The Central Limit Theorem

The previous discussion of the sampling distribution of the mean is the basis for one of the most important theorems in statistics—the **central limit theorem.**

In effect, the theorem states that when the size of the sample is large enough, the distribution of the sample means is normal, regardless of the shape of the population from which the sample is drawn. Most statisticians use a sample of 30 or more as being "large enough" to employ the central limit theorem.

f. Yes.

g. The range for the sampling distribution is $5.5 - 2.5 = 3.0$. The range for the population is $6 - 2 = 4.0$.

h. Nonnormal.

> **Central Limit Theorem** Regardless of the shape of the population, the distribution of the sample means approaches the normal probability distribution as the sample size increases.
>
> The bigger the Sample, the better.

The central limit theorem tells us that the distribution of sample means selected from any shape population is approximately normal, provided the sample size is sufficiently large.

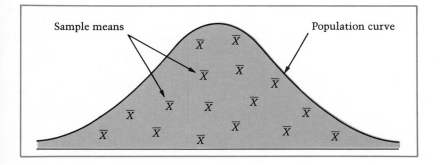

Confidence Intervals

The information just developed about the shape of the sampling distribution of $\overline{X}$ allows us to locate an interval that has a high probability of containing the population mean μ. For reasonably large samples the following can be said:

1. Ninety-five percent of the sample means selected from a population will lie within 1.96 standard deviations of the population mean μ.

2. Ninety-nine percent of the sample means will lie within 2.58 standard deviations of the population mean.

Intervals computed in this fashion are called the **95% confidence intervals** and the **99% confidence intervals.**

How are the values of 1.96 and 2.58 obtained? The 95% and 99% refer to the approximate percent of the time that similarly

Confidence Interval   A range within which the population parameter is expected to fall for a preselected level of confidence.

constructed intervals would include the parameter being estimated. The 95%, for example, refers to the middle 95% of the observations. Therefore, the remaining 5% is equally divided between the two tails. See the following diagram. The central limit theorem states that the distribution of the sample mean will be approximately normal; therefore Appendix C may be used to find the appropriate z values. Locate 0.4750 in the body of the table, then read the corresponding row and column total. It is 1.96; that is, the probability of being in the interval between $z = 0$ and $z = 1.96$ is 0.4750. Likewise, the probability of being in the interval between -1.96 and 0 is also 0.4750. Combining these two probabilities, the probability of being in the interval -1.96 to 1.96 is 0.95 (see the graph that follows). The z value corresponding to 0.99 is determined in a similar fashion.

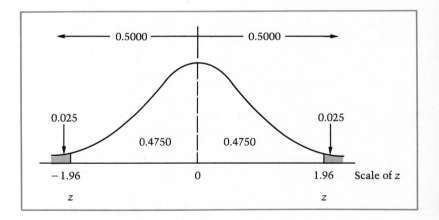

How do we construct the 95% confidence interval? To illustrate, assume our research involves the annual starting salary of graduates with a master's degree. We have computed the mean of the sample to be \$22,500 and the standard deviation of the sample means to be \$200. The 95% confidence interval is \$22,108 and \$22,892, found by \$22,500 $\pm$ 1.96(\$200). If 100 samples of the same size were selected from the population of interest and the corresponding 100 confidence intervals determined, one could expect to find the population mean in about 95 out of the 100 confidence intervals. Shown in a diagram for a few samples:

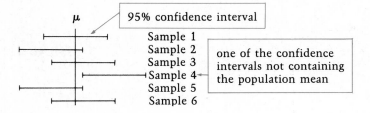

Standard Error of the Mean

In the previous section, the standard deviation of the sampling distribution was given as $200. It is called the **standard error of the mean,** often shortened to the **standard error.**

> **Standard Error of the Mean**   The standard deviation of the sampling distribution of the sample means.

The standard error is a measure of the variability of the sampling distribution. It is computed by

$$\sigma_{\overline{X}} = \frac{\sigma}{\sqrt{n}}$$

where

$\sigma_{\overline{X}}$ is the symbol for the standard error of the mean.

σ is the population standard deviation.

n is the sample size.

In most "real world" cases the population standard deviation is not known. However, if the sample size is 30 or more, the sample standard deviation s will closely approximate the population standard deviation σ. The formula for this approximation of the standard error of the mean then becomes:

$$\sigma_{\overline{x}} = \frac{s}{\sqrt{n}}$$

where s is the sample standard deviation.

Note how the standard error is affected by the size of the sample, which is recorded in the denominator. As the sample size n increases, the variability of the sample means decreases. This outcome is logical, because an estimate made with a larger sample should be subject to less variability.

Point Estimates and Interval Estimates

The data in Table 8-1 represented a population—the length of service of six social workers. The mean of this population was easily computed. However, in most cases, the very thing we are trying to estimate is a population parameter—the mean length of service of the social workers, for example. This parameter is unknown in practice, and we are trying to find its value. The single number with which we estimate a population parameter is called a **point estimate.**

> **Point Estimate**   The value, computed from a sample, which is used to estimate a population parameter.

A sample mean is a point estimate of the population mean. To estimate the mean age of purchasers, a distributor of stereo equipment records the age of a sample of 50 customers. The mean age of the sample is a point estimate of the mean age of the population of all purchasers.

However, a point estimate only tells part of the story. While we expect the point estimate to be close to the population parameter, we would like some way to measure how close it is. The **interval estimate** serves this purpose.

> **Interval Estimate**   A range of values within which we have some confidence that the population parameter lies.

$$\overline{x} \pm z \, \sigma_{\overline{x}} \qquad \sigma_{\overline{x}} = \frac{\sigma}{\sqrt{N}}$$

For example, we estimate the mean yearly income of a group of farmers to be $15,000. The range of that estimate could be from $14,000 to $16,000. We can describe how confident we are that the population parameter is in that interval by making a probability statement. The resulting confidence interval is an interval estimate of the population parameter. Confidence levels such as 95% and 99% are often used to indicate the degree of belief or credibility to be placed on a particular interval estimate of a population parameter. We might say, for example, that we are 90% sure that the mean yearly income is between $14,000 and $16,000.

Constructing the 95% and 99% Confidence Intervals

The 95% confidence interval is constructed by

$$\overline{X} \pm 1.96\, \sigma_{\overline{X}} = \overline{X} \pm 1.96\, \frac{s}{\sqrt{n}}$$

where $\overline{X}$ is the sample mean, s the sample standard deviation, and n the sample size. The 99% confidence interval is computed by

$$\overline{X} \pm 2.58\, \frac{s}{\sqrt{n}}$$

Problem

Construct a 95% confidence interval for the mean hourly wages of the geologists employed by the five top oil companies. For a sample of 50 geologists, $\overline{X} = \$15$, $s = \$3$.

Solution

The standard error of the sample mean is estimated to be $0.42, found by

$$\sigma_{\overline{X}} = \frac{s}{\sqrt{n}} = \frac{\$3}{\sqrt{50}} = \$0.42$$

Thus, the 95% confidence interval for μ is

$$\$15 \pm 1.96(\$0.42)$$
$$\$15 \pm \$0.82$$

This interval goes from $14.18 up to $15.82. If we were to repeat the sampling process many times, 95% of the intervals so constructed would include the population mean hourly wage. We feel quite confident (95%) that it is between $14.18 and $15.82.

Problem

We sampled 160 Hispanic families with respect to income, number of children, their ages, and so on. "What is the age of your youngest child?" was one of the questions asked. The sample mean was computed to be 6.7 years, the sample standard deviation was 2.5 years. Construct a 99% confidence interval for the mean age of the youngest child in all families.

Solution

To determine the 99% confidence interval, the standard error of the mean is needed. It is approximately 0.2 years, found by

$$\sigma_{\overline{x}} = \frac{\sigma}{\sqrt{n}} = \frac{s}{\sqrt{n}} = \frac{2.5}{\sqrt{160}} = 0.2 \text{ years}$$

Recall that 99% of the normal distribution is within 2.58 standard deviations of the population mean. Hence, the 99% confidence interval is

$$6.7 \pm 2.58(0.2)$$
$$6.7 \pm 0.5$$
$$6.2 \text{ years up to } 7.2 \text{ years}$$

Thus, we are 99% confident that the population mean age is between 6.2 years and 7.2 years. In other words, our maximum error is 0.5 years, which is one-half of the width of the interval.

In summary, the researcher selects the desired degree of confidence (such as 95% or 99%) and then proceeds to construct an interval corresponding to that percent. Note that only one confidence interval is calculated from a particular sample. The confidence interval either includes the population mean μ or it does not. Thus, probability statements can be made only before the sample is taken. We should be careful when interpreting a confidence interval. The level of confidence does *not* specify the probability that a sample mean is included in the interval, but rather the percent of times that similarly constructed intervals could be expected to include the population mean. Referring to the Problem-Solution, we can say that about 99% of the similarly constructed confidence intervals for the youngest Hispanic child would bracket the population mean. We cannot say that the probability is 0.99 that the population mean is in that interval.

1. A sample of 900 registered voters in the state were surveyed about age, political party, and so on. The mean of the sample was computed to be 42 years, the standard deviation 12 years. What is the 95% confidence interval for the population mean μ?

2. A sample of 312 grades on a mathematics test given nationwide reveals that the sample mean is 560 and the sample standard deviation is 120. What is the 99% confidence interval for the population mean?

Chapter Exercises

2. A simple random sample of size 300 is taken from a population of clerical workers in a particular city and the data on their hourly wages are recorded. The sample mean and the standard deviation are found to be $4.44 and $1.27, respectively. Construct a 90% confidence interval for the mean hourly wage of this population.

3. A simple random sample of 80 school children enrolled at Home Street school were asked the distance they travel to school. The sample mean and standard deviation are found to be 7.52 km. and 1.32 km., respectively. What is the 95% confidence interval for the population mean distance traveled to school?

Choosing an Appropriate Sample Size

A concern that usually arises when a statistical study is being designed is, "How many items should be in the sample?" If a sample is too large, money and effort are wasted collecting the data. Similarly, if the sample is too small, the resulting conclusions will be uncertain. The correct sample size depends upon three factors:

1. the level of confidence desired,
2. the variability in the population being studied, and
3. the maximum allowable error.

You, the researcher, select the level of confidence. As noted in the previous sections, confidence intervals of 95% and 99% are selected

Self-Review 8-5

1. $\sigma_{\bar{x}} = \dfrac{12}{\sqrt{900}} = \dfrac{12}{30} = 0.4$

Then

$42 \pm 1.96(0.4) = 42 \pm 0.784$
$= 41.216$ years
and 42.784
years

2. $\sigma_{\bar{x}} = \dfrac{120}{\sqrt{312}} = \dfrac{120}{17.66352} = 6.79$

Then

$560 \pm 2.58(6.79) = 560 \pm 17.52$
$= 542.48$ and 577.52
Would probably be rounded to 542 and 578.

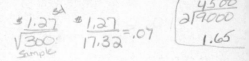

$\dfrac{\$1.27}{\sqrt{300}}$ sample $\dfrac{\$1.27}{17.32} = .07$

$\dfrac{4500}{2\overline{)9000}}$ 1.65

mean
$\$4.44 \oplus 1.65 (.07)\ 4.56$
$\$4.44 \ominus 1.65 (.07)\ 4.32$

most often. A 99% confidence level corresponds to a standard normal (z) value of ± 2.58 and a 95% confidence level yields $z = \pm 1.96$.

If the population is widely dispersed, a large sample is required. On the other hand, a small standard deviation (a homogeneous population) would not require as large a sample. The population variability is measured by the sample standard deviation s and the standard error of the sample mean is, of course, $s/\sqrt{n}$. Often a small "pilot" study is conducted in order to estimate s before the major study is undertaken.

Finally, the maximum allowable <u>error (E)</u> is the amount that was added and subtracted from the sample mean to obtain the limits of the confidence interval in previous sections. It is the amount of error the researcher is willing to tolerate. In general, it is one-half of the width of the corresponding confidence interval. A small allowable error will require a larger sample size, whereas a large allowable error will require smaller sample sizes.

We can express this interaction among these three factors and the sample size in the following formula:

$$z = \frac{E}{s/\sqrt{n}}$$

Solving this equation for n, the required sample is

$$n = \left[\frac{z \cdot s}{E}\right]^2 \qquad n = \frac{z \cdot \sigma}{E}$$

where

E is the maximum allowable error.

s is the estimate of the population standard deviation.

n is the size of the sample.

z is the standard normal value corresponding to the desired level of confidence.

Since the result of this computation is not always a whole number, the usual conservative practice is to round up any fractional result. For example, 72.1 would be "rounded up" to 73.

Problem

A study is to be conducted on the mean salary of mayors of cities with a population of less than 100,000. The error in estimating the mean is to be less than $100, and a confidence level of 95%

is desired. Suppose the standard deviation of the population is estimated to be $1,000. What is the required sample size?

Solution

The allowable error, $E,$ is $100. The value of z for a 95% level of confidence is 1.96. Substituting the values into the formula, the required sample size is determined to be

$$n = \left[\frac{(1.96)(\$1,000)}{\$100}\right]^2 = (19.6)^2 = 385$$

Thus, a sample of 385 is required. If a higher level of confidence were desired, say 99%, then a larger sample would also be required:

$$n = \left[\frac{(2.58)(\$1,000)}{\$100}\right]^2 = (25.8)^2 = 666$$

You are conducting a study designed to estimate the mean number of hours worked per week by suburban housewives. Your pilot study revealed that the population standard deviation is 2.7 hours. How large a sample should be selected if you wished to be 95% confident that the sample mean differs from the population mean by at most 0.2 hours?

Self-Review 8-6

$\frac{475}{95500} = 1.96$

$$n = \left[\frac{(1.96)(2.7)}{0.2}\right]^2$$

$n = 700$

Chapter Exercises

$\overline{X} \pm E.$

4. A company wishes to estimate the average starting salaries for security personnel in their plants. From a previous study they estimate that the standard deviation is $2.50. How large a sample should be selected in order to be 95% confident that the sample mean differs from the population mean by at most $0.50?

$$n = \left[\frac{(1.96)(2.50)}{0.50}\right]^2$$

$n = 96.04$

5. A meat packer is investigating the marked weight shown on links of summer sausage. A pilot study showed a mean weight of 11.8 pounds per link and a standard deviation of 0.7 pounds. How many links should be sampled in order to be 95% confident that the sample mean differs from the population mean by at most 0.2 pounds?

Research in the areas of criminology, medicine, aging, mental retardation, fish management, business, education, and others often involves taking a sample. Rarely does the investigation encompass a

Summary

study of all the criminals, all the mentally retarded, all the fish in the lake, or all senior citizens, collectively referred to as the population. Studying an entire population is too time-consuming, and the expense is usually prohibitive.

A small number (the sample) selected from the population (the entire group) is used to reason from specific instances to generalizations about the entire population. Care must be taken that the members of the sample are representative of the entire population. Chance must govern the selection. One method of ensuring that each member of the population has the same chance of being included in the sample is called random sampling. Another way of ensuring a representative sample is called systematic random sampling. As an example, if the population is defined as all inmates in Rahway Prison, a list of all the prisoners is secured. Then a starting point (say, the 12th prisoner on the list) is chosen at random. Starting with the 12th prisoner, every 10th prisoner on the list might constitute the sample. Prisoners numbered 12, 22, 32, 42, and so on would be interviewed.

For certain research problems, a stratified random sample might be the most appropriate sampling method. If the population of interest contains a number of distinct groups, it may be desirable to stratify the population into subgroups. Cluster sampling is used if the population is spread out over a large area. The purpose of this sampling technique is to save time and money.

It was pointed out that even though a sample is carefully selected from the population, by chance the mean of the sample will probably not be exactly the same as the population mean. In order to examine further the concept of this sampling error, all possible samples of a given size were selected from the population and the mean of each sample computed. When all the sample means are organized into a frequency distribution, the distribution is called the sampling distribution of the mean. It was noted that (1) the mean of the sampling distribution of the mean is the same as the population mean and (2) there is less spread in the distribution of sample means than in the distribution of the population values. Further, it was pointed out that

1. If the population from which the sample is drawn is normal, the sampling distribution of the means will also be normally distributed.

2. If the population from which the sample is drawn is not normally distributed, the distribution of the sample means will become more and more normal as the number of members in the sample becomes larger and larger. This is usually referred to as the central limit theorem. In essence, it states that when the size of the

sample is large enough, the distribution of the sample means is normal regardless of the shape of the population from which the sample is selected. A sample of 30 or more is usually considered "large enough."

Finally, based on the central limit theorem, a confidence interval can be set up for a sample mean. It specifies an interval that has a high probability of containing the population mean μ. For a normal distribution

1. Ninety-five percent of the sample means selected from that population will lie within 1.96 standard deviations of the population mean.
2. Ninety-nine percent of the sample means will lie within 2.58 standard deviations of μ.

Intervals computed in this manner are called the 95% confidence interval and the 99% confidence interval. The intervals are constructed in the following manner.

For the 95% confidence interval:

$$\overline{X} \pm 1.96\, \sigma_{\overline{X}} = \overline{X} \pm 1.96\, \frac{s}{\sqrt{n}}$$

For the 99% confidence interval:

$$\overline{X} \pm 2.58\, \sigma_{\overline{X}} = \overline{X} \pm 2.58\, \frac{s}{\sqrt{n}}$$

where s is the standard deviation of the sample and n the sample size.

Sampling Distributions and the Central Limit Theorem

I. **Sampling**
   A. Definition. A sample is a part or portion of the population.
   B. Objective. To gain information about the entire population.
II. **Methods of Probability Sampling**
   A. Characteristics. Each member of the population has a known probability of being selected for the sample.

B. Types of Probability Sampling
　1. *Random Sampling*
　　Procedure. One possible way of selecting a sample is to secure a list of the entire population. Write the name, or identifying number, of each of its members on a slip of paper. Scramble the slips and select the desired number for the sample.
　2. *Systematic Random Sampling*
　　Procedure. Select a random starting point from the population. Then systematically choose, say, every 40th name on the list.
　3. *Stratified Random Sampling*
　　Procedure. Subdivide the population into natural subgroups, called strata (for example, age, income, type of crime committed). Then, select a random sample from each stratum.

III. Distribution of Sample Means

A. Definition. The probability distribution of the means for all possible samples of the same size taken from the population.
B. Characteristics. The mean of the sample means is the same as the population mean, and the spread in the sampling distribution is less than the spread in the population. The standard error is computed by

$$\sigma_{\overline{X}} = \frac{\sigma}{\sqrt{n}} = \sigma_{\overline{X}} = \frac{s}{\sqrt{n}}$$

IV. Central Limit Theorem

The theorem states that if the population from which the sample is drawn is not normal, the distribution of the sample means will become more and more normal as the number of members in the sample becomes larger and larger. A sample of 30 or more is considered by most as being large enough.

V. Confidence Intervals

A. Definition. A range of values within which the population parameter is expected to fall with a preselected probability.
B. Objective. To determine the interval that has a high probability of including the population mean.
C. Construction
　1. Ninety-five percent confidence interval:

$$\overline{X} \pm 1.96 \frac{s}{\sqrt{n}}$$ where s is the sample standard deviation, n the size of the sample

2. Ninety-nine percent confidence interval:

$$\bar{X} \pm 2.58 \frac{s}{\sqrt{n}}$$

Chapter Exercises

6. You are studying voters' reactions in your state to a piece of statewide legislation. Describe how you would take a sample of these voters.

7. You wish to study the birth weights of newborn infants in your city. How would you go about obtaining the sample?

8. If you wished to estimate the percent of total working time a secretary spends on specific tasks: typing, answering the telephone, making copies, and so forth, how would you sample his or her activities?

9. A certain variety of flower grows to only three different heights: two inches, four inches, or six inches. If each of these heights is equally likely, draw a histogram of the probability distribution of this population. Find the mean and standard deviation. List all possible samples of size two that could be drawn from this population and calculate the corresponding sample average. Draw a histogram of the probability distribution of sample averages. Find its mean and standard deviation.

10. A study dealing with divorced couples gathered data on the length of time from marriage to separation. A random sample of 100 divorced couples had an average length of marriage of 5.9 years with a sample standard deviation of 2.0 years. Construct a 99% confidence interval for the mean length of time from marriage to separation for the population of all divorced couples.

11. A random sample of 80 terms of sentence for rape (first offense) showed a mean equal to 3.9 years with a standard deviation of 1.8 years. Construct a 95% confidence interval for the mean term of imprisonment for all such sentences.

12. Sick leave records obtained from a random sample of 200 social workers showed a mean number of days' sick leave equal to 25.6 last year. If the sample standard deviation is 5.1, construct a 90% confidence interval for the mean number of days' sick leave for all social workers last year.

13. The 1980 costs per pupil for seven city school systems in Ohio are listed on the following page:

City	Cost per Pupil
Akron	$1,375
Canton	1,320
Cincinnati	1,525
Cleveland	1,745
Columbus	1,480
Dayton	1,715
Toledo	1,395

   a. Using the random number table in Appendix B, select two school systems at random.

   b. Calculate the mean cost for your sample of two.

   c. Compare your sample mean with the population mean.

14. Can a sample be representative without being random? Explain.

15. On a fair die the numbers 1 through 6 each have a probability of 1/6 of occurring. Suppose two fair dice are thrown and the average number of points showing is calculated. What is the probability distribution of this average? Find its mean and standard deviation.

16. The California Department of Highways is analyzing traffic patterns on a busy section of Interstate 605, near Long Beach. They want to estimate the mean number of cars that pass this section each day. The requirements are that the estimate be within 20 cars per day of the population mean, and that the analysts be 95% confident of the results. A similar study showed the standard deviation to be 75 cars per day. How large a sample is required?

17. A consultant to a large Idaho ski resort wants to estimate the mean daily amount spent by its guests. How large a sample should the consultant select in order to estimate the mean daily amount spent to within $2.00 with a 99% level of confidence? A reasonable estimate of the standard deviation is $5.00.

Chapter
Achievement Test

Do all the problems. Then check your answers against those given in the Answer section of the book.

  I. True-False Questions. Indicate whether the statement is true or false. If it is false, make the needed corrections (4 points each).

    1. A sample is a list of every member in a group.

    2. Stratified random sampling is a nonprobability method of sampling.

3. Samples of size 30 or more are generally considered large.
4. A sampling distribution is a distribution that lists all the possible values of a sample statistic and their corresponding probability of occurrence.
5. All sampling distributions are symmetrical.
6. The mean of the sampling distribution of $\overline{X}$ is equal to the mean of the population.
7. The standard error of the mean is the standard deviation of the population from which the samples have been taken.
8. The standard error of the mean increases as the sample size increases.
9. The shape of the distribution of sample means is always normal.
10. The central limit theorem is true whenever the sample size is reasonably large.

II. Computation Problems (12 points each).

11. A population consists of the hourly wages of six employees. The wages are $10, $4, $12, $11, $9, and $8.
    a. Determine the population mean.
    b. A sample of size four is to be selected at random from the population. List all the possible samples of size four. (Hint: as a check for the total number of samples, solve $_nC_r = n!/[r!(n - r)!]$) Then compute the mean of each sample.
    c. Develop a distribution of the sample means and compute the mean.
    d. Portray the distribution of the sample means in the form of a histogram. Immediately below the histogram show the probability distribution for the population values.
    e. Draw conclusions regarding the two means (the population mean and the mean of the sample means). Also, make an observation with respect to the spread of the two probability distributions.
12. A population has a mean age of 50 years and a standard deviation of 10 years. Nine members of the population were selected at random. What is the standard error of this mean?
13. A sample of 817 truck drivers was given the Army General Classification Test. The sample mean is 96.2 with a sample standard deviation of 9.7. Construct a 95% confidence interval for the population mean score of all truck drivers. Interpret.
14. A study is made of how long inner-city families had lived at the current address. A random sample of 40 families revealed a mean of 35 months with a sample standard devia-

tion of 6.3 months. Construct a 99% confidence interval for the mean time inner-city families have lived at their current address. Interpret.

15. A study of the income of farming households in a particular state is undertaken. How large a sample is required to estimate the mean income within $200 with a 95% level of confidence? The standard deviation is estimated to be $3,000.

9
Hypothesis Tests: Large-Sample Methods

OBJECTIVES

When you have completed this chapter, you will be able to
- Describe the five-step hypothesis-testing procedure.
- Distinguish between a one-tailed and a two-tailed statistical test.
- Identify and describe possible errors in hypothesis testing.
- Conduct a hypothesis test about a population mean.
- Conduct a hypothesis test between two population means.

Introduction

In Chapter 6 we developed a discrete probability distribution. It described the possible outcomes of an experiment. Chapter 7 dealt with the normal probability distribution, a continuous distribution. We noted that it is symmetrical, bell shaped, and the tails taper off into infinity.

In Chapter 8, we used both the normal distribution and the notion of a probability distribution to develop a sampling distribution. In that chapter it was also pointed out that it is often too expensive, or impossible, to study an entire population. Instead, a part of the population of interest, called a sample, is examined. The express purpose of sampling, therefore, is to learn something about the population. For example, we might want to approximate the mean income or the mean age of the population. We showed that this could be accomplished by constructing confidence intervals within which the population mean might fall.

This chapter continues our study of the use of the normal distribution in sampling. Instead of building an interval in which a population parameter (such as the mean) is expected to fall, we test the validity of a statement about a population parameter.

The General Idea of Hypothesis Testing

To illustrate the concept of hypothesis testing, let us first examine a nonstatistical case. It concerns private eye Ms. Sharpe as she tries to unmask a mysterious murderer. Upon her arrival at the scene of the crime, Ms. Sharpe observed that the victim had been struck from above by a left-handed person whose size 7 shoes had been covered with mud.

What course of action will our detective pursue? Naturally, she will first suspect that the butler performed the foul deed. Then, she will examine each piece of evidence in succession to see if it is consistent with the presumption that the butler might be the murderer. If the butler is a short, right-handed man who wears size 11 shoes, it is highly unlikely that he committed the crime and he will be dismissed as a suspect.

As she considers the butler's innocence or guilt, what goes through the detective's mind? She realizes that she must either accuse or dismiss the prime suspect. Either way, the butler may in fact be either guilty or innocent. Thus, there are four possibilities that might occur when Ms. Sharpe finally reaches her decision. They are:

1. She can accuse the butler when the butler actually committed the crime—a correct decision.

2. She can dismiss the butler when the butler is innocent—again, a correct decision.

3. She can accuse the butler when the butler is actually innocent—an incorrect decision.

4. She can dismiss the butler when the butler is actually guilty—another incorrect decision.

These four possibilities facing our investigator can be diagrammed as follows:

	accuse the butler	dismiss the butler
the butler did it	correct	error!
the butler is innocent	error!	correct

If Ms. Sharpe's problem were one involving statistics, the first thing she would do is to set up a **null hypothesis.**

> **Null Hypothesis**   A claim about the value of a population parameter. H_0

The null hypothesis, designated H_0, in the murder case could be

$$H_0\text{: the butler is innocent}$$

The null hypothesis is a claim that is established for the purpose of testing. This claim is either rejected or not rejected. If the evidence is sufficient to reject the null hypothesis, then the **alternate hypothesis** is accepted.

> **Alternate Hypothesis**   A claim about the population parameter that is accepted if the null hypothesis is rejected.

In Ms. Sharpe's murder investigation, the alternate hypothesis, designated H_a, is

$$H_a\text{: the butler is guilty}$$

Note that the procedure is to test the null hypothesis. The alternate hypothesis is accepted if, and only if, the null hypothesis is rejected. The strategy is to make a decision first with respect to the null hypothesis.

After examining all the evidence, the detective might accept the null hypothesis (the butler is innocent) and release him. Or, she might reject the null hypothesis and have him arrested.

As noted before, the investigator could make two kinds of mistakes during her investigation. They are called **Type I** and **Type II** errors, respectively.

> **Type I Error**   An error that occurs when a true null hypothesis is rejected.

> **Type II Error**   An error that occurs when a false null hypothesis is not rejected.

If, for example, the butler had been arrested as a suspect when he was actually innocent of the murder, a Type I error would have been committed. But if a null hypothesis is accepted when it is actually not true, a Type II error is committed. That is, if the butler had been released when he actually did commit the murder, a Type II error would have been committed. To summarize:

	action	
	fail to reject H_0	H_0 rejected
H_0 is true		Type I error
H_0 is false	Type II error	

Of course, a Type I error could be avoided by never rejecting H_0, the null hypothesis. In the crime problem, if Ms. Sharpe never rejected the H_0 that a suspect was innocent, she would never make the mistake of arresting an innocent man. Clearly, this is a rather extreme way for a crime-fighter to avoid a Type I error! It also increases the chances that a Type II error will be committed.

A Type II error could always be avoided by rejecting the null hypothesis. If our private eye were always to reject the null hypothesis and accept the alternate hypothesis, she would never release a murderer. Either extreme position is unrealistic. As will be noted in the following sections, statistical theory deals with "decision rules" based on probability, which attempt to balance these two kinds of potential errors.

We will now leave Ms. Sharpe to her ongoing murder investigation. Many different types of statistical hypothesis-testing problems will now be considered in this and in subsequent chapters. As will become evident, however, the thinking and procedure developed in the nonstatistical murder investigation is very similar to research problems involving statistics. A systematic five-step approach will be applied in solving each problem. A brief discussion of these five steps follows. Later, as each of the tests is presented, the five steps will be explained in greater detail.

In statistics we employ a method of "proof by contradiction." Generally, we hope to prove something is true (the alternate hypothesis) by rejecting the claim in the null hypothesis. If the sample evidence contradicts the null hypothesis, we can reject it and "believe" the alternate hypothesis. On the other hand, if the sample information does not contradict the null hypothesis, you still have not proven that the null hypothesis is true. You have merely found that the null hypothesis cannot be rejected.

Step 1   The **null hypothesis** is stated. Along with the null hypothesis, a second statement called the **alternate hypothesis** is made. The alternate hypothesis is accepted if the null hypothesis is rejected.

Step 2   A **level of significance** is chosen. Usually, either the 0.05 or the 0.01 level is selected. The level of significance refers to the probability of making a Type I error.

Step 3   A **test statistic** is chosen. A test statistic is a quantity calculated from the sample information. Its value will be used in Steps 4 and 5 to arrive at a decision regarding the null hypothesis.

Step 4   A **decision rule** is set up based on the level of significance chosen in Step 2 and the sampling distribution of the test statistic (from Step 3).

Step 5   **One or more samples are selected.** Then, using the sample results, *the value of the test statistic is computed.* In this chapter, z will be the only test statistic employed. Finally, the decision rule in Step 4 is used to **make a decision**—either to reject the null hypothesis or not to reject the null hypothesis.

A Test Involving the Population Mean (Large Samples)

One type of hypothesis-testing problem involves checking whether a reported mean is reasonable. In order to perform the test, a random sample is taken from the population. It requires that the sample be "large." Samples of size 30 or greater are generally considered "large samples." It should be noted, however, that if you know the population is normally distributed, any size sample can be used.

Problem

According to the Bureau of the Census, the mean annual income of government employees during a recent year was $14,632. There was some doubt that this mean was representative of incomes of government employees living in the San Francisco Bay area during the same period. The question to be explored is: Is there sufficient evidence to conclude that the mean annual income of government employees living in the Bay area was different than the national average during that period?

Solution

Even before starting to collect data, the first step is to state the hypothesis to be tested. Recall from the murder investigation that we call this statement the null hypothesis, usually expressed symbolically as H_0. The null hypothesis, in this case, is that the population mean will not be affected—or, stated another way, that the mean in the Bay area is equal to $14,632. The mean is designated by the Greek lowercase letter μ. Symbolically, then, the null hypothesis is

$$H_0: \mu = \$14,632$$

The alternate hypothesis is that the mean annual income in the San Francisco Bay area is *not* equal to $14,632. It is written

$$H_a: \mu \neq \$14,632$$

Note that the equality condition appears in the null hypothesis. This will always be the case.

The second step in testing a hypothesis is to select the *level of significance*, which is the probability of a Type I error.

> **Level of Significance**   The probability of rejecting the null hypothesis when it is true.

The crucial question in this income problem is to determine when to reject the null hypothesis. The significance level will define more precisely when the sample mean is too far removed from the hypothesized value of $14,632 for the null hypothesis to be plausible. How do you decide what level of significance to select? You should consider the "cost" of being wrong. As an example of the "cost" of being wrong, it is a more serious error for meteorologists to predict sunshine and then have it rain than it is for them to predict rain and then have it sunny. The most common levels of significance used are 0.05 and 0.01, although any value is possible. For this illustration a 0.05 significance level might be selected. If rejecting a true hypothesis is relatively serious, then set the significance level quite low. If accepting a false null hypothesis is relatively more serious, then pick a high level of significance.

Suppose, for example, that a test had been devised to detect cancer. The null hypothesis might be that a patient is free of the disease (H_0: no disease). A Type I error would reject the null hypothesis when it is true—that is, telling a patient that he has cancer when in fact he does not. While such an error would cause much pain and suffering, in a case like this one it is not as serious or costly as a Type II error. Remember, a Type II error means that a false null hypothesis is accepted. In this example, it means you tell a patient he is healthy when in fact he has cancer. Most physicians would agree that the Type II error here is much more serious than the Type I error; consequently, a high level of significance should be selected for this test—say, 0.10 instead of 0.01.

The third step is to select the appropriate test statistic.

> **Test Statistic**   A quantity, calculated from the sample information, used as a basis for deciding whether or not to reject the null hypothesis.

Recall from Chapter 8 that, according to the central limit theorem, the sampling distribution of the sample mean is approximately normal, and the standard deviation of the sampling distribution of means ($\sigma_{\bar{x}}$) is $\sigma/\sqrt{n}$. Hence, the following test statistic is appropriate:

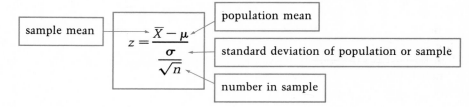

where

$\bar{X}$ is the sample mean.

μ is the population mean.

$\dfrac{\sigma}{\sqrt{n}}$ is the standard error of the mean.

n is the number in the sample.

Recall that σ is the standard deviation of the population from which the sample was drawn. For this test, we assume that either σ is already known based on prior studies, or that a good estimate of its value can be obtained from the sample data.

The fourth step is to formulate a decision rule.

> **Decision Rule**   A statement of the condition or conditions under which the null hypothesis is rejected.

In the study of Bay area wages, the decision rule is an objective statement that will allow us to test the null hypothesis.

The question we are exploring is: Does the mean annual income in the Bay area differ significantly from the national average? The mean in the Bay area could be larger or smaller than the national average. The decision rule is designed to accommodate both these possibilities. Thus, the test is called a **two-tailed test.**

As described in Chapter 8, the distribution of the means of all possible samples of the same size selected from a population is normally distributed. When the null hypothesis is true, the test statistic z will also be normally distributed. Recall from Chapter 7 that 95% of the area of the curve is between −1.96 and 1.96. Thus, if the significance level is 0.05, the region of rejection falls to the left of −1.96 and to the right of 1.96—that is, less than −1.96 and greater than 1.96. These two values are called the critical values.

> **Critical Value** A value or values that separate the region of rejection from the remaining values.

The decision rule, therefore, is: Do not reject the null hypothesis if the computed value of z is in the region from −1.96 to 1.96. When that is not the case, the null hypothesis is rejected, and the alternate hypothesis is accepted. The decision rule can be shown in the form of a diagram.

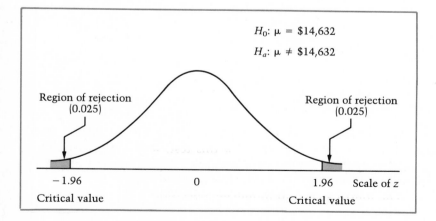

H_0: μ = \$14,632

H_a: μ ≠ \$14,632

Region of rejection (0.025)

Region of rejection (0.025)

−1.96   0   1.96   Scale of z

Critical value   Critical value

The final step is to take a sample of government employees in the Bay area, compute the sample mean $\overline{X}$, and determine the value of the test statistic z. Based on the computed value of z, the null hypothesis is either accepted or rejected. Suppose the sample consisted of 49 employees, the mean of the sample were $15,415, and the standard deviation of the sample were $1,827. Because the sample is reasonably large, the sample standard deviation (s) is a good estimate of the population standard deviation (σ).

$$z = \frac{\overline{X} - \mu}{\dfrac{\sigma}{\sqrt{n}}}$$

$$= \frac{\$15,415 - \$14,632}{\dfrac{\$1,827}{\sqrt{49}}} \quad \text{Square root} = 7$$

$$= \frac{\$783}{\$261}$$

$$= 3.0$$

Since 3.0 is in the rejection region, the null hypothesis is *rejected* at the 0.05 level of significance. The alternate hypothesis, which states that the mean annual income of government employees in the San Francisco Bay area is *not* equal to $14,632, is accepted. Such a large difference between the sample mean and the population mean cannot reasonably be attributed to chance.

    The correct course of action is either to reject or fail to reject the null hypothesis. By failing to reject, the investigator takes the position that the evidence is not sufficient to rule out—that is, reject—the null hypothesis. However, in practice often we think in terms of "accepting" the null hypothesis rather than "failing to reject" it. While this notion is technically incorrect, its use is common.

Self Review 9-1

A reminder: cover the answers in the margin.

In his team's first report, Alfred Kinsey indicated that the average frequency of sexual relations was 9.5 times per month. The standard deviation was 3.9 times per month. Twenty years later, *Redbook* conducted an inquiry to establish whether there had been a change in the frequency of sexual relations since publication of the Kinsey Report. The 0.01 level of significance was used.

a. $H_0 = 9.5$ $H_0 = M = 9.5$

    $H_a \neq 9.5$ $H_a = M = 9.5$

b. 0.01

c. $z = \dfrac{\overline{X} - \mu}{\dfrac{\sigma}{\sqrt{n}}}$

a. State the null hypothesis and alternate hypothesis.

b. What is the level of significance?

c. What is the appropriate test statistic? Give its formula.

d. Show the decision rule in the form of a diagram. *spine*
e. A sample of 18,000 responded to the *Redbook* questionnaire. *Redbook* reported that Americans had sexual relations an average of 9.2 times per month. Compute *z*. At the 0.01 level of significance, does this evidence indicate that there had been a change in the frequency of sexual relations in the twenty years since the Kinsey Report?

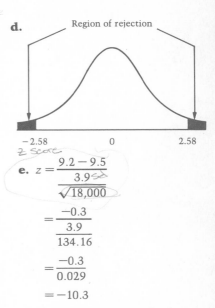

d. Region of rejection

−2.58   0   2.58

z score

e. $z = \dfrac{9.2 - 9.5}{\dfrac{3.95}{\sqrt{18,000}}}$

$= \dfrac{-0.3}{\dfrac{3.9}{134.16}}$

$= \dfrac{-0.3}{0.029}$

$= -10.3$

Reject H_0, accept H_a. There was a change.

$H_0 : M = 1,010$
$H_a : M \neq 1,010$

$Z = \dfrac{\bar{X} - \mu}{\dfrac{\sigma}{\sqrt{n}}}$

$\dfrac{\$1,090 - \$1,010}{\dfrac{300}{\sqrt{50}}}$

$\dfrac{80}{42.43}$

$\underline{1.88}$

not rejected

Chapter Exercises

1. The U.S. Public Health Service publishes the *Annual Data Tabulations, Continuous Air Monitoring Projects*, which recently indicated that a large midwestern city had an annual mean level of sulfur dioxide of 0.12 (concentration in parts per million). In order to change this concentration, many steel mills and other manufacturers installed antipollution equipment. Plans are to make about 36 random checks during the year to determine if there has been a change in the sulfur dioxide level. The 0.05 level is to be used.
   a. State both the null hypothesis and the alternate hypothesis.
   b. What is the level of significance? 0.05
   c. What is the appropriate test statistic? Give its formula.
   d. Show the decision rule in the form of a diagram.
   e. Thirty-six random checks were made throughout the year. It was found that the sample mean was 0.10 and the sample standard deviation 0.03. At the 0.05 level, does this evidence indicate that there has been a change in the sulfur dioxide level in the city?

2. During the past several years, frequent checks were made of the spending patterns of citizens returning from a vacation of 21 days or less to countries in Europe. Results indicated that travelers spent an average of $1,010 on items such as souvenirs, meals, film, gifts, and so on. A new survey is to be conducted to determine if there has been a change in the average amount spent. The 0.01 level is to be used.
   a. State the null hypothesis and alternate hypothesis in the form of H_0 and H_a.
   b. State the decision rule.
   c. A survey of 50 travelers has a sample mean of $1,090. The standard deviation of the sample was $300. At the 0.01 level, is there evidence that there has been a change in the mean amount spent abroad, or is the increase of $80 probably due to chance?

One-Tailed Tests

The earlier example that dealt with the annual income of government employees in the San Francisco Bay area required a *two-tailed* test. Before the sample of 49 employees was taken, there was no knowledge that the average income would be above or below the national average. For the 0.05 level there were two rejection regions, one to the right of 1.96, the other to the left of −1.96. (Consult the diagram on p. 265.)

Another problem might require the application of a *one-tailed* test. Suppose we suspected that federal employees in the heavily populated Bay area earn *more* than typical government employees. In this case, the only concern is whether the Bay area employees earn significantly *more than* typical government employees. To test our suspicions, our null hypothesis would state that the population mean is *equal to or less than* $14,632 or, stated symbolically:

$$H_0: \mu \leq \$14{,}632$$

The alternate hypothesis is that the mean income for Bay area government employees is *greater than* $14,632. It is written

$$H_a: \mu > \$14{,}632$$

Rejection of the null hypothesis and acceptance of the alternate hypothesis will allow us to conclude that Bay area government employees earn higher salaries than the national average.

Using the 0.05 level of significance, the critical value is 1.65. It is determined by referring to Appendix C again and searching the body of the table for 0.4500, or the value closest to it. The critical *z* value is in the margin. The decision rule states that the null hypothesis will not be rejected if the computed value of *z* is equal to or less than the critical value of 1.65. If the computed *z* is greater than 1.65, the null hypothesis will be rejected and the alternate hypothesis accepted. Shown in a diagram:

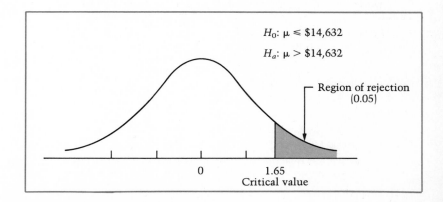

$H_0: \mu \leq \$14{,}632$

$H_a: \mu > \$14{,}632$

Region of rejection (0.05)

0   1.65
Critical value

Problem

Hyperactive children are often disruptive in the typical classroom setting because they find it difficult to remain seated for extended periods of time. Baseline data from a very large study show that the typical frequency of "out-of-seat behaviors" was 12.38 per 30-minute period with a standard deviation of 3.52. A treatment known as covert positive reinforcement was applied to a group of 30 hyperactive children. The average number of "out-of-seat behaviors" was reduced to 11.59 per 30-minute observation period. Using the 0.01 significance level, can we conclude that this decline in "out-of-seat behaviors" is significant?

$$\frac{3.52}{\sqrt{30}}$$

Solution

Step 1   The null hypothesis is: the mean is equal to or greater than 12.38. That is

$$H_0: \mu \geq 12.38$$

The alternate hypothesis is: the mean is less than 12.38.

$$H_a: \mu < 12.38$$

Step 2   State the level of significance. It is 0.01.

$$\begin{array}{r} 100\% \\ -0.01 \\ \hline .99 \end{array}$$

$$\frac{.99}{2} = .4950 = Z\text{-score of } 2.58$$

Step 3   Give the appropriate test statistic. It is

$$z = \frac{\overline{X} - \mu}{\dfrac{\sigma}{\sqrt{n}}}$$

Step 4   Since there is interest in demonstrating that the covert treatment *lowers* the mean value, a one-tailed test is appropriate. This requires a one-tailed test in the negative direction. Therefore, the critical value of z is on the left side of the curve. The critical value for the 0.01 significance level is −2.33 (found by referring to Appendix C and .4900).

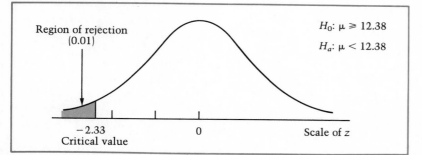

Region of rejection (0.01)

$H_0: \mu \geq 12.38$

$H_a: \mu < 12.38$

−2.33
Critical value

0

Scale of z

The decision rule is: reject the null hypothesis if the computed value of z is to the left of -2.33. Otherwise, accept the null hypothesis.

Note that the inequality sign in the alternate hypothesis (Step 1), "points" in the negative direction. Thus, the critical region will be in the left tail (the direction in which the inequality is pointing) and the critical value will have a negative sign. If a one-tailed test is employed, and the inequality sign in the alternate hypothesis points in the positive direction, the area of rejection will always appear in the positive tail and the sign of the critical value will be positive.

Step 5   Compute z and arrive at a decision.

$$z = \frac{\overline{X} - \mu}{\frac{\sigma}{\sqrt{n}}}$$

$$= \frac{11.59 - 12.38}{\frac{3.52}{\sqrt{30}}}$$

$$= -1.23$$

Since the computed z value of -1.23 is not in the rejection region, we fail to reject the null hypothesis at the 0.01 level. The difference between 11.59 and 12.38 "out-of-seat behaviors" can be attributed to sampling error. From a practical standpoint, it cannot be concluded that the covert positive reinforcement treatment reduced the "out-of-seat behaviors" of the hyperactive children.

Self-Review 9-2

a. $H_0: \mu \geq 6.0$
$H_a: \mu < 6.0$

b. Reject H_0 if computed z is to left of -1.65. Otherwise, accept H_0 at 0.05 level.

c. $z = \dfrac{5.8 - 6.0}{\dfrac{0.5}{\sqrt{100}}}$

$= -4.0$

Reject H_0. Mothers are staying less time.

Experience over a long period of time had shown that, on the average, a mother stayed 6.0 days in a certain hospital after childbirth. The standard deviation was 0.5 days. Hospital administrators, doctors, and other groups decided to make a joint effort to reduce the average time spent in the hospital. After this campaign, a sample of the files of 100 mothers revealed that the new mean length of stay was 5.8 days. Are mothers staying in the hospital less time, or could the difference between 6.0 and 5.8 be due to sampling error?

a. What is H_0 and H_a?
b. Using the 0.05 level, state the decision rule.
c. Arrive at a decision.

Chapter Exercises

3. A machine is set to fire 30.00 decigrams of chocolate pellets into a box of cake mix as it moves along the production line. Of course, there is some variation in the weight of the pellets. A sample of 36 boxes of mix revealed that the average weight of the chocolate pellets was 30.08 decigrams with a sample standard deviation of 0.50 decigrams. Is the increase in the weight of the pellets significant at the 0.05 level? Apply the usual five steps to be followed in hypothesis testing. (Hint: this involves a one-tailed test because there is interest in finding out whether there has been an increase in weight.)

4. The board of education of a suburban school district wants to consider a new academic program funded by the Department of Education. In order to be eligible for the federal grant, the arithmetic-mean income per household must not exceed $16,000. The board has hired a research firm to gather the required data. In its report, the firm has indicated that the arithmetic-mean income in the district is $17,000. Its survey included 75 households, and the standard deviation of the sample was $3,000. Using a one-tailed test and the 0.01 level of significance, can the board argue that the larger household income ($17,000) is due to chance?

5. A test was constructed to measure the degree of people's alienation. The mean score was 78 and the standard deviation of the population of scores was 16. The test was administered to 37 Vietnam veterans. Their mean score was 84. Test the hypothesis that Vietnam veterans are more alienated than the general population. Use the 0.05 significance level.

A Test for Two Population Means (Large Samples)

The previous section dealt with one large sample (30 or more). This section is concerned with *two population means*. The five-step hypothesis-testing procedure is the same one used for one-sample tests; however, the formula for the test statistic z is slightly different.

Problem

There have been complaints that resident physicians and nurses at the Las Palmas Hospital central desk respond slowly to emergency calls from senior citizens who are medical or surgical patients. It is claimed that other patients receive faster service. The 0.01 level of significance is to be used to test the hypothesis that the response times to emergency calls from senior citizens and from other patients are the same. The alternate hypothesis is that the response times for the senior citizens are *greater than* for other medical or surgical patients.

Solution

Unknown to the resident physicians and nurses, lengths of time it took them to respond to the calls of both senior citizens and other patients were recorded. The sample results are summarized as follows:

Patients	Sample Mean	Sample Standard Deviation	Number in Sample
Senior citizens	5.5 minutes	0.4 minutes	50
Other patients	5.3 minutes	0.3 minutes	100

The important question to be tested is whether the mean time to answer senior citizens' calls (5.5 minutes) really differs from that of the mean time for other patients (5.3 minutes). As in all problems involving sampling, we know that there is a distinct possibility that the difference between 5.5 minutes and 5.3 minutes is due to chance.

The null and alternate hypotheses are

$$H_0: \mu_1 \le \mu_2$$
$$H_a: \mu_1 > \mu_2$$

where μ_1 is the population-mean time for the senior citizens and μ_2 the population-mean time for other patients. The way the alternate hypothesis is stated (the mean time for senior citizens is *greater* than for other patients) reveals to us that this is to be a one-tailed test.

The test statistic z is

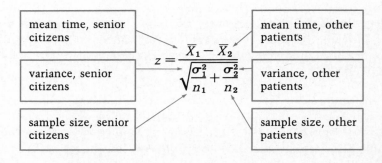

The decision rule for the 0.01 level is: do not reject the null hypothesis if the computed value of z is equal to or less than 2.33. Otherwise, reject the null hypothesis and accept the alternate hypothesis. A diagram illustrating the decision rule follows.

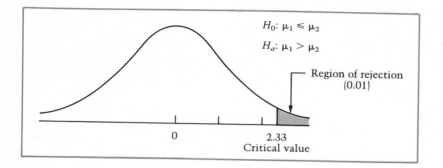

$H_0: \mu_1 \leq \mu_2$

$H_a: \mu_1 > \mu_2$

Region of rejection (0.01)

0

2.33
Critical value

Computing z:

$$z = \frac{\overline{X}_1 - \overline{X}_2}{\sqrt{\dfrac{\sigma_1^2}{n_1} + \dfrac{\sigma_2^2}{n_2}}}$$

$$= \frac{5.5 - 5.3}{\sqrt{\dfrac{(0.4)^2}{50} + \dfrac{(0.3)^2}{100}}}$$

$$= \frac{0.2}{0.064}$$

$$= 3.13$$

The computed value of 3.13 falls in the tail to the right of 2.33. Therefore, the null hypothesis is rejected at the 0.01 level of significance. There *is* a difference in the length of time it takes physicians and nurses to answer calls from senior citizens as compared to calls from other patients. The probability that the difference of 0.2 minutes between the two means $(5.5 - 5.3)$ is due to chance (sampling error) can be dismissed.

It should be noted that in order to use the testing procedure for two means presented in the preceding paragraphs two conditions must be met.

1. n_1 and n_2 must be 30 or more. In the preceding problem, 50 and 100 exceed the minimum number of 30. This restriction can be ignored, however, if the two populations are normally distributed and their variances are known.

2. The samples are *independent.* This means that the samples are unrelated. For example, if senior citizen Smith were chosen in the sample of senior citizens, her selection in no way affects the selection of any other patient, either in the senior citizen group or the other patient group.

Self-Review 9-3

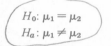
Test question

a. A two-tailed test.

$$H_0: \mu_1 = \mu_2$$
$$H_a: \mu_1 \neq \mu_2$$

b. Accept H_0 if computed z is between -1.96 and $+1.96$.

c. $z = \dfrac{90 - 94}{\sqrt{\dfrac{(12)^2}{40} + \dfrac{(15)^2}{50}}}$

$= \dfrac{-4.000}{2.846}$

$= -1.41$

Do not reject H_0. There is no difference between the two groups.

The admissions officer at a university wants to investigate whether there is any difference between the scores on the mathematics placement test of students who attend primarily during the day and the scores of students who work during the day and attend evening classes.

a. Is this a one-tailed or a two-tailed test? Symbolically, what are the null hypothesis and alternate hypothesis?
b. The 0.05 level of significance is to be used. State the decision rule.
c. The records of both day and evening students revealed:

Student	Mean Score	Standard Deviation	Number in Sample
Day	90	12	40
Evening	94	15	50

Compute z and arrive at a decision.

Chapter Exercises

6. Two machines fill bottles with a cough syrup. The performance of each is checked periodically by testing bottles removed at random from the production line. The sample statistics for machine 1 are: mean weight of the contents of 40 bottles was 202.6 milligrams, and the standard deviation of the sample was 3.3 milligrams. For machine 2: mean weight of 50 bottles was 200.0 milligrams, and the standard deviation was 2.0 milligrams. Using a two-tailed test and the 0.01 level of significance, test whether there is any difference in the performance of the two machines.
  a. State both the null hypothesis and the alternate hypothesis.
  b. State the decision rule.
  c. Compute z and arrive at a decision.

7. You are conducting a study with respect to the annual incomes of probation officers in metropolitan areas of less than 100,000 population and in metropolitan areas having more than 500,000 population. Your sample data are

	Population less than 100,000	Population more than 500,000
Sample size	25	60
Sample mean	$14,290	$14,330
Sample standard deviation	$135	$142

Test the hypothesis that the annual incomes of probation officers in areas having more than 500,000 population are significantly greater than those paid in areas of less than 100,000 population. Use the 0.05 level of significance. (Hint: note that the problem includes the words "greater than.") Use a systematic approach by stating the null hypothesis, the alternate hypothesis, and so on.

8. A sociologist asserts that the average length of courtship is longer before a second marriage than before a first. He bases this claim on the observation of (1) 800 first marriages where the average is 265 days with a sample standard deviation of 60 days, and (2) 600 second marriages where the average is 268 days with a sample standard deviation of 50 days. Test his assertion by the five-step hypothesis-testing procedure. Use a 0.01 significance level.

This chapter has presented the general idea of hypothesis testing. A five-step testing procedure was developed. The steps followed are

1. A null hypothesis and an alternate hypothesis are stated.

2. A level of significance is chosen. The most common levels are 0.05 and 0.01.

3. An appropriate test statistic is selected. The test statistic in both types of problems in this chapter is z.

4. A decision rule is set up. The objective statement found in the decision rule gives the researcher a basis on which to make a decision. The test can be one-tailed or two-tailed, depending on how the alternate hypothesis is phrased.

5. Take one or more samples, depending on the problem, and then *compute z*. Based on the computed value of z and the critical values stated in the decision rule, either accept the null hypothesis or reject it.

Two types of errors are possible when sampling is employed. These potential errors are called Type I and Type II errors. A Type I error is the level of significance chosen in Step 2. Suppose the null hypothesis in a problem is rejected. Thus, it might be said, for example, that "the probability that a Type I error has been committed is 0.05." This indicates that the probability is 0.05 that the null hypothesis was rejected when it should have been accepted because the null hypothesis was true. A Type II error is the probability of accepting the null hypothesis when it should have been rejected.

Two statistical hypothesis tests were presented. Both are based on "large samples"—30 or more. The first, a one-sample test, was designed to determine if a reported mean is reasonable. The test statistic used is

$$z = \frac{\bar{X} - \mu}{\dfrac{\sigma}{\sqrt{n}}}$$

The second test compared two samples to find out if the two population means are equal. The test statistic used is

$$z = \frac{\bar{X}_1 - \bar{X}_2}{\sqrt{\dfrac{\sigma_1^2}{n_1} + \dfrac{\sigma_2^2}{n_2}}}$$

Chapter Outline

Hypothesis Tests: Large-Sample Methods

I. General Approach to Hypothesis Testing

A. Objective. To illustrate a formal "rules of evidence" approach to making statistical decisions based on objective scientific procedures.

B. Steps in hypothesis testing
  1. State the null hypothesis and alternate hypothesis.
  2. Choose a level of significance. 0.05 and 0.01 are most commonly used.
  3. Select the appropriate test statistic. It is designated as z for hypothesis tests discussed in this chapter.
  4. Find one or more critical values and state the decision rule.
  5. Take one or more samples, compute z, and arrive at a decision to accept or reject the null hypothesis.

II. One-Sample Testing Involving the Population Mean (Large Samples)

A. Criterion for a large sample: 30 observations or more.

B. Illustrative problem and procedure. Problem: test whether the mean age of males receiving social security benefits is 70.

Step 1 The null and alternate hypotheses are H_0: $\mu = 70$. For a two-tailed test, H_a is $\mu \neq 70$. For a one-tailed test, H_a is either $\mu < 70$ or $\mu > 70$.

Step 2 Level of significance arbitrarily chosen to be 0.01.

Step 3 Test statistic is computed by

$$z = \frac{\overline{X} - \mu}{\dfrac{\sigma}{\sqrt{n}}}$$

Step 4 If test is two-tailed, the decision rule is: accept null hypothesis if computed z is between -2.58 and 2.58; otherwise, reject. (Critical values found by referring to Appendix C, finding .4950, and reading 2.58 in the margin.)

Step 5 Take a sample, compute z, and reach a decision. 400 males on social security are surveyed. $n = 400$; $\overline{X} = 70.2$; standard deviation = 2 years. Computing z:

$$z = \frac{70.2 - 70.0}{\dfrac{2}{\sqrt{400}}} = 2.0$$

Accept null hypothesis because 2.0 is between -2.58 and 2.58. Conclusion: mean age is 70.0. The difference of 0.2 years between sample mean and population mean can be attributed to sampling error.

III. Hypothesis Tests Involving Two Sample Means (Large Samples)

A. Criteria for large samples: n_1 and n_2 must be 30 or more, and the samples must be independent.

B. Illustrative problem and procedure. Problem: test whether the mean age of males and mean age of females receiving social security are equal. H_0: $\mu_1 = \mu_2$. Two-tailed alternate is $H_a: \mu_1 \neq \mu_2$. One-tailed alternate is either $H_a: \mu_1 < \mu_2$ or $\mu_1 > \mu_2$. Appropriate test statistic is

$$z = \frac{\overline{X}_1 - \overline{X}_2}{\sqrt{\dfrac{\sigma_1^2}{n_1} + \dfrac{\sigma_2^2}{n_2}}}$$

If the 0.05 level of significance and a two-tailed test are used, the decision rule is to accept the null hypothesis if the computed value of z is between -1.96 and 1.96; otherwise, reject. The final step is to take a sample from each group, compute z, and make a decision based on the decision rule.

Chapter Exercises

9. The Department of Health and Human Services reported that the mean number of years of school completed by adults in the United States was 11.5 years in 1980. The Tennessee Department of Employment Security wants to determine if the level of education of their employees is less than the national average.
   a. State the null and alternate hypotheses both in words and using the letters H_0 and H_a.
   b. The level of significance is 0.05.
   c. Give the test statistic z.
   d. State the decision rule in words. Show it in a diagram.
   e. A sample of 100 employees revealed that the mean is 13 years with a standard deviation of 1.1 years. Compute z and arrive at a decision regarding the null hypothesis.

10. A psychologist wants to evaluate whether windowless schools lead to anxiety in the psychological development of children. A sample of children who are in windowless classrooms is selected and given an anxiety test. The same procedure is followed for children in classrooms with windows. The results are:

	Windowless Schools	Schools with Windows
Mean	94	90
Standard deviation	8	10
Sample size	100	80

Using the usual five-step hypothesis-testing procedure, the 0.01 level, and a one-tailed test, arrive at a conclusion regarding the anxiety levels of the two groups.

11. A medical researcher contends that lung capacity varies significantly between smokers and nonsmokers. The mean capacity of a sample of 30 nonsmokers is 5.0 liters with a sample standard deviation of 0.3 liters. Of the 40 smokers in the sample, the mean lung capacity is 4.5 liters with a standard deviation of 0.4 liters. At the 0.01 significance level, is there sufficient evidence to conclude that lung capacity is larger among nonsmokers?

12. Patients entering a hospital have complained that it takes 30 minutes to fill out the forms required for admittance. As a result of their complaints, the forms and procedure have been revised. A recent analysis of a random sample of 40 incoming patients reveals that the mean time to fill out the forms now is 28.5 minutes with a standard deviation of 5 minutes. Is there sufficient evidence at the 0.02 level of significance to show that the new system is an improvement?

First answer all the questions. Then check your answers against those given in the Answer section of the book.

Chapter Achievement Test

I. Multiple-Choice Questions. Select the response that provides the best answer (5 points each).

Questions 1 through 5 refer to the following information:

$$H_0: \mu = 200 \qquad \sigma = 20$$
$$H_a: \mu \neq 200 \qquad n = 36$$
$$\overline{X} = 210$$

1. Which of the following sentences is the correct statement of the null hypothesis?
   a. the mean of the population is not 200
   b. the sample mean is 210
   c. the population mean is 200
   d. the mean of the sample is 200
   e. none of the above are correct
2. The alternate hypothesis indicates
   a. the test is two-tailed
   b. the null hypothesis should be accepted when $\overline{X}$ is 210
   c. the test is one-tailed
   d. the alternate hypothesis should be accepted when $\overline{X} \neq$ 210
   e. none of these are correct

3. The value of the appropriate test statistic is computed to be
   a. 1.93
   b. 2.00
   c. 2.24
   d. 3.00
   e. 8.20

4. If the level of significance selected is 0.05, the critical values are
   a. ±1.65
   b. ±1.96
   c. ±2.33
   d. ±2.58
   e. can't tell from the information given

5. The decision is
   a. reject the null hypothesis and accept the alternate hypothesis
   b. the sample mean is not 210
   c. fail to reject the null hypothesis
   d. the population mean is 210
   e. cannot reach a decision based on the information given

6. The critical value for a one-tailed z test with a 0.05 level of significance is
   a. −1.96
   b. −1.65
   c. 1.65
   d. 1.96
   e. it depends on the alternate hypothesis

7. It is decided to use a two-tailed test with 0.10 as the level of significance in a large-sample test between two sample means. If, in fact, the null hypothesis is true, what is the probability of making a Type I error?
   a. 0
   b. 0.025
   c. 0.05
   d. 0.10
   e. can't tell from the information given

8. Which of the following is an assumption necessary for the correct use of the two-sample test involving two means?
   a. the population is 100 or more
   b. each sample is less than 30
   c. the samples are taken from two normally distributed populations
   d. the population standard deviations are known
   e. all of the above are required assumptions

Questions 9 and 10 refer to the following data:

$$\overline{X}_1 = 100 \qquad \overline{X}_2 = 90$$
$$\sigma_1 = 12 \qquad \sigma_2 = 12$$
$$n_1 = 100 \qquad n_2 = 100$$

9. The computed test statistic for the null hypothesis $\mu_1 = \mu_2$ is
   a. 1.00
   b. 1.77
   c. 2.50
   d. none of these are correct

10. The critical values for a 0.01 level of significance and a two-tailed test would be
    a. ±1.65
    b. ±1.96
    c. ±2.33
    d. ±2.58
    e. none of these are correct

II. Work out in detail each of the following computation problems (25 points each).

11. Planned Parenthood is investigating the differences between families in the Midwest and those on the Atlantic Coast. A particular study concerns the age of mothers at the birth of their last child. A random sample of 36 women from the Midwest had a sample mean of 32.9 years and a sample standard deviation of 5.7 years. A second random sample of 49 women from the Atlantic Coast had a mean of 29.6 years with a sample standard deviation of 5.5 years. Does this represent a significant difference between the two geographic regions? Use the 0.05 level of significance.

12. A West Coast study of a very large population showed the mean time devoted to volunteer political and civic activities was 32 minutes per month with a population standard deviation of 4.5 minutes. A smaller group of 49 similar individuals who lived in the South reported a mean of 34 minutes per month. Is this difference possibly due to sampling error, or does it represent a significant difference at the 0.05 level?

10
Testing a Hypothesis About Proportions

OBJECTIVES

When you have completed this chapter, you will be able to
- Define a proportion.
- Test a hypothesis about a population proportion.
- Test a hypothesis about two population proportions.

Introduction

Chapter 9 introduced statistical hypothesis testing. Hypothesis testing is a scientific statistical procedure designed to establish a set of rules for decision making. The testing techniques allow the researcher to distinguish between actual differences and differences that are likely to occur by chance. The five-step procedure introduced in that chapter is:

1. The null hypothesis and alternate hypothesis are stated.
2. A level of significance is selected.
3. A test statistic is chosen.
4. A decision rule is formulated.
5. One or more samples are selected, the test statistic computed, and a decision made either to accept or reject the null hypothesis.

Chapter 9 dealt with testing a hypothesis about one population mean or two population means. The data were interval-scale measurements. In this chapter we will discuss hypothesis testing dealing with data that are of nominal scale—that is, information that is merely classified into categories.

A Test Dealing with Population Proportions

Examples of the types of questions to be explored are: Is the probability 0.80 that a first offender placed on probation will commit a second crime? Do 20% of inner-city families live on a subsistence income? Is there a difference between the **proportion** of rural voters and urban voters who plan to vote for the incumbent governor in the forthcoming election?

X

Proportion   A fraction, ratio, percent, or probability that indicates what part of the sample or population has a particular trait.

Often, claims are made that a proportion, or fraction, of a population possesses a certain characteristic. The claim might be that 25% of nonworking adults watch daytime soap operas regularly. Another claim might be that 30% of children under ten do not brush their teeth daily. A test to verify such claims follows. This test is appropriate when *both np* and $n(1 - p)$ are *5 or more*. By meeting this requirement, the normal approximation to the binomial, discussed in Chapter 7, can be used. The test statistic follows the z distribution. The test statistic is

$$z = \frac{\dfrac{X}{n} - p}{\sqrt{\dfrac{p(1 - p)}{n}}}$$

where

X is the observed number in the sample possessing the trait.

n is the size of the sample.

p is the population proportion possessing the characteristic.

To employ this test statistic, the items in the sample must be *selected independently* and each must have the *same chance of being selected*.

Problem

We are investigating the question: Is the proportion of inner-city families living on a subsistence income 20%, or 0.20?

Solution

Using the same five-step hypothesis-testing procedure discussed in detail in the previous chapter:

Step 1   State the null hypothesis. Here, it is: the proportion of all inner-city families living on a subsistence income is 0.20. Symbolically:

$$H_0: p = 0.20$$

The alternate hypothesis is: the proportion of inner-city families living on a subsistence income is not 0.20. Symbolically:

$$H_a: p \neq 0.20$$

Since the alternate hypothesis does not state a direction, the test is two-tailed.

Step 2   Decide on the level of significance. The significance level determines when the sample measurement is too far away from the hypothesized proportion to be believable. The 0.05 level has been chosen for the inner-city problem.

Step 3   Select the appropriate test statistic. For this problem a sample of 200 inner-city families will be studied. $n = 200$ and $p = 0.20$, so np is $200(0.20) = 40$. And, $n(1 - p)$ is $200(1 - 0.20) = 160$. Both np and $n(1 - p)$ are greater than 5. Therefore, the appropriate test statistic is z.

$$z = \frac{\dfrac{X}{n} - p}{\sqrt{\dfrac{p(1 - p)}{n}}}$$

Step 4   Develop a decision rule. Referring to the standard normal distribution in Appendix C, the critical value for the 0.05 level of significance is 1.96. It is found by locating 0.4750 in the body of the table and then moving to the margin to read the critical z value of 1.96. The decision rule states that the null hypothesis will not be rejected if the computed value of z is in the interval between -1.96 and 1.96. Otherwise, it will be rejected and the alternate hypothesis accepted. Shown schematically:

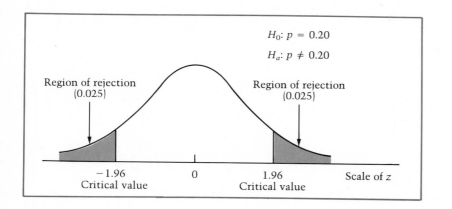

H_0: $p = 0.20$

H_a: $p \neq 0.20$

Region of rejection (0.025)

Region of rejection (0.025)

-1.96
Critical value

0

1.96
Critical value

Scale of z

Step 5   You select a sample from the population of all inner-city families, compute z, and then decide whether to reject H_0. A sample of 200 inner-city families revealed that 38 had incomes at the subsistence level. Computing z:

$$z = \frac{\frac{X}{n} - p}{\sqrt{\frac{p(1-p)}{n}}}$$

$$= \frac{\frac{38}{200} - 0.20}{\sqrt{\frac{0.20(1-0.20)}{200}}}$$

$$= \frac{-0.01}{\sqrt{0.0008}}$$

$$= -0.35$$

Your decision is not to reject the null hypothesis (that $p = 0.20$) because -0.35 is between -1.96 and 1.96. Although there is a difference between the sample proportion of 0.19, found by 38/200, and

0.20, the apparent difference can be attributed to chance (sampling error).

Self-Review 10-1

a. One-tailed.
b.

Region of rejection

−2.33   0
Critical value

c.

$$z = \frac{\frac{75}{100} - 0.80}{\sqrt{\frac{0.80(1-0.80)}{100}}}$$

.05
.04

$$= -1.25$$

Accept the null hypothesis. The rehabilitation program is not effective.

(#1) $Z = \begin{matrix} 1.96 \\ 1.64 \end{matrix}$

$$\frac{\frac{308}{350} - 0.90}{\sqrt{\frac{0.90(1-0.90)}{350}}}$$ reject

$$\frac{-.02}{.016} = -1.25$$

$H_0 : P \geq .90$

$H_a : P < .90$

A reminder: cover the answers in the margin.

National figures reveal that the probability a youthful offender on probation will commit another crime is 0.80. A special rehabilitation program was conducted for youthful offenders. A sample of 100 enrolled in the program showed that 75 subsequently committed another crime. The null hypothesis to be tested was $p \geq 0.80$. The alternate hypothesis was $p < 0.80$.

a. Based on the way the alternate hypothesis is stated, is this a one-tailed test or a two-tailed test?
b. Using the 0.01 significance level, show the decision rule graphically. $Z = 2.33$
c. Arrive at a decision.

Chapter Exercises $Z = \dfrac{\frac{x}{n} - P}{\sqrt{\frac{P \cdot (1-P)}{n}}}$

1. You believe that 90% of the population would continue to work even if they inherited sufficient wealth to live comfortably without working. This assertion is questioned by a social research team. A list of persons who recently inherited more than $400,000 is obtained, and 350 persons are selected from the list. Out of the 350 selected, 308 indicate that they are still working. At the 0.05 level, is this sufficient evidence to doubt your claim? (State the null hypothesis and other steps.)

2. The local newspaper conducted a study of the welfare claims in the area. It reported that "at least 20% of the present welfare recipients are ineligible and should not be receiving monthly payments." You are hired by the Welfare Department to investigate the newspaper's claim that some welfare recipients should not be receiving monthly payments. You randomly select 400 recipients from the files and carefully investigate each one and you find that 60 should not be receiving payments because they either filed false claims, failed to report that they work, or committed other fraudulent acts. At the 0.01 level of significance, should the newspaper's claim be rejected? Back your decision with statistical evidence.

A Test Dealing with Two Population Proportions

Some problems involve testing whether two population proportions are equal. In order to conduct such a test, first we select a random sample from each population. Then, if *both* samples meet the requirement that np and $n(1-p)$ are *5 or more*, we use the two-sample test that follows. The test statistic z follows the standard normal distribution and is computed by

$$z = \frac{\dfrac{X_1}{n_1} - \dfrac{X_2}{n_2}}{\sqrt{\bar{p}(1-\bar{p})\left(\dfrac{1}{n_1} + \dfrac{1}{n_2}\right)}}$$

where

X_1 is the number possessing the trait in the first sample.

X_2 is the number possessing the trait in the second sample.

n_1 is the number in the first sample.

n_2 is the number in the second sample.

$\bar{p}$ is the proportion possessing the trait in the combined samples. It is referred to as the *pooled estimate* of the proportion.

$$\bar{p} = \frac{\text{total number of successes}}{\text{total number samples}} = \frac{X_1 + X_2}{n_1 + n_2}$$

Problem
You know that a large group of homosexuals live in a certain district of a large city. An in-depth study is to be made of this group. One of your objectives is to determine if there is a difference in the respective proportions of white homosexuals to the total white population and of black homosexuals to the total black population living in that district.

Solution

Step 1 State the null and alternate hypotheses. The null hypothesis is: there is no difference between the two population proportions—that is, the two proportions are equal. Symbolically:

$$H_0\!: p_1 = p_2$$

The alternate hypothesis is that the two population proportions are not equal. Symbolically:

$$H_a: p_1 \neq p_2$$

The problem does not imply a direction; therefore, a two-tailed test is used.

Step 2   Give the level of significance. You decided on 0.05.

Step 3   The test statistic for problems involving two proportions is z. The formula is

$$z = \frac{\dfrac{X_1}{n_2} - \dfrac{X_2}{n_2}}{\sqrt{\bar{p}(1 - \bar{p})\left(\dfrac{1}{n_1} + \dfrac{1}{n_2}\right)}}$$

Step 4   The decision rule is: do not reject the null hypothesis if the value of the computed z is between -1.96 and 1.96. Otherwise, reject the null hypothesis and accept the alternate hypothesis. A graphic presentation of the rule is

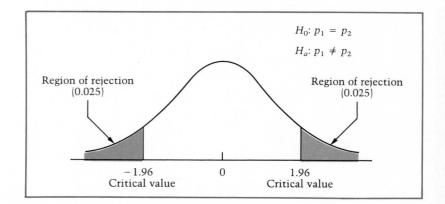

Step 5   You take a sample from the white population living in the district. Determine the number of male homosexuals in the sample. Then you take a sample from the black population living

in the area and determine the number of male homosexuals in the sample. The results are

	Number in Sample	Number of Male Homosexuals	Proportion of Male Homosexuals in Population
White	804	575	0.715, found by 575/804
Black	175	111	0.634, found by 111/175

Repeating the formula for z:

$$z = \frac{\dfrac{X_1}{n_1} - \dfrac{X_2}{n_2}}{\sqrt{\bar{p}(1-\bar{p})\left(\dfrac{1}{n_1} + \dfrac{1}{n_2}\right)}}$$

The pooled estimate of the proportion designated by $\bar{p}$ is 0.70, found by

$$\bar{p} = \frac{X_1 + X_2}{n_1 + n_2} = \frac{575 + 111}{804 + 175} = \frac{686}{979} = 0.70$$

Computing z:

$$z = \frac{\dfrac{575}{804} - \dfrac{111}{175}}{\sqrt{0.70(1-0.70)\left(\dfrac{1}{804} + \dfrac{1}{175}\right)}}$$

$$= \frac{0.715 - 0.634}{\sqrt{0.001461}}$$

$$= \frac{0.081}{0.038}$$

$$= 2.13$$

Since 2.13 is greater than the upper critical value of 1.96, the decision is to reject the null hypothesis. The conclusion is that there is a significant difference between the two population proportions. It is unlikely that the difference of 8.1% between the two sample proportions is due to chance (sampling error).

Self-Review 10-2

The effectiveness of a newly developed allergy-relief capsule is to be compared with that of one which has been on the market for a number of years. A sample of 250 persons using the new capsule revealed that 150 received satisfactory relief. Out of a group of 400 using the older capsule, 232 received satisfactory relief. Using the 0.02 level of significance, test the null hypothesis that the proportion receiving relief from the new capsule is equal to the proportion receiving relief from the old capsule. Use a two-tailed test.

a. $H_0: p_1 = p_2$
   $H_a: p_1 \neq p_2$

b.

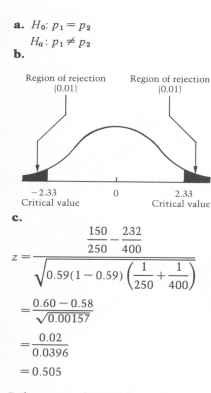

Region of rejection (0.01)   Region of rejection (0.01)

−2.33   0   2.33
Critical value   Critical value

a. State the null and alternate hypotheses, using the letters H_0 and H_a.
b. Show the decision rule graphically.
c. Compute z and arrive at a decision.

c.

$$z = \frac{\dfrac{150}{250} - \dfrac{232}{400}}{\sqrt{0.59(1-0.59)\left(\dfrac{1}{250} + \dfrac{1}{400}\right)}}$$

$$= \frac{0.60 - 0.58}{\sqrt{0.00157}}$$

$$= \frac{0.02}{0.0396}$$

$$= 0.505$$

Fail to reject the null hypothesis.

Problem

As another illustration of testing a hypothesis involving two proportions, suppose the incumbent governor comes from a large urban area. We suspect that a larger proportion of urban voters, in contrast to rural voters, will support the governor in her bid for reelection. Random samples of 500 urban and 400 rural voters are selected. The results are

	Urban	**Rural**
Plan to vote for the incumbent governor	$X_1 = 300$	$X_2 = 200$
Number in sample	$n_1 = 500$	$n_2 = 400$
Sample proportions	$\dfrac{X_1}{n_1} = \dfrac{300}{500} = 0.60$	$\dfrac{X_2}{n_2} = \dfrac{200}{400} = 0.50$

Notice that 60% of urban voters planned to support the governor and only 50% of the rural voters planned to support her. At the 0.05 level, does this demonstrate that a greater proportion of urban voters plan to vote for the incumbent governor?

Solution

The null hypothesis, written $H_0: p_1 \leq p_2$, is that there is no difference in the two proportions. The alternate hypothesis, written $H_a: p_1 > p_2$, states that the proportion of urban voters is *greater than* the proportion of rural voters. Therefore, the test will be one-tailed. The region of rejection will be in the right tail.

The decision rule is determined by referring to Appendix C,

the normal distribution. Locate 0.4500 in the body of the table, then move to the margin and read the value of z. It is 1.65. Hence, H_0 is rejected if the computed value of z is greater than 1.65. Shown schematically:

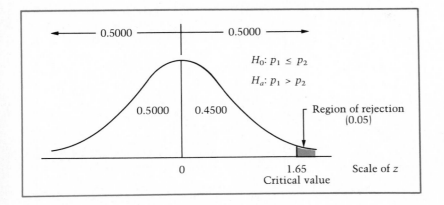

Computing z:

The pooled estimate of the proportion is

$$\bar{p} = \frac{X_1 + X_2}{n_1 + n_2} = \frac{300 + 200}{500 + 400} = 0.56$$

$$z = \frac{\dfrac{X_1}{n_1} - \dfrac{X_2}{n_2}}{\sqrt{\bar{p}(1 - \bar{p})\left(\dfrac{1}{n_1} + \dfrac{1}{n_2}\right)}}$$

$$= \frac{\dfrac{300}{500} - \dfrac{200}{400}}{\sqrt{0.56(1 - 0.56)\left(\dfrac{1}{500} + \dfrac{1}{400}\right)}} = 3.00$$

The null hypothesis is rejected at the 0.05 level because the computed z value of 3.0 is greater than the critical value of 1.65. The alternate hypothesis, namely, $p_1 > p_2$, is accepted. This indicates that a larger proportion of urban voters (0.60) than rural voters (0.50) plans to vote for the incumbent governor. At the 0.05 level of significance, the difference of 0.10 (found by 0.60 − 0.50) is too large to have reasonably occurred by chance.

Self-Review 10-3

a. $H_0: p_1 \geq p_2$
   $H_a: p_1 < p_2$
b. One-tailed.
c. -1.28
d.

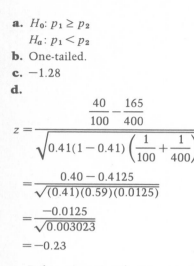

$$z = \frac{\dfrac{40}{100} - \dfrac{165}{400}}{\sqrt{0.41(1-0.41)\left(\dfrac{1}{100} + \dfrac{1}{400}\right)}}$$

$$= \frac{0.40 - 0.4125}{\sqrt{(0.41)(0.59)(0.0125)}}$$

$$= \frac{-0.0125}{\sqrt{0.003023}}$$

$$= -0.23$$

e. Fail to reject H_0. There is not sufficient evidence to reject the hypothesis that the two proportions are equal. The difference of 0.0125 could be the result of chance $(0.40 - 0.4125)$.

It is claimed that, compared with non-senior citizens, the proportion of senior citizens who regularly watch major golf tournaments on television is smaller. As an experiment, 100 senior citizens were selected at random and it was reported that 40 of them watch the tournaments regularly. Out of 400 non-senior citizens surveyed, 165 watch the tournaments regularly. At the 0.10 level, does this indicate the proportion of senior citizens watching the tournaments on television is smaller than the proportion of non-senior citizens?

a. State the null hypothesis and the alternate hypothesis symbolically.
b. Is this a one-tailed test or a two-tailed test?
c. What is the critical value?
d. Compute z.
e. Arrive at a decision.

Chapter Exercises

3. Subsalicylate bismuth is the active ingredient in several nonprescription drugs for the relief of stomach upsets. Only 14 out of 62 persons sampled who took the bismuth during their three weeks' vacation away from home suffered from the gastric and intestinal discomfort that often affect travelers. Of 66 persons who did not take bismuth, 40 suffered from stomach upsets. Is the difference in the two proportions significant at the 0.05 level? Use a two-tailed test.

4. A major automobile insurance company claims that a larger proportion of younger drivers take high-risk chances when driving compared with other drivers. To investigate this claim, you give a driving-simulator test to a sample of young drivers and a sample of other drivers. Out of 100 young drivers tested, 30 took high-risk chances. Out of 200 other drivers tested, 55 took high-risk chances. Do you think there is sufficient evidence to support the insurance company's contention? Use the 0.05 level of significance.

Summary

This chapter dealt with proportions. A proportion is the ratio of the number of persons or items possessing a trait relative to the total number in the sample or population.

Two hypothesis tests were presented. One test compares a sample proportion and a hypothesized population proportion. The test statistic is

$$z = \frac{\dfrac{X}{n} - p}{\sqrt{\dfrac{p(1-p)}{n}}}$$

To apply this test, np and $n(1-p)$ must be at least 5.

The other test of hypothesis compares two population proportions. The test statistic is

$$z = \frac{\dfrac{X_1}{n_1} - \dfrac{X_2}{n_2}}{\sqrt{\bar{p}(1-\bar{p})\left(\dfrac{1}{n_1} + \dfrac{1}{n_2}\right)}}$$

To apply this test, both samples must meet the requirement that np and $n(1-p)$ be 5 or more.

The usual five-step hypothesis-testing procedure is employed in both tests.

Testing a Hypothesis About Proportions

I. Hypothesis Tests Involving One Proportion

A. Criterion. Both np and $n(1-p)$ are 5 or more.

B. Illustrative problem and procedure. H_0: the proportion of high-school males in suburban Chicago who are drug addicts is 0.15.

1. The null hypothesis is H_0: $p = 0.15$. For a two-tailed test the alternate hypothesis is H_a: $p \neq 0.15$. For a one-tailed test the alternate hypothesis is either $p < 0.15$ or $p > 0.15$.

2. The test statistic is

$$z = \frac{\dfrac{X}{n} - p}{\sqrt{\dfrac{p(1-p)}{n}}}$$

3. Procedures for the decision rule, computing z, and making a decision are the same as for previous tests.

II. Hypothesis Tests Involving Two Proportions

A. Criterion. Both samples must meet the requirement that np and $n(1 - p)$ be 5 or more.

B. Illustrative problem and procedure. Is the proportion of male prisoners who are habitual offenders equal to the proportion of female prisoners who are habitual offenders?

   1. $H_0: p_1 = p_2$ Alternate, two-tailed $p_1 \neq p_2$
                                 Alternate, one-tailed $p_1 < p_2$
                                            or $p_1 > p_2$

   2. The test statistic is

$$z = \frac{\dfrac{X_1}{n_1} - \dfrac{X_2}{n_2}}{\sqrt{\bar{p}(1 - \bar{p})\left(\dfrac{1}{n_1} + \dfrac{1}{n_2}\right)}}$$

   3. Procedures for the decision rule, computing z, and arriving at a decision are the same as for previous tests.

Chapter Exercises

5. Past elections indicate that 40% of the central-city votes are needed to pass any amendment that will increase taxes in the entire county. A preelection poll revealed a tax increase is favored by 300 out of 1,000 central-city voters. Is it likely that the difference between the sample proportion (0.30) and the proportion needed (0.40) is due to chance (sampling error)? Use the 0.01 level.

· **6.** Suppose that a random sample of 1,000 American-born citizens had revealed that 200 favor resumption of full diplomatic relations with Cuba. Similarly, 110 out of a sample of 500 foreign-born citizens favor it. Test at the 0.05 level the H_0 that there is no significant difference in the two proportions. Use a two-tailed test.

7. Past experience at a university indicates that 50% of the students change their major area of study after their first year of college work. At the end of the past year, a sampling of 100 students revealed that 48 of them had changed their major at the end of their first year. Has there been a significant decrease in the proportion of students who change their major, or can the difference between the expected proportion (50%) and the sample proportion (48%) be attributed to chance (sampling error)? Use the 0.05 level of significance.

8. A survey dealing with car pooling in metropolitan Philadelphia found that during a particular October, 83 out of 420 vehicles had more than one occupant. Five years later, a similar sample of 423 cars showed 146 had more than one occupant. Is this increase significant at the 0.01 level?

9. Preliminary research suggests that the proportion of young women who use a particular method of family planning is 0.30. A detailed study conducted in a small Indiana town showed a total of 46 users in a random sample of 200 young women. Are these results consistent with the preliminary research? Use the 0.10 level of significance.

First answer all the questions. Then check your answers against those given in the Answer section of the book.

Chapter Achievement Test

I. Multiple-Choice Questions. Select the response that best answers each of the questions (5 points each).

For questions 1 through 4 use the information that follows. An urban planner claims that, nationally, 20% of all families renting condominiums move during the year. A random sample of 200 families renting condominiums in a large development revealed that 56 had moved during the past year. Does the evidence demonstrate that this development has a greater proportion of movers than the national average?

1. Symbolically, the null hypothesis would be stated as
   a. $\mu = 0.20$
   b. $p \geq 0.20$
   c. $p < 0.20$
   d. $\mu < 0.20$

2. If a 0.05 level of significance is desired, the critical value is
   a. -1.96
   b. -1.65
   c. 1.28
   d. 1.65
   e. 1.96

3. The value of the test statistic is
   a. 1.28
   b. 2.82
   c. 5.66
   d. 16
   e. none of these

4. The null hypothesis is
   a. not rejected
   b. rejected and the alternate hypothesis accepted
   c. neither of these is correct

For questions 5 through 8 use the following information. A noted medical researcher has suggested that a heart attack is less likely to occur among men who actively participate in athletics. A random sample of 300 men is obtained. Of that total, 100 are found to be athletically active. Within this group, 10 had suffered heart attacks; among the 200 athletically inactive men, 25 had suffered heart attacks. Test the hypothesis that the proportion of men who are active and suffered heart attacks is equal to the proportion of men who are not active and suffered heart attacks. The z statistic is computed to be -0.63.

5. If p_1 refers to those who participate in athletics, the following is appropriate for stating the null hypothesis:
   a. $\mu_1 = \mu_2$
   b. $p_1 < p_2$
   c. $p_1 \geq p_2$
   d. $\mu_1 \neq \mu_2$

6. If a 0.05 significance level is used, the critical value is
   a. -1.96
   b. -1.65
   c. 1.28
   d. 1.65
   e. 1.96

7. The pooled estimate $\bar{p}$ is
   a. 0.10
   b. 0.125
   c. 35/300
   d. 0.025
   e. -0.025

8. The null hypothesis is
   a. accepted
   b. rejected and the alternate hypothesis accepted
   c. neither of these is correct.

II. Problems (30 points each).

9. Research into the use of physical punishment (spanking) in child rearing yielded this information: out of 100 children aged 3 to 9 years, 82 had been spanked by their parents during the previous month. Similarly, 40 out of 60 children aged 10 to 14 years had been spanked. Do parents use spankings significantly less often on older children? Test at the 0.01 level of significance.

10. A drug manufacturer has developed a drug that claims to cure postnatal depression in 85% of the cases. A random sample of 150 women who gave birth at the Colorado General Hospital and who used the drug in a two-year period revealed that 120 of them found it effective. Does this result contradict the manufacturer's claim? Use a 0.02 level of significance.

HIGHLIGHTS
From Chapters 8, 9, and 10

The last three chapters dealt with the fundamental concepts of sampling. They described (1) how sampling can be used to develop a range of values within which a population parameter is likely to occur, and (2) the procedure for testing a hypothesis about a population parameter.

Key Concepts

1. **Sampling** is important because complete knowledge regarding the population is seldom available. The basic purpose of sampling is to provide an estimate about a population parameter based on sample evidence.

2. Sampling is necessary because
   a. It may be impossible to check the entire population.
   b. The cost to study the entire population may be prohibitive.
   c. To contact the entire population would be too time-consuming.
   d. The tests may destroy the product.

3. A **confidence interval** is a range within which the population parameter is expected to fall for a preselected level of confidence.

4. The **sampling distribution of the mean** is a probability distribution that describes all possible sample means of a given size selected from a population.

5. The **central limit theorem** states that the distribution of the sample means is approximately normal regardless of the shape of the population.

6. **Hypothesis testing** is an extention of the concept of interval estimation and offers a strategy for choosing among alternative courses of action. For the purpose of testing, two claims are made about a population parameter.

Key Terms

Probability sampling
Random sample
Systematic random sample
Stratified random sample
Cluster sample
Sampling error
Sampling distribution of the mean
Central limit theorem
Confidence interval
Standard error of the mean
Point estimate

Interval estimate
Null hypothesis
Alternate hypothesis
Type I error
Type II error
One-tailed test
Two-tailed test
Level of significance
Test statistic
Decision rule
Critical value
Proportion

Key Symbols

$\sigma_{\overline{x}}$ The standard error of the mean.
E Maximum allowable error.
H_0 The null hypothesis.
H_a The alternate hypothesis.

p The proportion of objects or things in the population that possess a particular trait.

$\overline{p}$ The pooled estimate of the population proportion based on two samples.

Review Problems

1. A special ten-question examination was given to a population of five students. The number of correct responses by each student was as follows:

Student	Number Count
Taylor	5
Tigranian	6
Sims	7
Sobsak	9
Hilix	8

a. How many different samples of size two are possible from this population?
b. List the possible samples of size two.
c. Organize the sample means into a probability distribution.
d. Compare the mean of the sampling distribution to the mean of the population.
e. Compare the spread of the sampling distribution to that of the population.

2. The Department of Mathematics consists of seven faculty members. Their salaries are as follows:

Professor	Salary (000)
Hause	$23.0
Delgado	26.0
Gallaghar	29.0
Jackson	21.0
Ramsdall	20.0
Tang	19.0
Elsass	23.0

a. How many different samples of size two are possible?
b. List the possible samples of size two.
c. Organize the sample means into a probability distribution.
d. Compare the mean of the sampling distribution to the mean of the population.
e. Compare the spread of the sampling distribution to that of the population.

3. A study has been conducted on the moviegoing habits of young adults. A random sample of 50 reveals the mean number of movie-viewing hours per month to be 9.0 hours. The standard deviation is 2.8 hours. Develop a 95% confidence interval for the mean number of hours per month of movies viewed.

4. The Levision Brothers Department Store recently made an analysis of its delinquent accounts. A random sample of 90 delinquent accounts showed a mean delinquent amount of $78.65 with a standard deviation of $36.51. Construct a 90% confidence interval for the population mean.

5. A large health insurer is considering raising its rates. However, the decision is based on the average annual family expenditure for medical

care. The ensurer wants to be 99% confident that the sample is within $70 of the correct mean spent for medical care. The standard deviation is estimated to be $250. How large a sample is required?

6. The manager of Clem's Supermarket wants to estimate the mean amount spent in his store per customer. The manager estimates the population standard deviation to be $25.00, and he would like to be within $4 of the population mean. Assume a 99% level of confidence. How large a sample is required?

7. Coffee cans are filled to a net weight of 32 ounces, but there is some variability. A random sample of 36 cans revealed an average weight of 31.8 ounces with a standard deviation of 0.4 ounces. Test the hypothesis, at the 1% level of significance, that the net weight of the coffee is actually 32 ounces.

8. Records maintained by Orange Cross Insurance Company indicate that the average length of hospital confinement for a routine appendectomy is 4.0 days. St. Mark's Memorial Hospital selected a sample of 40 patients who had a routine appendectomy and found the mean length of confinement was 3.5 days with a standard deviation of 1.2 days. At the 0.05 significance level, can it be concluded that St. Mark's patients are released significantly earlier than insurance records suggest?

9. The president of Wattsburg Community College reported in a meeting with students: "When I came here four years ago the typical student traveled only 7 miles to class." He further stated that "a recent survey by the Admissions Office staff showed that in a sample of 35 students the mean distance traveled to class was 9.2 miles, and the standard deviation was 6.0 miles." Does this recent study show that students are now driving significantly farther to attend class? Use a 0.01 level of significance.

10. The student newspaper interviewed a random sample of 200 male students and found that 150 were in favor of having a homecoming dance. A random sample of 400 girls revealed 312 were in favor of the dance. Does the sample evidence show (at the 0.10 significance level) that there is a difference between the proportion of boys and of girls favoring a dance?

11. Shoplifting has become a national problem. One report indicates that 10% of the customers that enter a large discount store will steal something. The store manager of the Jamesway Discount Store randomly selects 150 customers and observes them behind a one-way mirror. Twenty of these shoppers attempted to steal something. Do these data suggest that shoplifting is more of a problem for Jamesway than for the typical large discount store? Use a 0.05 significance level.

12. A random sample of 40 male students and 42 female students revealed the following information about the number of bottles of beer (12-ounce size) they consumed last week.

	Male Students	Female Students
Mean	8.75	7.25
Standard deviation	2.30	1.95
Sample size	40	42

At the 0.05 significance level, can it be shown that the male students drink more beer?

13. A machine is set to turn out ball bearings having a radius of 1.5 cm. A sample of 36 produced by the machine had a radius of 1.504 cm, with a sample standard deviation of 0.009 cm. Is there reason to believe that the machine is producing ball bearings with a mean radius greater than 1.5 cm? Use the 0.05 significance level.

Case Analysis

The McCoy's Market Case

(Data for these two cases can be found in the first Chapter Highlights, pp. 126 - 130.)

1. Assume the 65 measurements provided on pp. 127 - 128 represent the sample of all shoppers. Develop a 95% confidence interval for the mean amount spent by all shoppers. How large a sample is required if the manager wishes to be 99% confident that the sample mean is within $2 of the population mean?

2. Test the hypothesis at the 0.05 level of significance that the mean number of items purchased is 20.

3. Industry sources claim the typical customer spends $27.75. Test the hypothesis at the 0.01 level of significance that McCoy's Supermarket customers spend more.

4. At the 0.10 significance level, do the data show a significant difference between the mean number of items purchased by males and by females?

5. At the 0.05 level, do significantly more males shop on weekends rather than midweek?

6. Test the hypothesis that there are as many male as female customers.

The St. Mary's Emergency Room Case

1. Test the hypothesis, at the 0.01 level of significance, that the proportion admitted during shift 2 is equal to the proportion admitted during shift 3.

2. Construct a 90% confidence level for the mean cost of service per patient.

3. If the administrator wished to be 98% confident she had estimated the mean cost within $10, how large a sample should she take?

4. Test the hypothesis, at the 0.05 level of significance, that the care of patients over 60 years of age requires more staff than the care of younger patients (under 30 years of age).

11
Hypothesis Tests: Small-Sample Methods

Test equasions

$$\sigma_{\bar{x}} = \frac{\sigma}{\sqrt{n}} \qquad M \pm Z\,\sigma_{\bar{x}}$$

$$t = \frac{\bar{x} - M}{\frac{S}{\sqrt{n}}}$$

OBJECTIVES

When you have completed this chapter, you will be able to
- Describe the characteristics of the t distribution.
- Conduct a hypothesis test for a population mean, given that the sample size is smaller than 30 and the population standard deviation is unknown.
- Conduct a hypothesis test for the difference between two population means, given that each sample is smaller than 30 and the population standard deviations are unknown.
- Conduct a hypothesis test for the difference between paired observations, given that the sample size is smaller than 30.

Introduction

In Chapter 9, hypothesis tests for normally distributed interval-level data were considered. The population standard deviation σ was either known, or estimated from a "large" sample—generally considered by most statisticians to be of size 30 or larger. In many situations, however, there is interest in testing a hypothesis about a normally distributed interval-level variable, but the population standard deviation is not known and the sample size is small—that is, under 30. In such cases the z statistic is not appropriate because σ is unknown or cannot be accurately estimated based on only a few observations. In its place, Student's t distribution is used. This distribution first appeared in the literature in 1908 when W. S. Gosset, an Irish brewery employee, published a paper about the distribution under the pseudonym "Student." In his paper, Gosset assumed the samples were taken from normal populations. Approximate results are also obtained in practice when sampling from nonnormal populations.

Characteristics of the t Distribution

The t distribution is similar to the z distribution in some respects, but quite different in others. It has the following major characteristics:

1. It is a continuous distribution like the _z_ distribution used in Chapter 9.

2. It is bell shaped and symmetrical, again similar to the _z_ distribution.

3. There is only one standard normal distribution _z_, but there is a "family" of _t_ distributions. That is, each time the size of the sample changes, a new _t_ distribution is created.

4. The _t_ distribution is more spread out at the center (that is, "flatter") than the normal distribution.

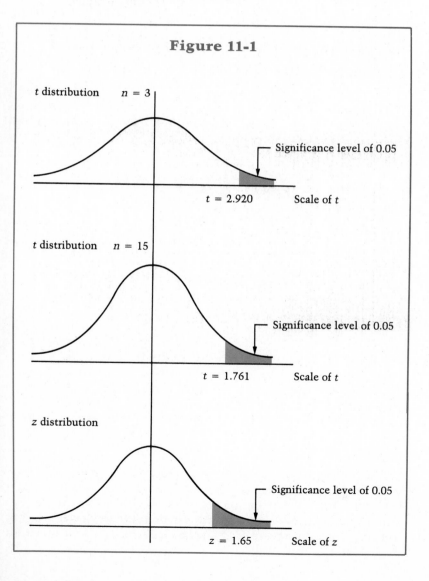

Figure 11-1

Because the t distribution has a greater spread than the z distribution, the critical values of t for a particular significance level will be greater in magnitude than the corresponding critical values of z. As the size of the sample increases, the value of t approaches the value of z for a particular significance level. For example, the critical values of a one-tailed test with a significance level of 0.05 are shown in Figure 11-1 on page 307. The sample sizes of 3, 15, and the z value corresponding to a one-tailed test are shown. Note that the t value for a sample size of 3 is larger than that for 15: for an n of 3 t is 2.920, for 15 it is 1.761.

Testing a Hypothesis About a Population Mean

The t distribution is used to test a hypothesis about a population mean when the population standard deviation is not known and the sample size is small. The test statistic is

$$t = \frac{\overline{X} - \mu}{\dfrac{s}{\sqrt{n}}}$$

where

   $\overline{X}$ is the sample mean.
   μ is the hypothesized population mean.
   s is the sample standard deviation.
   n is the number of items in the sample.

Note in the formula for t that the sample standard deviation (s) is used instead of the population standard deviation (σ). The usual five-step hypothesis-testing procedure is followed.

Problem

Strong Muffler Company advertises extensively that they can "mufflerize" any car in 30 minutes or less, on the average. The Consumer Protection Agency decides to investigate this claim. Ten of its fleet of cars, all unmarked and in need of new mufflers, have been dispatched to nearby Strong Muffler shops. The time it has taken for

$$H_o: M \leq 30$$
$$H_a: M > 30$$

each of the ten mufflers to be installed is 26, 32, 24, 37, 28, 29, 33, 31, 34, and 36 minutes, respectively. Does this evidence suggest that Strong Muffler is unable to meet its advertised claim of being able to change mufflers in 30 minutes or less, on the average?

$$t = \frac{31 - 30}{\frac{10}{\sqrt{10}}}$$

Solution

The five-step hypothesis-testing procedure employed by the Consumer Protection Agency is:

✱Step 1   State the null hypothesis. The null hypothesis is that the average time to mufflerize a car is 30 minutes. The alternate hypothesis is that it is more than 30 minutes. Symbolically:

$$H_0: \mu \leq 30$$
$$H_a: \mu > 30$$

✱The way the alternate hypothesis is stated dictates a one-tailed test.

Step 2   The level of significance is selected. The 0.05 level of significance was chosen for this problem.

Step 3   The appropriate test statistic is chosen. The t distribution is used because the population standard deviation is not known and the sample size is small (ten). The t distribution formula is

$$t = \frac{\overline{X} - \mu}{\frac{s}{\sqrt{n}}}$$

Step 4   The decision rule is stated. A critical value separates the region of acceptance from the region of rejection. Appendix D gives the critical values for Student's t distribution. In order to use the table, the number of _degrees of freedom_ (df) must be determined. The number of degrees of freedom is equal to the sample size minus the number of samples. One sample of size ten is used in this problem. Thus, the degrees of freedom is $n - 1 = 10 - 1 = 9$. An example is presented to show why 1 is subtracted from the sample size. Suppose a sample of four measurements is obtained and the values are 4, 6, 8, and 10. The mean of the sample is 7. If any three of the values are changed then the fourth value is automatically fixed so that the mean is still 7. Suppose that the first three values were changed to 3, 5, and 9. The last number must be 11 for the four numbers to have a mean of 7, that is, $(3 + 5 + 9 + 11)/4 = 28/4 = 7$. Thus, only three out of the four numbers are free to vary, and it is said that there are 3 "degrees of freedom." For this example, the degrees of freedom could be determined by $n - 1 = 4 - 1 = 3$ df.

Let us now return to the critical values for Student's t distribution as given in Appendix D. The following table (Table 11-1) excerpts a few of the critical values from Appendix D. Note that both two-tailed and one-tailed values are given. To locate the critical value of t, go down the left margin in Table 11-1, labelled "degrees of freedom" (df), until $n - 1$ degrees of freedom is located. There are ten cars being mufflerized in the sample, so $10 - 1 = 9$ degrees of freedom. Then move to the right and read the value given under the "One-Tailed Value" column headed by 0.05. The critical value for 9 df and a one-tailed test is 1.833 for the 0.05 level of significance.

Table 11-1

Portion of the t Distribution Table

One-tailed Value   **Two-tailed Value**

Appendix D
Student's t Distribution

DEGREES OF FREEDOM (df)	ONE-TAILED VALUE					
	0.25	0.10	0.05	0.025	0.01	0.005
	TWO-TAILED VALUE					
	0.50	0.20	0.10	0.05	0.02	0.01
1	1.000	3.078	6.314	12.706	31.821	63.657
2	0.816	1.886	2.920	4.303	6.965	9.925
3	.765	1.638	2.353	3.182	4.541	5.841
4	.741	1.533	2.132	2.776	3.747	4.604
5	.727	1.476	2.015	2.571	3.365	4.032
6	.718	1.440	1.943	2.447	3.143	3.707
7	.711	1.415	1.895	2.365	2.998	3.499
8	.706	1.397	1.860	2.306	2.896	3.355
9	.703	1.383	1.833	2.262	2.821	3.250
10	.700	1.372	1.812	2.228	2.764	3.169

Thus, the decision rule is: reject H_0 if the computed value of t is greater than 1.833.

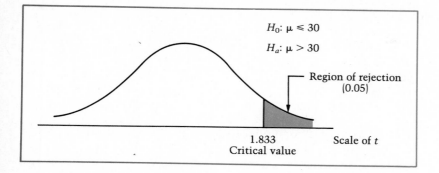

$H_0: \mu \leqslant 30$

$H_a: \mu > 30$

Region of rejection (0.05)

1.833
Critical value

Scale of t

Step 5 Make the statistical decision. The mean length of time Strong Muffler took to replace the mufflers on the Consumer Protection Agency cars is 31 minutes, and the standard deviation of the sample of ten is 4.24 minutes, found by using either of these equations from Chapter 4:

$$s = \sqrt{\frac{\Sigma(X - \bar{X})^2}{n - 1}} \quad \text{or} \quad \sqrt{\frac{\Sigma X^2 - \frac{(\Sigma X)^2}{n}}{n - 1}}$$

The essential calculations are shown in Table 11-2.

Table 11-2

Essential Calculations for the Sample Standard Deviation

Time (in minutes) X	X²	
26	676	
32	1,024	$s = \sqrt{\dfrac{\Sigma X^2 - \dfrac{(\Sigma X)^2}{n}}{n - 1}}$
24	576	
37	1,369	
28	784	$s = \sqrt{\dfrac{9772 - \dfrac{(310)^2}{10}}{10 - 1}}$
29	841	
33	1,089	
31	961	$s = 4.24$
34	1,156	
36	1,296	$\bar{X} = \dfrac{310}{10} = 31$
310	9,772	

The computed value of t is 0.745, found by

$$t = \frac{\overline{X} - \mu}{\dfrac{s}{\sqrt{n}}}$$

$$= \frac{31 - 30}{\dfrac{4.24}{\sqrt{10}}}$$

$$= 0.745$$

Comparing the computed t value of 0.745 with the critical value of 1.833, we fail to reject the null hypothesis. Although there is a difference between the hypothesized mean (30) and the sample mean (31), it can be attributed to sampling error. The Consumer Protection Agency does not have sufficient statistical evidence to disprove that Strong Muffler is able to mufflerize a car in 30 minutes or less, on the average, as claimed.

Self-Review 11-1

A reminder: cover the answers in the margin.

The American Association of Retired Persons (AARP) reported that the typical (average) senior citizens living at high altitudes claim their systolic blood pressure is lower than the average of 160. To test this claim, 16 senior citizens living at high altitudes were selected at random and their blood pressures checked. Their mean systolic pressure was 151, the standard deviation of the sample was 12. The question to be explored is: Do senior citizens living at high altitudes have significantly *lower* systolic blood pressure? Use the 0.05 level.

a. State the null hypothesis and the alternate hypothesis symbolically.

b. Give the formula for the test statistic.

c. How many degrees of freedom are there?

d. Give the decision rule.

e. Compute t.

f. Arrive at a decision.

a. $H_0: \mu \geq 160$
$H_a: \mu < 160$

b. $t = \dfrac{\overline{X} - \mu}{\dfrac{s}{\sqrt{n}}}$

c. 15, found by $16 - 1$.

d. Reject H_0 if computed t falls to the left of -1.753. Otherwise accept H_0.

e. $t = \dfrac{151 - 160}{\dfrac{12}{\sqrt{16}}}$

$= -3$

f. Reject H_0. Senior citizens living in high altitudes do have a lower systolic blood pressure than the national average.

Chapter Exercises

1. A car manufacturer asserts that with the new collapsible bumper system, the average body repair cost for the damages sustained in a collision impact of 15 miles per hour does not exceed $400. To test the validity of this claim, six cars are crashed into a barrier at 15 miles per hour and their repair costs recorded. The mean and the standard deviation are found to be $458 and $48, respectively. At the 0.05 level of significance, do the test data contradict the manufacturer's claim that the repair cost does not exceed $400?

2. It is claimed that a new treatment for prolonging the lives of cancer patients is more effective than the standard one. Records of earlier research show the mean survival period to have been 4.3 years with the standard treatment. The new treatment is administered to 20 patients and the duration of their survival is recorded. The sample mean is 4.6 years, and the standard deviation is 1.2 years. Is the claimed effectiveness of the new method supported at the 1% level of significance?

A Computer Example

The Statistical Package for the Social Sciences (SPSS) was introduced in Chapter 4 as an efficient way of doing routine calculations by computers. A similar package, called MINITAB, was developed at Pennsylvania State University. A MINITAB illustration using the Consumer Protection Agency problem follows. The first step is for you to enter, or input the times needed to mufflerize the ten cars (see DATA on the next-to-last line of the printout).

```
MINTAB     04:00 PM      22-Mar-82

Minitab release 81.1 *** Copyright Penn State Univ 1981
MARCH 22, 1982  University of Toledo version PDP-11  81.1
Type NEWS or HELP for information on MINITAB

Storage available     3995 @    36056

MTB >SET DATA INTO C1
DATA>26,32,24,37,28,29,33,31,34,36
DATA>END
```

Next, you enter TTEST, the null hypothesis that $\mu = 30$, and the alternate hypothesis. Note that t is the same as we calculated before and that we cannot reject the null hypothesis at the 0.05 level.

```
MTB >TTEST OF MU=30 VS. ALTERNATE=1, ON DATA IN C1
    C1        N =   10      MEAN =      31.000      ST.DEV. =       4.24

    TEST OF MU =     30.0000 VS. MU G.T.     30.0000
    T =   0.745
    THE TEST IS SIGNIFICANT AT  0.2375
    CANNOT REJECT AT ALPHA = 0.05
```

Self-Review 11-2

a. $H_0: \mu \le 20$
   $H_a: \mu > 20$
b. One-tailed.
c. σ is unknown, sample is small.

d. $s = \sqrt{\dfrac{2218 - \dfrac{(104)^2}{5}}{5-1}}$

   $= 3.70$

   $t = \dfrac{20.8 - 20.0}{\dfrac{3.70}{\sqrt{5}}}$

   $= 0.48$

e. Fail to reject H_0
f. Although the sample mean of 20.8 is greater than 20, the difference could be due to sampling error. The product should not be marketed.

A new toy has been developed and the manufacturer hopes to market it for the coming Christmas season. Before going into full production, a large number of toys were handcrafted and sent to five test market areas. The manufacturer plans to start full production if the monthly sales in the test markets average more than $20,000 during the one-month trial period. The results, in thousands of dollars, were $20, $16, $25, $19, and $24. Using the 0.05 level of significance, can it be shown that the mean monthly sales volume is greater than $20,000?

a. State the null and alternate hypotheses.
b. Is this a one-tailed or a two-tailed test?
c. Why is Student's t being used?
d. Calculate the value of the test statistic.
e. What is your decision regarding the null hypothesis?
f. Interpret your results for the toy manufacturer.

Chapter Exercises

3. A city health department wishes to determine if the mean bacteria count per unit-volume of water at Siesta Lake beach is below the safety level of 200. Researchers have collected ten water samples and have found the bacteria counts per unit-volume to be 175, 180, 215, 188, 194, 207, 211, 195, 198, 190, respectively. Do the data warrant cause for concern? Use the 0.10 level of significance.

4. Sorenson Pharmaceutical has been conducting restricted studies on small groups of people to determine the effectiveness of a measles vaccine. The following measurements are readings on the antibody strength for five individuals injected with the vaccine: 1.2, 2.5, 1.9, 3.0, and 2.4. Use the sample data to test the hypothesis at the 0.01 level of significance that the mean antibody strength for individuals vaccinated with the new drug is 1.6 or more.

Comparing Two Population Means

A t test can be used to compare two population means. We assume

1. Each population is approximately normally distributed,
2. Their population standard deviations are equal but unknown, and
3. The two samples are unrelated (independent).

The t statistic for the two-sample case is similar to that employed for the two-sample z statistic in Chapter 9. The additional calculation required is that the two computed sample standard deviations are pooled to form a single estimate of the population standard deviation. This test is normally employed under the conditions where both samples have less than 30 observations. The formula for t is

$$t = \frac{\overline{X}_1 - \overline{X}_2}{s_p \sqrt{\dfrac{1}{n_1} + \dfrac{1}{n_2}}}$$

where

$\overline{X}_1$ is the mean of the first sample.
$\overline{X}_2$ is the mean of the second sample.
n_1 is the number in the first sample.
n_2 is the number in the second sample.
s_p is a pooled estimate of the population standard deviation. Its formula is:

$$s_p = \sqrt{\frac{(n_1 - 1)(s_1^2) + (n_2 - 1)(s_2^2)}{n_1 + n_2 - 2}}$$

where

s_1 is the standard deviation of the first sample.

s_2 is the standard deviation of the second sample.

Recall that the number of degrees of freedom is equal to the total number of items sampled minus the number of samples. The sample sizes for this problem are n_1 and n_2, and there are two samples. Hence, there are $n_1 + n_2 - 2$ degrees of freedom; this appears in the denominator in the pooled standard deviation.

Problem

You are conducting an experiment involving the third-grade class of eleven children at the Toth Elementary School to determine if there is a difference in student comprehension using two different teaching methods. You randomly assign the eleven children to two groups. Then, they are taught the basic concepts of multiplication by the same teacher, but using two different methods. Finally, after the first week of instruction, you administer the same ten-question examination to both groups. The number of correct answers out of ten are listed in Table 11-3.

Table 11-3

Number Correct on a Ten-Question Examination

	Teaching Method I Score		Teaching Method II Score
Sally	4	Olga	9
James	3	Orville	6
Abdul	5	Peter	8
Jackson	7	Rachel	4
Noreen	6	Susan	7
		Andrew	5

The question being investigated is: Is there a significant difference in the performance under the two teaching methods? Use the 0.01 significance level.

Solution

The null hypothesis states that there is no difference in the mean scores of the two groups. The alternate hypothesis states that there is a difference in the mean scores. Symbolically:

$$H_0: \mu_1 = \mu_2$$
$$H_a: \mu_1 \neq \mu_2$$

The alternate hypothesis indicates that a two-tailed test is called for.

Again, the decision rule depends on the number of degrees of freedom. In this case, that number is equal to the combined number of observations in the two samples minus the number of samples. This is expressed as $n_1 + n_2 - 2$. For this problem, $n_1 + n_2 - 2 = 5 + 6 - 2 = 9$ degrees of freedom.

The critical value for the 0.01 level, two-tailed test is 3.250 (from Appendix D). The decision rule is that you fail to reject the null hypothesis if the computed value of t is in the interval from -3.250 to 3.250. If it is outside this interval, you reject the null hypothesis. Shown schematically:

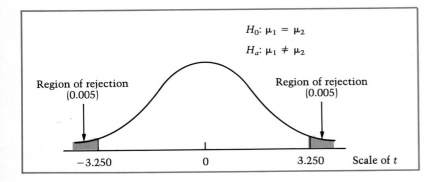

The calculation of Student's t can be accomplished in three stages. First, calculate the standard deviation of each sample. Second, "pool" these standard deviations into a single estimate of the population standard deviation. Third, calculate t.

Stage 1   Calculate the sample standard deviations.

Method 1		Method 2	
X_1	X_1^2	X_2	X_2^2
4	16	9	81
3	9	6	36
5	25	8	64
7	49	4	16
6	36	7	49
		5	25
25	135	39	271

$$s_1 = \sqrt{\frac{\Sigma X_1^2 - \frac{(\Sigma X_1)^2}{n_1}}{n_1 - 1}} \qquad s_2 = \sqrt{\frac{\Sigma X_2^2 - \frac{(\Sigma X_2)^2}{n_2}}{n_2 - 1}}$$

$$= \sqrt{\frac{135 - \frac{(25)^2}{5}}{5 - 1}} \qquad = \sqrt{\frac{271 - \frac{(39)^2}{6}}{6 - 1}}$$

$$= 1.58 \qquad\qquad\qquad = 1.87$$

Stage 2   Pool the standard deviations.

$$s_p = \sqrt{\frac{(n_1 - 1)(s_1^2) + (n_2 - 1)(s_2^2)}{n_1 + n_2 - 2}}$$

$$= \sqrt{\frac{(5 - 1)(1.58)^2 + (6 - 1)(1.87)^2}{5 + 6 - 2}}$$

$$= \sqrt{3.05}$$

$$= 1.746$$

Stage 3   Calculate t.

$$t = \frac{\overline{X}_1 - \overline{X}_2}{s_p \sqrt{\frac{1}{n_1} + \frac{1}{n_2}}}$$

$$= \frac{5.0 - 6.5}{1.746\sqrt{\frac{1}{5} + \frac{1}{6}}}$$

$$= \frac{-1.5}{1.058}$$

$$= -1.417$$

Since -1.417 falls within the interval between -3.250 and 3.250, the null hypothesis cannot be rejected. You conclude that there is no significant difference in the mean scores of the children taught under the two methods. The difference in comprehension evidenced in the two samples could be attributed to sampling error.

A Computer Example

If you use MINITAB to solve the Toth Elementary School problem, the data are entered first (line 2).

```
MTB >SET DATA INTO C10
DATA>4,3,5,7,6
DATA>SET DATA INTO C11
DATA>9,6,8,4,7,5
DATA>PRINT C10,C11
COLUMN          C10             C11
COUNT            5               6
ROW
  1              4.              9.
  2              3.              6.
  3              5.              8.
  4              7.              4.
  5              6.              7.
  6                              5.
```

Next, the instruction POOLED T causes the computer to perform the necessary calculations.

```
MTB >POOLED T, ALTERNATE 0,DATA IN C10,AND C11
   C10       N =   5    MEAN =    5.0000    ST.DEV. =    1.58
   C11       N =   6    MEAN =    6.5000    ST.DEV. =    1.87

   DEGREES OF FREEDOM =   9

   A 95.00   PERCENT C.I. FOR MU1-MU2 IS (    -3.8951,       0.8951)

   TEST OF MU1 = MU2 VS. MU1 N.E. MU2
   T = -1.417

   THE TEST IS SIGNIFICANT AT   0.1901
   CANNOT REJECT AT ALPHA = 0.05
```

calculated value of t

Self-Review 11-3

a. $H_0: \mu_1 \geq \mu_2$
   $H_a: \mu_1 < \mu_2$

b. df $= 7 + 5 - 2 = 10$

c. $s_1 = \sqrt{\dfrac{976 - \dfrac{(82)^2}{7}}{6}} = 1.60$

$s_2 = \sqrt{\dfrac{1621 - \dfrac{(89)^2}{5}}{4}} = 3.03$

d. $s_p = \sqrt{\dfrac{(6)(1.60)^2 + (4)(3.03)^2}{7 + 5 - 2}}$

   $s_p = 2.28$

e. Accept H_0 if t is less than -1.812.

f. $t = \dfrac{11.714 - 17.80}{2.28\sqrt{\dfrac{1}{7} + \dfrac{1}{6}}} = -4.55$

g. Reject H_0.

h. Interceptive treatment is effective in reducing the time in braces.

An orthodontist wants to investigate the effectiveness of the interceptive treatment she prescribes for some of her patients. (Interceptive treatment is dental work performed on relatively young patients in hopes of forestalling more extensive treatment later.) She compares the length of time a sample of interceptive patients must wear braces with a random sample of noninterceptive patients. Results follow (time is in months):

| | Time Wearing Braces | | |
Interceptive		Noninterceptive	
Joseph	12	Sally	16
Karen	13	George	22
Pam	11	Enrico	14
Peter	12	Aldine	18
Nickie	14	Jenny	19
Rosa	9		
Kurt	11		

The research question to be explored is: Is there sufficient evidence to indicate that interceptive patients spend less time in braces? Use the 0.05 significance level.

a. State the null and alternate hypotheses. Use the subscript $_1$ to refer to the interceptive group.

b. How many degrees of freedom are there in this problem?

c. Calculate the two sample standard deviations.

d. Compute the pooled estimate of the standard deviation.

e. What is the decision rule?

f. Calculate the value of the test statistic.

g. What is your decision regarding the null hypothesis?

h. Interpret the results.

Chapter Exercises

5. The peak oxygen intake per unit of body weight, called the "aerobic capacity," of an individual performing a strenuous activity is a measure of work capacity. For a comparative study, measurements of aerobic capacities are recorded for a group of 20 Peruvian highland natives and for a group of 10 U.S. lowlanders acclimatized to high altitudes as adults. The following summary statistics were obtained from the data:

	Peruvians	Americans
Mean	46.3	38.5
Standard deviation	5.0	5.8

Do the data provide a strong indication (at the 0.05 level of significance) of a difference in the mean aerobic capacities?

6. To compare the effectiveness of isometric and isotonic exercise methods in abdominal reduction, 20 overweight business executives are included in an experiment. Ten use each type of exercise and after ten weeks the reductions in abdomen measurements are recorded in centimeters.

	Isometric	Isotonic
Mean	2.5	3.1
Standard deviation	0.8	1.0

At the 0.01 level of significance, do these data support the claim that the isotonic method is more effective?

A Test of Paired Observations

In some experiments, the investigator is concerned with the *difference* in a pair of related observations. For example, those enrolled in a physical fitness class are weighed both before the course starts and after it is completed. The purpose of the experiment is to examine the effectiveness of the fitness program. Therefore, the t test focuses on the weight loss of each person and not on the means of the two populations. In such cases, the test is based on the difference in each pair of observations, instead of on the value of the individual observations. The distribution of this population of differences is assumed to be normal with an unknown standard deviation. The mean of this population of differences is designated D. As pointed out earlier, it is often impossible to study the entire population of differences. Therefore, a sample is selected. The symbol d is used to designate a particular observed difference and $\bar{d}$ the mean difference. The formula for t is

$$t = \frac{\bar{d}}{\dfrac{s_d}{\sqrt{n}}}$$

where

$\bar{d}$ is the average difference between the paired observations.

s_d is the standard deviation of the difference between the paired observations.

n is the number of paired observations.

Problem

A company is considering the purchase of a new Colton typewriter for use in the typing pool. Colton claims that their new-model type-writer will increase typing efficiency. To back up this claim, Colton loans the company ten new typewriters. Ten typists are randomly selected from the typing pool and assigned to test the new model. The results are

Typist	Current Typewriter	Proposed Typewriter
Woodstock	55	61
LaRoche	54	60
Hayes	47	56
Kootz	59	63
Kinney	51	56
Markas	61	63
Chung	57	59
Jaynes	54	56
Shue	63	62
Keller	58	61

Using a 0.01 level of significance, how would you analyze the data?

Solution

Before making the change to the new model, the company would like to know that the typists can type *faster* on the proposed model. The difference d is $X - Y$, where X is the speed on the current typewriter model and Y the speed on the proposed Colton model. The null hypothesis is that the typists will perform at least as well on the currently used typewriter model. Colton hopes the null hypothesis will be rejected. Thus a one-tailed test is necessary. The null and alternate hypotheses are

$$H_0: D \geq 0$$
$$H_a: D < 0$$

To reject H_0 and accept H_a will indicate that the difference is not zero or some larger number. The sample size n is equal to the number of paired observations (10). In this example, there are 9 (10 − 1) degrees of freedom. The decision rule is to reject H_0 if the computed value of t is to the left of −2.821 (see Appendix D).

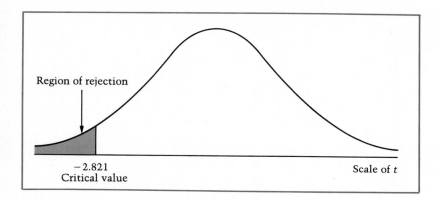

The sample results necessary to the test are shown in Table 11-4, followed by the computations.

Table 11-4.

Comparative Results of Efficiency

Typist	Current Typewriter X	Proposed Typewriter Y	X − Y = d	d²
Woodstock	55	61	−6	36
LaRoche	54	60	−6	36
Hayes	47	56	−9	81
Kootz	59	63	−4	16
Kinney	51	56	−5	25
Markas	61	63	−2	4
Chung	57	59	−2	4
Jaynes	54	56	−2	4
Shue	63	62	+1	1
Keller	58	61	−3	9
			−38	216

$$\bar{d} = \frac{\Sigma d}{n} = \frac{-38}{10} = -3.8$$

$$s_d = \sqrt{\frac{\Sigma d^2 - \dfrac{(\Sigma d)^2}{n}}{n-1}}$$

$$= \sqrt{\frac{216 - \dfrac{(-38)^2}{10}}{10-1}}$$

$$= 2.82$$

Computing the test statistic t:

$$t = \frac{\bar{d}}{\dfrac{s_d}{\sqrt{n}}}$$

$$= \frac{-3.8}{\dfrac{2.82}{\sqrt{10}}}$$

$$= -4.26$$

Since -4.26 is to the left of -2.821, the null hypothesis is rejected. The alternate hypothesis is accepted $(D < 0)$. Therefore, the conclusion is that the typists do not perform as effectively on the current model as on the proposed model.

Note that in Table 11-4 X represents the current model and Y the proposed model. The difference d is $X - Y$, where Y is expected to be larger than X. This leads to the negative direction of the alternate hypothesis. The entire test could have been reversed if d were found by $Y - X$. In this case the d values would have been positive and the direction of the alternate hypothesis would have been positive.

Self-Review 11-4

Advertisements for the Sylph Physical Fitness Center claim that completion of their course will result in a loss of weight. A random sample of recent students revealed the following body weights before and after completion. At the 0.01 level, can it be concluded that the course will result in a significant weight loss?

Name	Before	After
Wellman	155	154
Gersten	228	207
Tamayo	141	147
Miller	162	157
Ringman	211	196
Garbe	185	180
Monk	164	150
Heilbrunn	172	165

a. State the null and alternate hypotheses.
b. What is the critical value of t?
c. Calculate the value of the test statistic.
d. What is your decision regarding the null hypothesis?
e. Interpret your results.

a. $H_0: D \leq 0$
$H_a: D > 0$
b. $t = 2.998$
c. $t = \dfrac{7.75}{\dfrac{8.60}{\sqrt{8}}}$
$= 2.55$

d. H_0 is accepted.
e. The program cannot be shown to result in a significant weight loss.

Chapter Exercises

7. Two methods of memorizing difficult material are being tested. Nine pairs of students are matched according to I.Q. and background and then assigned to one of the two methods at random. A test is finally given to all the students, with the following results:

	Pair								
	1	2	3	4	5	6	7	8	9
Method A	90	86	72	65	44	52	46	38	43
Method B	85	87	70	62	44	53	42	35	46

Using the 0.05 level of significance, test to determine if there is a difference in the effectiveness of the two methods.

8. Measurements of the left-handed and right-handed gripping strengths of ten left-handed persons are recorded:

	1	2	3	4	5	6	7	8	9	10
Left hand	140	90	125	130	95	121	85	97	131	110
Right hand	138	87	110	132	96	120	86	90	129	100

Do these data provide evidence, at the 0.01 level of significance, that those tested have greater gripping strength in their dominant hand?

Summary

The hypothesis-testing methods in this chapter assume that the populations are normally distributed, that the standard deviation is unknown, and that the sample size is small (under 30). Under these conditions, the appropriate test statistic is Student's t distribution. The t distribution is based on the number of degrees of freedom. (The number of degrees of freedom is equal to the sample size minus the number of samples.) The critical value for any hypothesis test can be obtained from Appendix D.

Three hypothesis-testing situations were considered:

1. Comparing a single sample against some population mean.
2. Comparing two sample means, where the samples are independent and have the same standard deviation.
3. Comparing the difference between two related pairs of observations.

In each situation, the usual five-step hypothesis-testing procedure is used.

Chapter Outline

Hypothesis Tests: Small-Sample Methods

I. Small-sample versus population mean
  A. Objective. To compare the sample mean with a population mean.
  B. Assumptions:
    1. Normal population;
    2. Standard deviation is unknown;
    3. Sample size is less than 30.
  C. Test statistic

$$t = \frac{\overline{X} - \mu}{\dfrac{s}{\sqrt{n}}}$$

II. Comparing two independent sample means
  A. Objective. To test two sample means for equality.
  B. Assumptions:
    1. Each population is normal;
    2. Standard deviations are equal but unknown;
    3. Each sample is less than 30;
    4. Samples are unrelated (independent).

C. Test statistic

$$t = \frac{\bar{X}_1 - \bar{X}_2}{s_p \sqrt{\dfrac{1}{n_1} + \dfrac{1}{n_2}}}$$

where s_p is a pooled estimate of the common standard deviation, found by

$$s_p = \sqrt{\frac{(n_1 - 1)(s_1^2) + (n_2 - 1)(s_2^2)}{n_1 + n_2 - 2}}$$

III. Differences between related pairs

A. Objective. To test for a significant difference between observed pairs.

B. Assumptions:
   1. Two samples are paired or related;
   2. Sample sizes are each less than 30;
   3. A difference is computed and its distribution is normal with unknown standard deviation.

C. Test statistic

$$t = \frac{\bar{d}}{\dfrac{s_d}{\sqrt{n}}}$$

Chapter Exercises

In each of the following situations:
   a. State the appropriate null and alternate hypotheses.
   b. Determine the decision rule.
   c. Calculate the value of the test statistic.
   d. Interpret the decision.

9. The university library is interested in determining whether the average number of books checked out per visit has increased. In the past, the average was 3.0 books per student visit. A random sample of ten students revealed a mean of 4.1 books with a standard deviation of 2.0 books. At the 0.05 level of significance, does this information provide sufficient evidence to show that students are now checking out more books per visit?

10. A recent newspaper article claimed that the average American was 20 pounds overweight. To test this claim, 15 randomly

selected persons were weighed. The result averaged 18 pounds overweight, with a standard deviation of 5 pounds. At the 0.05 significance level, is there sufficient evidence to reject the newspaper's claim?

11. An association of college-textbook publishers recently reported the average retail cost of its members' books to be $15.00. A group of students lobbying for increased state support to students because of higher education costs has challenged this claim. A random sample of 20 books is selected. Calculations indicate that the average cost is $15.80 with a standard deviation of $3.80. At the 0.05 level of significance, is there sufficient evidence to reject the claim of the publishers' association? Can the students assert that the average cost is higher?

12. Your new car has an EPA rating of 26.0 miles per gallon. The mileage figures actually obtained on six trips were 24.3, 25.2, 24.9, 24.8, 25.6, and 25.4. Is there sufficient evidence, at the 0.01 level of significance, to conclude that the car performs below the EPA specifications?

13. A toothpaste manufacturer claims that children brushing their teeth with Bianca will have fewer cavities than those brushing with Sparkle. In a carefully supervised study, the number of cavities using the two brands is compared. Is there sufficient evidence, at the 0.01 level of significance, to support the manufacturer's claim?

	Cavities
Bianca	1,2,3,4,2,0,2
Sparkle	4,5,4,2,1,2,4

14. The National Weather Service reports the high temperature on July 1 is as likely to be above 25°C as below. Test this claim at the 0.10 significance level, given the following July 1 temperature readings over the past 16 years: 22, 26, 28, 24, 27, 20, 29, 32, 28, 21, 25, 27, 26, 28, 30, 22.

15. The following data are the weight gains, measured in pounds, of babies from birth to age six months. All babies in the sample weighed between seven and eight pounds at birth. One group of babies was breast-fed and the other fed a specific formula. Is there evidence that the weight gains are different among the two groups? Use a 0.01 significance level.

Breast-Fed	Formula-Fed
7	9
8	10
6	8
10	6
9	7
8	8
9	

16. A manufacturer of wrist supports for bowlers maintains that use of this support will improve a bowler's average. A sample of 12 bowlers who have a league-sanctioned average of over 150 roll two complete games, one with the wrist support and one without it. Does the evidence support the manufacturer's claim? Assume a significance level of 0.01.

Bowler	With Wrist Band	Without Wrist Band
Clarke	230	217
Redenback	225	198
Simmons	223	208
Pelton	216	222
Griffin	229	223
Farthy	201	214
Hawkins	205	187
Nugent	193	187
Bryan	177	178
Hucklebury	201	195
Baker	178	169
Berry	207	194

17. Six junior executives were sent to a class to improve their verbal skills. In order to test the quality of the program, the students were tested before and after taking the class, with the following results:

	Before	After
Levin	18	30
Baker	38	70
Craft	8	20
Denfrey	10	4
Longhi	12	10
Foster	12	20

Do these records indicate a significant improvement in verbal skills at the 0.10 significance level?

18. To measure the effectiveness of his sales training program, a car dealer selects at random eight sales representatives to take the course. The monthly sales volume of each sales representative is shown below. At the 0.05 significance level, can it be concluded that the new program is effective in increasing sales?

Sales Rep.	Gross Sales After the Course (000)	Gross Sales Before the Course (000)
1	$14.0	$13.5
2	10.7	11.4
3	12.4	10.7
4	11.1	11.1
5	10.9	9.8
6	10.5	9.6
7	10.8	10.7
8	13.0	11.7

Chapter Achievement Test

First answer all the questions. Then check your answers against those given in the Answer section of the book.

I. Multiple-Choice Questions. Select the response that best answers each of the questions that follow (5 points each).
  1. The t test for the difference between the means of two independent samples assumes that
     a. the samples were obtained from normal populations
     b. the population standard deviations are equal
     c. the samples were obtained from independent populations
     d. all of the above
  2. If we are testing for the *difference* between the means of two related samples with $n = 15$, the number of degrees of freedom is equal to
     a. 28
     b. 30
     c. 15
     d. 14
     e. none of the above
  3. The t distribution approaches which distribution as the sample size increases?
     a. binomial
     b. normal
     c. normal approximation to the binomial
     d. all of the above
     e. none of the above

4. A random sample of 16 is selected from a normal population. The population standard deviation is unknown. If a two-tailed test of significance is to be used at the 0.01 significance level, the null hypothesis is not rejected if
   a. z is between -2.58 and 2.58
   b. t is between -2.921 and 2.921
   c. t is between -2.947 and 2.947
   d. t is less than 2.602
   e. none of the above
5. The two sample t tests and the t test for paired observations will always yield the same results.
   a. true
   b. false

For questions 6 through 11 use the following information.

   A U.S. congressman claims that the average enrollment in public institutions of higher learning is less than 4,500 students. To test this claim, a random sample of six schools is selected. The enrollments at the selected schools are

Central State	3,206
Edison CC	1,721
Lacy State	5,634
Northside Tech	1,457
Shawnee State	1,728
Washington CC	727

6. The appropriate null and alternate hypotheses are
   a. $H_0: \overline{X}_1 \geq \overline{X}_2$
      $H_a: \overline{X}_1 < \overline{X}_2$
   b. $H_0: \mu \geq 4,500$
      $H_a: \mu < 4,500$
   c. $H_0: \mu = 4,500$
      $H_a: \mu \neq 4,500$
   d. $H_0: \overline{X} \geq 4,500$
      $H_a: \overline{X} < 4,500$
7. The appropriate test statistic is t because (assume the population is approximately normal)
   a. the population standard deviation is not known
   b. the sample size is small
   c. the enrollments are about equal
   d. a. and b. are both correct
   e. none of the above

8. The population standard deviation must be estimated from the sample information. The best estimate is
   a. 1,772.7
   b. 1,618.70
   c. 50,642,135
   d. 3,144,239.5
   e. none of the above

9. If the congressman assumed a 0.05 significance level, the null hypothesis would be rejected if
   a. t is to the left of -1.476
   b. t is to the left of -2.015
   c. t is outside the interval -2.015 and 2.015
   d. z is to the left of -1.65
   e. none of the above

10. The computed value of the test statistic is
    a. $t = -2.63$
    b. $z = -2.12$
    c. $t = 3.00$
    d. $t = -2.12$
    e. none of the above

11. Which of the following would be a correct conclusion?
    a. enrollment is not less than 4,500
    b. enrollment is at least 4,500
    c. enrollment is less than 4,500
    d. cannot be determined
    e. none of the above

II. For each of the following computation problems use the five-step hypothesis-testing procedure (15 points each).

12. The Food and Drug Administration is conducting tests on a certain drug to determine if it has the undesirable side effect of reducing the body's temperature. It is known that the average human temperature is 98.6°F. The new drug is administered to 25 patients and the patients' mean temperature drops to 98.3°F with a standard deviation of 0.64°. At the 0.05 significance level, is there sufficient reason to conclude that the drug reduces body temperature?

13. A home builder claims that the addition of a heat pump will reduce electric bills in all-electric homes. To support his claim he tests the electric bill for the month of January for two consecutive years, one before the heat pump was installed and one after. Is there sufficient evidence to show that heating bills were reduced at the 0.01 level?

Customer	Before	After
Garcia	180	160
Huffman	156	164
Johnson	188	172
Palmer	132	130
Kerby	208	200
Beard	196	190
Sauve	190	184

14. The fire chief for Slocum County is evaluating two policies for the location of emergency medical equipment. One plan calls for supplies to be kept near the engines most often used by paramedical personnel. The second one calls for storage near the crew's sleeping quarters. In order to decide objectively if one location is better than the other, the chief tries each of the two locations, clocking the time it takes paramedics to collect their equipment under emergency conditions. The results, shown in seconds are listed below. Use a 0.05 significance level.

Plan 1	Plan 2
10	15
55	9
30	47
30	3
53	34
	41
	30
	29

12
Analysis of Variance

OBJECTIVES

When you have completed this chapter, you will be able to
- Describe the F distribution.
- Construct an analysis-of-variance table for one criterion of classification.
- Test for a difference among two or more population means.

Introduction

In this chapter we continue the discussion of hypothesis testing started in Chapter 9. Recall that Chapter 9 developed the general theory of hypothesis testing and applied it to the means of two normally distributed populations using the z statistic. Chapter 10 tested the difference between two proportions, again employing the z statistic. Chapter 11 analyzed differences between means of two normally distributed populations, but using Student's t statistic instead of the z statistic.

This chapter compares several population means. The data are presumed to be of interval scale and normally distributed. We will describe a technique that allows us to compare *two or more population means* simultaneously. This technique is called **analysis of variance,** abbreviated **ANOVA.**

Underlying Assumptions for ANOVA

To use the ANOVA technique, the following conditions are assumed:

1. The populations being studied are normally distributed.
2. The populations have equal standard deviations (σ).
3. The samples selected from those populations are independent and random.

When those conditions are met, the F statistic is used (instead of z or t) to test if the means of the populations are equal. There is a "family" of F distributions; each F distribution is positively skewed and its values cannot be negative. Whenever the assumptions about the normality of the population distribution and equal standard deviations cannot be met, an analysis-of-variance technique developed by Kruskal and Wallis, to be discussed in a later chapter, may be used.

Because analysis of variance had its beginnings in agriculture, the term **treatment** is generally used to identify the different populations being examined.

Treatment A specific source or cause of variation in a set of data.

Two illustrations will help to clarify the term "treatment" and to demonstrate the application of the ANOVA technique.

1. Wheat yields. A farmer wants to use the brand of fertilizer that will produce the maximum yield per acre of wheat. Assume three different commercial brands, namely, Prothro, Scotts, and Anderson are to be applied. As an experiment, the farmer divides his field into 15 plots of equal size. The 15 areas are planted in the same manner and at the same time, but he randomly assigns Prothro to five plots, Scotts to five plots, and Anderson to five plots. At the end of the growing season the number of bushels of wheat produced by each plot is recorded. The results are

Prothro	Anderson	Scotts
40	72	51
45	71	55
47	68	60
50	75	57
47	66	54

Do the treatments differ? In this case, "treatment" refers to the different fertilizers that are applied. Are the mean yields of wheat different among the three populations? Figure 12-1 is an illustration of how the means would appear if the yields were *different*.

Suppose that the means were, in fact, identical. From a practical standpoint, this would indicate that the fertilizers all produced the same yield of wheat. Figure 12-2 illustrates three yields with equal

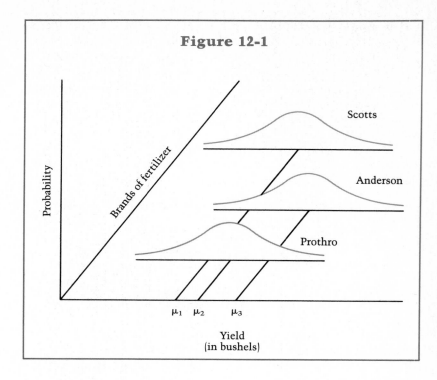

Figure 12-1

population means ($\mu_1 = \mu_2 = \mu_3$). Note that the distributions are approximately normal and that the dispersion of each is about the same.

2. Instructional modes. The instructor of a nursing arts course wants to know if the extent of learning in the course differs with the type of instruction. Four methods (treatments) are proposed: (1) lecture, (2) movie, (3) lecture and experience, and (4) movie and experience. (The experience is to be obtained by having the students follow the particular nursing procedure with the instructor's guidance.) As an experiment, the 19 students in the current class were randomly assigned to the four groups. After the course, the same final examination was administered to all students. The results were

Lecture	Movie	Lecture and Experience	Movie and Experience
80	59	85	81
72	65	84	84
69	68	77	76
75	61	69	71
	70	73	72

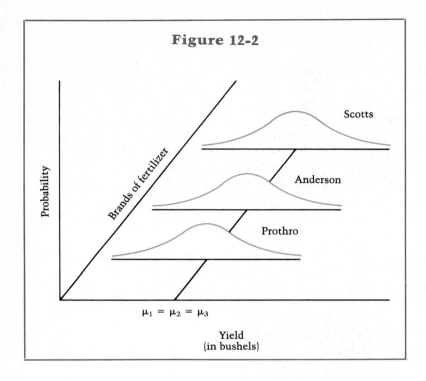

Figure 12-2

Do these data demonstrate to the instructor that there is a difference in the extent of learning achieved with different types of instruction?

The ANOVA Test

The purpose of this section is to explore some of the reasoning behind ANOVA. First, ANOVA breaks down the total variation into two parts. One part measures the variation between sample means (treatments). The other part measures the variation of the observations from their treatment mean, that is, it measures the variability within each of the sampled populations. Next, this relationship between the two sources of variation is compared by forming the F ratio in the following manner:

$$F = \frac{\text{estimate of the population variance based on differences between sample means}}{\text{estimate of the population variance based on variation within samples}}$$

The F distribution has three major characteristics:

1. Both the numerator and denominator of F are variances. Since variances are always positive, F cannot be negative.

2. The *F* distribution's curve is positively skewed, and the values range from zero to infinity.

3. There is a "family" of *F* distributions. This graph, in which three *F* distributions are shown schematically, illustrates the point. There is one *F* distribution for the combination of 32 degrees of freedom in the numerator and 30 degrees of freedom in the denominator. There is a different *F* distribution for the pair of 20 degrees of freedom in the numerator and 8 degrees of freedom in the denominator. The third *F* distribution (8 df in the denominator, 8 df in the numerator) is different yet.

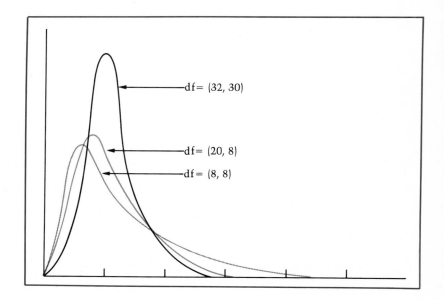

The following problem provides additional insight into the analysis-of-variance technique.

Problem

A clinical psychologist plans to study the age at which people become psychologically independent of their families. It is suspected that religious orientation may be one factor affecting this variable (age). As an experiment, the psychologist selects random samples of young people of Protestant, Catholic, Jewish, and other faiths. A tally of religious affiliations and of the respective ages when those sampled became independent of their families reveals the information given in Table 12 - 1.

Table 12-1

Comparison of Religious Affiliation and Age at
which those Sampled Became Independent

Jewish	Religious Affiliation Catholic	Protestant	Other
22	27	20	18
19	25	18	16
13	22	21	24
19	27	21	19
23	19	16	22
15	23	17	22
16	21	20	24
18	28	18	
20	23	17	
20	25	19	
	27	18	

The question to be explored is: Is there a significant statistical difference among the mean ages at which independence was achieved by people in each of these four categories?

Solution

The usual five-step hypothesis-testing procedure is followed.

Step 1 The null hypothesis is that the mean age is the same for each faith.

$$H_0: \mu_1 = \mu_2 = \mu_3 = \mu_4$$

The alternate hypothesis is that the means are not all equal.

$$H_a: \mu_1 \neq \mu_2 \neq \mu_3 \neq \mu_4$$

Step 2 The 0.01 significance level is chosen.

Step 3 The appropriate test is based on the F statistic.

Step 4 The decision rule is formulated. Remember that in order to arrive at a decision rule we need to identify the *critical value*. The critical values for the F statistic can be found in Appendix E. You will find the critical values for the 0.05 significance level on the first page of that appendix and the values for the 0.01 significance level on the second. To use the table you need to know two numbers: the degrees of freedom in the numerator and the degrees of freedom in the denominator. The degrees of freedom in the numerator refer to the number of treatments, designated k, minus 1. That is, the degrees of freedom are found by $k - 1$. The degrees of freedom in the denominator refer to the total number of observations, designated N, minus the number of treatments. It is written $N - k$. For this problem there are four treatments and a total of 39 observations. Thus:

Degrees of freedom in numerator $= k - 1 = 4 - 1 = 3$
Degrees of freedom in denominator $= N - k = 39 - 4 = 35$

Refer to Appendix E and the 0.01 level of significance. Move horizontally at the top of the table to 3 degrees of freedom in the numerator. Then move down that column to the critical value opposite 40 degrees of freedom for the denominator. The critical value is approximately 4.31. The decision rule, therefore, is to accept the null hypothesis if the computed value of F is less than or equal to 4.31. If the computed value is greater than 4.31, the null hypothesis is rejected. Recall that the F distribution is positively skewed. The decision rule, shown diagrammatically, is

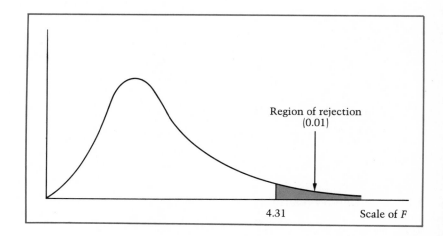

Region of rejection
(0.01)

4.31   Scale of F

Step 5 Compute F and make a decision. It is often convenient to record the computations for the F statistic in a table. The general form is:

ANOVA Table

Source of Variation (1)	Sum of Squares (2)	Degrees of Freedom (3)	Mean Square (4)
Treatments	SST	$k-1$	$\dfrac{SST}{k-1}$
Within	SSE	$N-k$	$\dfrac{SSE}{N-k}$
Total	SS total		

Referring to the general format for the analysis of variance, note that three totals, called **sums of squares,** are needed to compute F—namely, SST, SSE, and SS total.

1. SST. It is the abbreviation for the sum of squares due to the treatment effect and is found by

$$SST = \Sigma\left(\frac{T_c^2}{n_c}\right) - \frac{(\Sigma X)^2}{N}$$

where

T_c is the column total for all observations in the treatment.

n_c is the number of observations (sample size) for each respective treatment.

ΣX is the sum of all observations.

k is the number of treatments.

N is the total number of observations.

2. SSE. It is the abbreviation for the sum of squares within (error). SSE is computed by

$$SSE = \Sigma(X^2) - \Sigma\left(\frac{T_c^2}{n_c}\right)$$

3. SS total. It is the total variation. That is, it is the sum of SST and SSE.

$$SS \text{ total} = SST + SSE$$

As a check, SS total is calculated by

$$SS \text{ total} = \Sigma(X^2) - \frac{(\Sigma X)^2}{N}$$

To compute F:

$$F = \frac{\dfrac{SST}{k-1}}{\dfrac{SSE}{N-k}}$$

The calculations for F are shown in Table 12-2.

Table 12-2.

Calculations Needed for Computed F

	Jewish	Catholic	Protestant	Other	Grand Totals	
Religious Affiliation						
Ages (X)						
	22	27	20	18		
	19	25	18	16		
	13	22	21	24		
	19	27	21	19		
	23	19	16	22		
	15	23	17	22		
	16	21	20	24		
	18	28	18			
	20	23	17			
	20	25	19			
	—	27	18	—		
Column totals (T_c)	185	267	205	145	802	ΣX
Sample size (n_c)	10	11	11	7	39	
Squared totals (ΣX^2)	3,509	6,565	3,849	3,061	16,984	$\Sigma(X^2)$

The entries for the ANOVA table are computed:

$$SST = \Sigma \frac{T_c^2}{n_c} - \frac{(\Sigma X)^2}{N}$$

$$= \frac{(185)^2}{10} + \frac{(267)^2}{11} + \frac{(205)^2}{11} + \frac{(145)^2}{7} - \frac{(802)^2}{39}$$

$$= 234.93$$

$$SSE = \Sigma(X^2) - \Sigma\left(\frac{T_c^2}{n_c}\right)$$

$$= (22^2 + 19^2 + \cdots + 24^2)$$

$$- \left[\frac{(185)^2}{10} + \frac{(267)^2}{11} + \frac{(205)^2}{11} + \frac{(145)^2}{7}\right]$$

$$= 256.66$$

$$SS \text{ total} = SST + SSE$$
$$= 234.93 + 256.66$$
$$= 491.59$$

As a check:

$$SS \text{ total} = \Sigma(X^2) - \frac{(\Sigma X)^2}{N}$$

Inserting these values into an ANOVA table:

Source of Variation	Sum of Squares	Degrees of Freedom	Mean Square
Treatments	234.93	3	78.31
Within	256.66	35	7.33
Total	491.59		

Now, to compute F:

$$F = \frac{\dfrac{SST}{k-1}}{\dfrac{SSE}{N-k}}$$

$$= \frac{78.31}{7.33} = 10.68$$

Since the computed F value of 10.68 is greater than the critical F value of 4.31 (determined in Step 4), the null hypothesis is rejected at the 0.01 level. This indicates that it is quite unlikely that the differences in the four means could have occurred by chance. Thus, we conclude that the means are not all equal.

A Computer Example

The statistical package called MINITAB, mentioned in the previous chapters, has an ANOVA routine. The ANOVA table for the problem involving the age at which a person becomes psychologically independent of his or her family is used to illustrate the typical output. (In the output, MINITAB uses the term FACTOR for "Treatment" and ERROR for "Within.")

```
MTB >ONEWAY C1 C2

ANALYSIS OF VARIANCE

DUE TO        DF            SS       MS=SS/DF      F-RATIO
FACTOR         3        234.93         78.31        10.68
ERROR         35        256.66          7.33
TOTAL         38        491.59
```

Problem

The head nurse at University Hospital has the responsibility of assigning personnel to the emergency room. Present policy calls for the same number of registered nurses to be assigned to all three shifts. The head nurse, however, thinks that the number of emergencies handled may not be the same for each shift. It was decided to use the ANOVA technique to investigate whether the same number of emergencies are handled on each shift.

A random sample of five days from each shift is selected. The results are shown. The "treatment" in this problem is the shift. Note that the sample sizes are equal. ANOVA follows the same hypothesis-testing procedure outlined in Chapter 9.

Number of Emergency Cases Reported per Shift

	Day	Afternoon	Night
	44	33	39
	53	42	24
	56	15	30
	49	30	27
	38	45	30
Mean	48	33	30

Solution

Repeating this table with all necessary calculations:

	Day X	Day X²	Afternoon X	Afternoon X²	Night X	Night X²	Grand Totals
	44	1,936	33	1,089	39	1,521	
	53	2,809	42	1,764	24	576	
	56	3,136	15	225	30	900	
	49	2,401	30	900	27	729	
	38	1,444	45	2,025	30	900	
Column totals (T_c)	240		165		150		555 ⟵—ΣX
Sample size (n_c)	5		5		5		15
Sum of squares (ΣX^2)		11,726		6,003		4,626	22,355 ⟵—$\Sigma(X^2)$

Step 1 The null hypothesis is that the mean number of emergencies is the same for each shift. That is, $\mu_1 = \mu_2 = \mu_3$. The alternate hypothesis is that the means are not all equal.

Step 2 The level of significance: the 0.05 level is selected.

Step 3 The statistical test: the appropriate test is the F test.

Step 4 Formulate the decision rule.

Step 5 Calculate F. The F statistic is the ratio:

$$F = \frac{\dfrac{SST}{k-1}}{\dfrac{SSE}{N-k}}$$

where k is the number of treatments and N the total number of items sampled.

Calculate SST (the abbreviation for the sum of squares due to the treatment effect):

$$SST = \Sigma \left(\frac{T_c^2}{n_c}\right) - \frac{(\Sigma X)^2}{N}$$

where

T_c is the column total for all observations in the treatment.

n_c is the number of observations (sample size) for each respective treatment. There are five observations on each of the three shifts in the example.

ΣX is the sum of all observations (total number of emergencies). It is 555 in the example.

k is the number of treatments (shifts). There are three in the example.

N is the total number of observations. There are 15.

Hence, here

$$SST = \Sigma \left(\frac{T_c^2}{n_c}\right) - \frac{(\Sigma X)^2}{N}$$

$$= \frac{(240)^2}{5} + \frac{(165)^2}{5} + \frac{(150)^2}{5} - \frac{(555)^2}{15}$$

$$= 930$$

Next, calculate SSE, the abbreviation for sum of squares within (error).

$$SSE = \Sigma(X^2) - \Sigma \left(\frac{T_c^2}{n_c}\right)$$

$$= 22{,}355 - \left(\frac{(240)^2}{5} + \frac{(165)^2}{5} + \frac{(150)^2}{5}\right)$$

$$= 22{,}355 - 21{,}465$$

$$= 890$$

The F statistic can now be determined:

$$F = \frac{\dfrac{930}{3-1}}{\dfrac{890}{15-3}}$$

$$= 6.27$$

The decision rule is to reject the null hypothesis if the computed F value exceeds 3.89. Since 6.27 is greater than 3.89, the null hypothesis is rejected at the 0.05 level of significance. To put it another way, the differences in mean number of emergencies handled per shift (48, 33, and 30) cannot be attributed to chance. From a practical standpoint, it may be concluded that the number of emergency cases handled on the three shifts is not the same.

A reminder: cover the answers in the margin.

Energy shortages have caused many schools to turn down the heat in the classroom. The principal at Penn Street Elementary School is concerned that this may have an effect on achievement. To investigate this question further, students in the fifth-grade mathematics class were randomly assigned to one of three groups. The three groups were then separated and placed in rooms having different temperatures. Each group received televised instruction in long division. At the end of the lesson, the same ten-question examination was given to all three groups. The results, scoring the number of correct answers out of ten, were

	Temperature	
65°F	**72°F**	**78°F**
3	7	4
5	6	6
4	8	5
3	9	7
4	6	6
	8	5
	8	4
		3

The question to be explored is: Is there any significant difference in the mean scores? Use the 0.05 significance level.

a. Compute SST.
b. Compute SSE.
c. Compute SS total.
d. Arrange the values in an ANOVA table and compute F.
e. Arrive at a decision.

Chapter Exercises

1. Deals-on-Wheels, a manufacturer of mobile homes, is interested in the ages of buyers of each of five available floor designs. He suspects that certain designs tend to appeal to younger buyers more than others. A random sample of 29 records is selected from last year's sales files, and the respective ages of the principal buyers are recorded. The outcome is:

Self-Review 12-1

a.

3	7	4	
5	6	6	
4	8	5	
3	9	7	
4	6	6	
	8	5	
	8	4	
		3	
19	52	40	$\Sigma X = 111$

Then:

$$\frac{(19)^2}{5} + \frac{(52)^2}{7} + \frac{(40)^2}{8} - \frac{(111)^2}{20}$$

$$= 42.44$$

b. $3^2 + 5^2 + 4^2 + \cdots + 3^2 -$

$$\frac{(19)^2}{5} + \frac{(52)^2}{7} + \frac{(40)^2}{8}$$

$$= 681.00 - 658.49 = 22.51$$

c. $42.44 + 22.51 = 64.95$

d.

Source	Sum of Squares	df	Mean Square
Between	42.44	2	21.22
Within	21.51	17	1.32
Total	64.95		

$$F = \frac{21.22}{1.32} = 16.08$$

e. The critical value of F is 3.59, found by 0.05 level, 2 df in numerator, 17df in denominator. Since $16.08 > 3.59$, H_0 is rejected. The means are not all equal.

Floor Design

A	B	C	D	E
30	48	54	52	44
32	52	60	50	48
31	45	56	43	50
41	38	50	42	41
35	34	47		35
34	42	57		
		63		
		57		

a. State the null and alternate hypotheses.
b. How many degrees of freedom are there in the numerator? In the denominator?
c. Using the 0.05 level of significance, what is the critical value of F?
d. Compute F.
e. State your decision.
f. Interpret the results.

2. A social researcher wants to evaluate the ethical behavior of attorneys in four major regions of the United States. Samples are selected and an index of ethical behavior computed for each attorney in the study.

West	South	North Central	East
12	19	34	19
16	20	29	21
12	18	31	17
14	9	19	24
26	22	26	
	19		

At the 0.01 level of significance, is there a statistically significant difference in ethical behavior among the four regions? Use the usual five-step testing procedure.

A final note: when doing an analysis of variance, the null hypothesis is that *the means are all equal.* The alternate hypothesis is that *at least one of the means is different.* Rejection of the null hypothesis and acceptance of the alternate hypothesis leads to the conclusion that there is a significant difference between at least one pair of means. Rejecting the null hypothesis does not pinpoint which pairs—or how many pairs—of means differ significantly, but it does indicate that there is a significant difference between at least one pair of means. Multiple-comparison tests, discussed in more advanced texts, may be used to identify the treatment that differs significantly.

The analysis of variance (ANOVA) technique is used to test simultaneously whether two or more population means are equal. It assumes that each of the populations are normally distributed with equal standard deviations. The samples are also presumed to be independent.

The F statistic, which is the ratio of two variances, is employed as test statistic. This F distribution is positively skewed and nonnegative. Its critical values are generally located in the right-hand tail of the distribution when doing an ANOVA test. F is computed by

$$\frac{SST/k - 1}{SSE/N - k}$$

where

$$SST = \Sigma \left[\frac{T_c^2}{n_c} \right] - \frac{(\Sigma X)^2}{N}$$

and

$$SSE = \Sigma(X^2) - \Sigma \left[\frac{T_c^2}{n_c} \right]$$

Summary

Analysis of Variance

Chapter Outline

I. Objective. To determine if several populations have identical means.

II. Procedure.

A. As usual, state H_0 and H_a

H_0: $\mu_1 = \mu_2 = \mu_3 = \mu_4$

H_a: The means are not all equal

B. Select a level of significance, usually 0.05 or 0.01.

C. Formulate a decision rule based on the F test.

The rationale is that if H_0 is true, the two variances will be equal and F will be 1. If H_0 is not true, the variation between the sample means (numerator) will be significantly larger than the variation within the samples (denominator). Thus, the resulting value of F will be equal to or greater than the critical value of F and the null hypothesis will be rejected in favor of H_a.

D. Computing F.

1. Design an ANOVA table.

Source of Variation	Sum of Squares	Degrees of Freedom	Mean Square
Treatments (between columns)	SST	$k - 1$	$\dfrac{SST}{k-1}$
Error (within columns)	SSE	$N - k$	$\dfrac{SSE}{N-k}$
Total	SS total		

2. Compute F:

$$F = \frac{SST/k - 1}{SSE/N - k}$$

where

$$SST = \Sigma \left[\frac{T_c^2}{n_c} \right] - \frac{(\Sigma X)^2}{N}$$

$$SSE = \Sigma(X^2) - \Sigma \left[\frac{T_c^2}{n_c} \right]$$

$$SS \text{ total} = \Sigma(X^2) - \frac{(\Sigma X)^2}{N}$$

N is the total number of observations, and n_c is the number of observations for each respective treatment.

3. If the computed value of F is smaller than the critical value of F (from Appendix E), accept the null hypothesis. Otherwise, reject it and accept H_a.

Chapter Exercises

3. A physician randomly selects 18 patients among those she is treating for high blood pressure. These patients are randomly assigned to three groups and treated with three different drugs, all designed to reduce blood pressure. The amount of reduction, in millimeters of mercury, is shown. At the 0.01 level of significance, is there sufficient evidence to show that the drugs act differently?

Drug A	Drug B	Drug C
10	13	9
10	14	8
9	11	6
10	10	10
7	9	10
6	10	7

4. There are four different instructors for the history course, Introduction to Western Civilization, taught at Scandia Tech. Following is the number of pages of reading assigned by each instructor every week for the first five weeks of the course. At the 0.05 level of significance, is there sufficient evidence to show a difference in the average length of the readings assigned by the four instructors?

Mr. Barr	Dr. Sedwick	Dr. Reading	Dr. Faust
25	35	30	28
29	20	27	32
30	20	18	33
42	17	19	35
35	30	26	24

5. The merchandising manager for Food Mart grocery stores is analyzing the effect of various placements of candy displays within the company's stores. He decides to conduct an experiment by locating the display in different areas in each of four Food Mart outlets. The amount in pounds sold in the various locations each week is recorded. Is there sufficient evidence to indicate that there is a difference in sales at the various locations? Use a 5% significance level.

Front of Store, Near Bread	Top Shelf, Near Cookies	End of Aisle, Near Meat	Third Shelf, Near Soda Pop
76	73	89	96
75	70	82	92
83	81	85	104
87	78	79	89
81	76	80	94

6. A psychologist wants to investigate the effect of social background on the time (in minutes) it takes freshmen to solve a puzzle. A random sample of students from different backgrounds is selected, resulting in the following data. Use the 0.05 level of significance to test the hypothesis that social background has no effect on the time required to solve the puzzle.

Inner City	Urban	Suburban	Rural
16.5	10.9	18.6	14.2
5.2	5.2	8.1	24.5
12.1	10.8	6.4	14.8
14.3	8.9		24.9
	16.1		5.1

7. A wholesaler is interested in comparing the weights in ounces of grapefruit from Florida, Texas, and California:

Florida	Texas	California
12.6	12.8	16.0
13.8	13.2	15.1
14.0	12.4	13.9
	13.2	14.3
		15.0

a. What are the null and alternate hypotheses?
b. Fill in an ANOVA table.
c. What is the critical value of F, assuming a 0.01 level of significance?
d. What decision should the wholesaler make?

8. Theft in student rooms is a major problem at a large university. In an effort to reduce the problem, the university conducted an experiment wherein 21 dormitories were randomly assigned to one of three groups. In the first group of eight dorms all students were involved in structured discussion groups about the theft problem. Informal "peer" group meetings were held for the students in seven other dorms. The six dorms in the third group were not subjected to any changes. The number of reported thefts in each dorm after one semester is shown by group:

Cluster A Structured	Cluster B Informal	Cluster C Control
35	18	14
20	10	3
32	21	16
27	14	10
23	13	11
18	12	12
29	15	
26		

Do these data present sufficient evidence to indicate a difference among the three methods? Use a 1% level of significance.

9. An oncologist—a physician who specializes in the treatment of cancer—has 24 patients with advanced lung cancer. He is aware of three treatments, reported in medical journals, that may gain remission for his patients. To assess the effect of the treatments, the doctor randomly assigns patients to each treat-

ment, then keeps careful records on the number of days the patients live after treatment starts. At the 0.05 level, can it be concluded that there is any difference in the effect of the treatments?

Laetrile	Chemotherapy	Radiation
75	80	64
88	82	90
62	64	58
97	45	64
62	67	82
81	84	71
93	55	59
	39	66
	60	

10. (It is suggested that a programmable calculator or a computer be used to solve the problem that follows.) In 1976, Mark "The Bird" Fidrych arrived on the major-league baseball scene. Not only did he have a great year on the field for the Detroit Tigers with a record of 19 - 9, but he was also a box-office attraction. Fidrych was the announced starter in 18 games at Tiger Stadium; the attendance figures are listed in the following table along with the 1976 attendance figures for three other Tiger starting pitchers. Determine if there are any differences in the attendance figures when "The Bird" is the starting pitcher. Use a 5% level of significance.

Fidrych	Ruhle	Roberts	Bare
14,583	11,802	8,317	51,650
17,894	19,909	13,252	24,856
36,377	9,283	14,923	8,141
21,659	24,038	10,866	13,426
47,855	30,110	24,824	9,896
51,032	22,553	27,630	3,616
51,041	14,924	14,835	
45,905	10,303	19,079	
44,068	12,156	8,204	
35,395		8,949	
36,523			
51,822			
34,760			
39,884			
32,951			
16,410			
20,371			
7,147			

Chapter Achievement Test

First answer all the questions. Then check your answers against those given in the Answer section of the book.

I. Multiple-Choice Questions. Select the response that best answers each of the questions that follow (5 points each).

Questions 1 through 4 refer to the following problem. Mr. Tourtellotte can drive to work along one of three different routes. The following data show the number of minutes it takes to make the trip on five different occasions for each route. Is there sufficient evidence to indicate that there is a difference in the average time it takes to drive the three routes?

Via Expressway	Via Downtown	Past the University
33	22	14
35	26	21
34	17	24
32	18	25
38	20	15

1. The number of treatments is
   a. 3
   b. 2
   c. 12
   d. 15
   e. 14
2. The number of degrees of freedom for the within-samples (error) estimate of the variance is
   a. 3
   b. 2
   c. 12
   d. 15
   e. 14
3. If a 0.05 significance level is used, the computed value of F that would cause the null hypothesis to be rejected is
   a. 19.4
   b. 3.49
   c. 3.89
   d. 6.93
   e. 1.96
4. The value of F was computed to be 23.07. The correct conclusion is
   a. accept H_0

b. reject H_a

c. reject H_0 and accept H_a

d. the samples are too small to be appropriate

5. The ANOVA technique was used to compare three population means based on samples of size nine drawn from each population. It was found that $SST = 90$ and $SSE = 200$. At the 0.05 level of significance the correct conclusion is

a. $F = 0.50$, accept H_0

b. $F = 2.22$, accept H_0

c. $F = 5.40$, reject H_0

d. $F = 12$, reject H_0

e. the value of F cannot be determined from the information given

6. In a particular ANOVA test the calculated value of F is between 0 and the value of the F table. The correct conclusion is

a. accept H_0 and conclude that the treatment means being tested are not significantly different

b. accept H_0 and conclude that the treatment means being tested differ significantly

c. reject H_0, accept H_a, and conclude that the treatment means are not significantly different

d. reject H_0, accept H_a, and conclude that the treatment means are significantly different

7. If the population means for each of the treatment groups were identical, the value of the F statistic would be

a. equal to 1.00

b. 0

c. infinite

d. a negative number

e. a number between 0 and 1.00

8. The F distribution

a. cannot be negative

b. is positively skewed for small samples

c. is determined by two parameters

d. all of the above are correct

9. It is possible for the within-samples (error) estimate of the variance to be negative.

a. true

b. false

10. If H_0 is rejected, it indicates that there is a significant difference between at least one pair of means.

a. true

b. false

II. Computation Problems (12 points each).

Questions 11 through 14 refer to the following problem. A manufacturer of automobiles is testing a new design of brakes. The Director of Engineering reports that test data for both the existing design and for two proposed designs are available. Test results were obtained by measuring the stopping distance, in feet, of the cars traveling at a speed of 15 miles per hour.

Existing Design	First Proposal	Second Proposal
5	5	8
7	5	4
6	8	5
	7	9
	6	

11. What are the null and alternate hypotheses?
12. Fill in an ANOVA table.
13. What is the critical value of F, assuming a 0.05 level of significance?
14. Can it be concluded that there is a difference in the mean stopping distances?

HIGHLIGHTS
From Chapters 11 and 12

In the last two chapters we examined two hypothesis-testing situations. The first deals with either one or two small samples. Generally, we consider a ''small sample'' to be one that consists of fewer than 30 observations. For such problems, **Student's t** is used as test statistic. The second hypothesis-testing procedure is used when comparing more than two population means to determine if they are equal. The technique is known as **analysis of variance,** or **ANOVA.** It uses F as test statistic.

Key Concepts

1. The **t distribution** is used to test hypotheses about the mean of a single population, the means of two populations, or the difference in related observations. In each of these instances the population is normal with an unknown standard deviation. Usually, the sample size is smaller than 30. The t distribution has the following characteristics:
   a. It is a continuous distribution.
   b. It is somewhat bell shaped and symmetrical.
   c. There is a ''family'' of t distributions. That is, each time the sample size changes, the t distribution changes.
   d. As the sample size increases, the t distribution approaches the normal distribution. Often, when $n > 30$ the normal distribution is used instead of t, because the values of these two statistics are close.
   e. The t distribution is more spread out than the normal distribution.

2. When testing the means of more than two populations simultaneously, the **analysis of variance (ANOVA)** is used. The test statistic follows the **F distribution,** which has the following characteristics:
   a. It is either 0 or positive (it cannot be negative).
   b. It is a continuous distribution.
   c. It is positively skewed and may range from 0 to infinity.

Key Terms

t distribution
Degrees of freedom
Pooled variance estimate
Independent samples
Dependent samples
F distribution

Treatment
Sum of squares total
Sum of squares error
Sum of squares treatment
Analysis of variance (ANOVA)

Key Symbols

t Depending on its usage, t refers either to the t test itself, the computed value of t, or the critical value of t.

df   Degrees of freedom.

s_p Pooled estimate of the population standard deviation.

s_d Standard deviation of the paired differences.

F The F probability distribution, which is the test statistic for ANOVA.

μ The population mean.
$\overline{X}$ The sample mean.
k The number of treatments.

SST Sum of squares treatment (between columns).

SSE Sum of squares error (within columns).

SS total   Sum of squares total.

Review Problems

1. In order to service her customers more effectively, Ms. Dodd, owner of Damschroders, an exclusive women's boutique, wants to know their mean age. A random sample of 15 customers revealed the mean to be 46.3 years with a standard deviation of 10.6 years. When Ms. Dodd

opened her boutique, the manager of the mall where the shop is located had reported the mean age of shoppers to be 40. Can Ms. Dodd conclude that her clients are older? Use the 0.05 significance level.

2. A government testing agency routinely tests various foods to ensure that they meet label requirements. A random sample of ten one-liter bottles of a soft drink actually contained the following amounts when tested: 0.93, 0.97, 0.96, 1.02, 1.05, 1.01, 1.02, 0.97, 0.98, and 0.97. At the 0.05 significance level, can the testing agency show that the soft-drink manufacturer is underfilling the product?

3. The personnel manager of a large corporation believes that nowadays people are retiring at a later age than ever before. A sample of 20 employees who retired in 1970 revealed that their mean age at retirement was 63.7 years with a standard deviation of 3.2 years. A sample of 15 employees who retired last year revealed the mean to be 66.5 years and the standard deviation 4.3 years. At the 0.01 significance level, can the manager conclude that the mean age at retirement has increased?

4. The dean of Northern University believes that student grade point averages have increased in recent years. She obtains a sample of seven student grade point averages for 1980 and eight for this year. Based on these data, can the dean conclude that grades have increased? Use the 0.01 significance level.

Student Grade Point Averages

1980	This Year
2.90	2.70
2.85	3.35
2.67	3.60
1.98	2.75
3.20	2.35
2.65	2.90
2.35	2.98
	3.01

5. Eight upper-middle-class families were surveyed to determine the extent of their medical expenses. The mean amount spent was $960 with a sample standard deviation of $135. Can it be concluded that the mean of the population's medical expenses is greater than $900 per year? Use the 0.05 significance level.

6. A study was designed to determine if drinking affects reaction time while driving. A random sample of 12 people were given a driving-simulator test. Each person was then asked to drink two ounces of whiskey and to repeat the test. The "errors" made in each test were tabulated. At the 0.05 significance level, can it be concluded that people make more errors after having drunk two ounces of whiskey?

Subject	Before Drinking	After Drinking
A	8	9
B	9	12
C	10	13
D	8	14
E	11	15
F	6	11
G	12	12
H	15	14
I	10	13
J	7	12
K	8	13
L	10	19

7. A large university is concerned about salary discrimination on the basis of sex. To investigate, a sample of 11 female instructors has been obtained and their salaries are determined. For each female selected, a male instructor with similar tenure status, academic rank, discipline, and so on is obtained and his salary is paired with that of the female instructor. The data are shown in the table that follows. Can it be concluded that female instructors earn significantly less? Use the 0.05 significance level.

Faculty Pair	Male	Female
1	$26.0	$25.9
2	22.9	23.5
3	23.9	22.7
4	27.8	21.7
5	25.5	24.7
6	24.3	23.0
7	20.7	20.8
8	25.4	24.8
9	36.4	29.5
10	31.0	27.2
11	18.3	18.1

8. The manager of Sally's Hamburger World suspects that the average number of customers between 4 P.M. and 7 P.M. differs by day of the week. The data show the number of customers during that period for randomly selected days of the week. At the 0.01 significance level, can it be concluded that the number of customers differs by day of the week?

Monday	Tuesday	Wednesday	Thursday	Friday
86	77	69	78	84
96	102	91	77	88
78	54	86	90	94
66	98	74	84	102
100		82	72	96
		78	74	
		84		

9. The Plumber's Union has gathered data on the hourly wages of random samples of plumbers in four southwestern cities. At the 0.05 significance level, can it be shown that the mean hourly wage differs in the four cities?

City			
A	**B**	**C**	**D**
$14.20	$14.40	$16.20	$15.40
15.40	15.20	15.40	15.80
14.80	15.80	16.00	16.60
13.80	16.20	16.00	14.80
	16.00	15.80	
	15.60		

10. A health spa has two one-week programs for grossly overweight persons. A client may select either of two available programs. In order to assess the effectiveness of the two programs, a sample of 300-pound persons were selected and their weight losses recorded. Seven were in Program A and nine in Program B.

Weight Losses	
Program A	**Program B**
40	39
38	21
41	29
52	42
27	43
32	28
40	29
	28
	46

Use the 0.05 level of significance to test whether there is a statistically significant difference between the two mean weight losses.

11. The health spa also has three programs designed to lower tension. An incoming group of patients were randomly assigned to the three programs. The reduction in tension after the week-long programs is shown (in percent).

Program A	**Program B**	**Program C**
27	17	22
21	26	31
18	33	16
32	26	18
26		27
		32

Test at the 0.01 level the hypothesis that there is no difference in the effectiveness of the three tension-reducing programs.

12. A fertilizer-mixing machine is set to give 10 pounds of nitrate for every 100 pounds of fertilizer. Eight 100-pound bags were examined and the pounds of nitrate were 8, 10, 9, 11, 7, 10, 9, and 10. Is there reason to conclude that the mean is not equal to 10 pounds? Use a 0.01 significance level.

13. A random sample of 12 women not employed outside the home were asked to estimate the selling prices of two 25-inch color TV sets. Their estimates (in dollars) are shown.

Homemaker	Model A	Model B
1	715	810
2	830	650
3	815	620
4	770	760
5	650	830
6	680	720
7	770	800
8	760	830
9	990	830
10	550	900
11	670	620
12	760	630

Is there sufficient evidence, at the 0.05 level of significance, to claim that homemakers perceive Model B to be more expensive than Model A?

Case Analysis

The McCoy's Market Case

(Data for these two cases can be found in the first Chapter Highlights, pp. 126 - 130.)

Among other things, the management of McCoy's Markets is asking the following questions:

1. Has the population mean amount spent increased significantly over last year's mean of $35.43? Use the 0.05 level of significance.
2. On the average, do midweek shoppers purchase fewer items than weekend shoppers? Use the 0.01 level of significance.
3. Is there a significant difference in the mean amount spent by men and the mean amount spent by women? Use the 0.10 level of significance.

The St. Mary's Emergency Room Case

The management of St. Mary's Medical Center would like to know:

1. Is there a difference between the mean ages of those who are admitted and those who are not admitted? Use the 0.10 level of significance.
2. Comparing the three different shifts, is there a significant difference in the mean cost of care for the patients seen by the hospital staff during different shifts? Use the 0.05 level of significance.

13
Correlation Analysis

```
┌──────────────────────────────────────────────────────┐
│                                                        │
│                   OBJECTIVES                           │
│                                                        │
│   When you have completed this chapter, you will be able to │
│     • Describe the relationship between two variables. │
│     • Compute Pearson's coefficient of correlation.    │
│     • Compute the coefficients of determination and non-│
│       determination.                                   │
│     • Test the statistical significance of the coefficient of │
│       correlation.                                     │
│     • Compute the Spearman coefficient of rank correlation. │
│                                                        │
└──────────────────────────────────────────────────────┘
```

Introduction

In Chapters 9 - 12 we dealt with hypothesis tests involving means and proportions. The techniques concentrated on only a *single* feature of the sampled item, such as income. This chapter begins a study of the relationship between *two or more* variables. We may want to determine if there is any relationship between the number of years of education completed by federal government employees and their incomes. Or, we may want to explore one of these questions: Does the crime rate in inner cities vary with the unemployment rate in those cities? Is there a relationship between the amount of money spent advertising a product such as a toothpaste and its sales? Is there any relationship between a student's grade point average in high school and that same student's corresponding grade point average in college? Note that in each case there are two separate characteristics—years of education and income, for example.

Correlation Analysis

The study of potential relationships between two variables is called correlation analysis.

The basic objective of correlation analysis is to determine the degree of correlation (relationship) between variables, from zero (no

> **Correlation Analysis**   The statistical techniques used to determine the strength of the relationship between two variables.

correlation) to perfect (complete) correlation. Our attention will focus first on the correlation between two interval-scaled variables. Then we will examine the relationship between two ordinal-scaled variables.

Scatter Diagram

One tool that is very useful for visualizing the relationship between two variables is the scatter diagram.

> **Scatter Diagram**   A graphic tool that visually portrays the relationship between two variables.

Problem

Suppose we were interested in studying the potential relationship between a student's grade point average in high school and his or her grade point average in college. For this illustration, we will consider a sample of only five students, using a 0 - 10 grading scale. Further, we will assume that high-school grade point averages range from a low of 1 to a high of 5 and college GPAs from 0 to 10. The paired data for the five students is shown in Table 13-1. Construct a scatter diagram for the paired data in Table 13-1.

Table 13-1

High-School and College Grade Point Averages

Student	High-School GPA X	College GPA Y
Janet Artz	3	6
Lee Jackson	5	9
James Coble	2	5
Sue Brown	3	7
Marita Estrada	4	8

Solution

Because we suspect that high-school grade point averages are good predictors of college grade point averages, it is implied that college GPA somehow is determined by, or *depends* upon, high-school GPA. Hence, we call college grade point average the *dependent* variable and high-school grade point average the *independent* variable.

It is traditional to put the dependent variable on the vertical axis (Y) and the independent variable on the horizontal axis (X). The paired data for Janet Artz ($X = 3$, $Y = 6$) are plotted by moving to 3 on the X-axis and then moving vertically to a position opposite 6 on the Y-axis and placing a dot at that intersection (see the scatter diagram that follows). This process is continued for the scores of all students in the sample. The completed scatter diagram is shown in Figure 13-1.

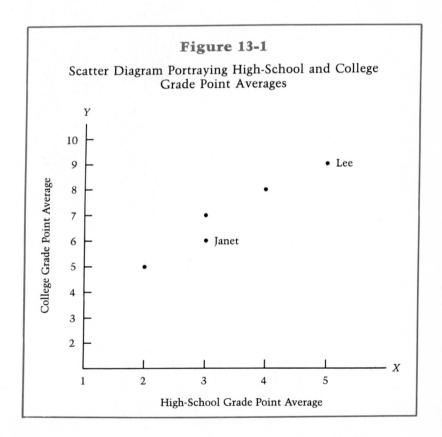

Figure 13-1

Scatter Diagram Portraying High-School and College Grade Point Averages

As will be explained in detail on pp. 373-376, the scatter diagram in Figure 13-1 indicates that there is indeed a very strong relationship (correlation) between performance in high school and performance in college. Students like Lee Jackson, who do exceptional work in high school, also tend to have high grade point averages in college.

A reminder: cover the answers in the margin.

An astute college recruiter noticed that the enrollment figures at the local campus of the state university seemed to fluctuate with the unemployment rate for the region. To establish whether or not his suspicions were justified, the recruiter collected relevant unemployment and enrollment figures for the same time periods. His findings were

Unemployment (in percent) X	State University Campus Enrollment Y
3.4	13,500
3.3	14,500
4.6	15,200
5.1	14,900
4.5	14,700
3.5	14,300
5.7	15,700
7.6	17,100

a. Construct a scatter diagram for the paired data.
b. As unemployment increases, does enrollment appear to increase, decrease, or remain the same?

Self-Review 13-1

a.

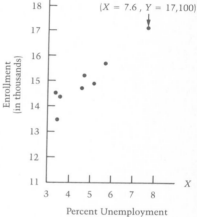

b. As unemployment in the region increases, enrollment at the state university campus also increases.

Chapter Exercises

1. A Peace Corps agronomist is studying the relationship between the average temperature and the yield in bushels per acre for a crop harvested in late fall. He has collected the following information for several regions:

Region	Temperature (C) X	Yield (in bushels per acre) Y
1	4	1
2	8	9
3	10	7
4	9	11
5	11	13
6	6	7

a. Construct a scatter diagram for the paired data.

b. As temperature increases, does yield appear to increase, decrease, or remain the same?

2. Suppose you are a personnel trainee. You believe there is a relationship between the number of years of employment with the company and the number of days a year an employee is absent from work. To explore this relationship further, you obtain the following information from the company records of six employees picked at random:

Employee	Length of Employment (in years) X	Absences Last Year (in days) Y
Phuong	1	8
Fadale	5	1
Sasser	2	7
Dilone	4	3
Thelin	4	2
Tortelli	3	4

a. Construct a scatter diagram for the paired data.

b. As the years of employment increase, do absences appear to increase, decrease, or remain the same?

3. A retailer of men's apparel wants to learn if there is a correlation between the number of suits sold and the number of salespersons covering the floor. A random sample of the company's files resulted in the following paired information:

Salespersons on Duty X	Suits Sold Y
3	7
1	5
2	6
3	6
4	9
5	10
1	4

Plot a scatter diagram with the number of salespersons on the X-axis and the number of suits sold on the Y-axis. Does it appear that more suits are sold when more salespersons are on duty?

4. A pair of dice are rolled five times and the number of spots appearing faceup noted. The results are

First Die X	Second Die Y
6	4
4	4
5	1
6	2
1	3

Develop a scatter diagram, scaling the first die on the X-axis and the second die on the Y-axis. What comments can you make? Would you expect to find a relationship?

The Coefficient of Correlation

About 1900 Karl Pearson, who made significant contributions to the science of statistics, developed a measure that describes the relationship between two sets of *interval-scaled variables*. It is called the coefficient of correlation and is designated by the letter *r*.

> **Coefficient of Correlation** A measure of the strength of the association between two variables.

Pearson's *r* is also known as the product-moment correlation coefficient, partly to distinguish it from other correlation coefficients. It is a valid measure of correlation if the relationship between the variables is *linear*. As illustrated by the scatter diagram in Figure 13-2, if all data points lie on a straight line, the correlation coefficient is either +1.00 or −1.00, depending on the direction of the slope of the line. Coefficients of +1.00 or −1.00 describe a *perfect correlation*.

If there is no relationship between X and Y, r will be 0, and the dots on the corresponding scatter diagram will be randomly scattered. This situation and several others are portrayed in Figure 13-3.

The strength and direction of the coefficient of correlation is summarized in the following diagram. Negative numerical values

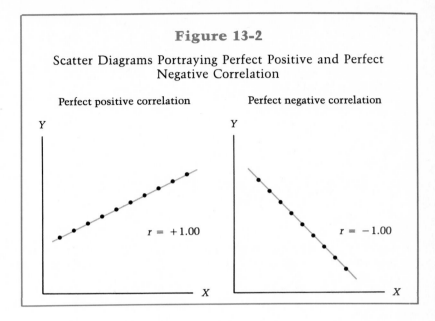

Figure 13-2

Scatter Diagrams Portraying Perfect Positive and Perfect Negative Correlation

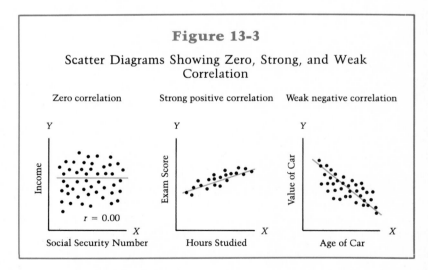

Figure 13-3

Scatter Diagrams Showing Zero, Strong, and Weak Correlation

such as −0.92 or −0.48 signify inverse correlation, whereas positive numerical values such as +0.83 and +0.46 indicate direct correlation. The closer Pearson's r is to 1.00 in either direction, the greater the strength of the correlation. *The strength of the correlation is not dependent on the direction.* Therefore, −0.12 and +0.12 are equal in strength (both weak). Coefficients of +0.94 and −0.94 are also equal in strength (both very strong).

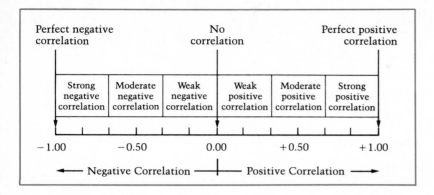

The coefficient of correlation r is computed by

$$r = \frac{n(\Sigma XY) - (\Sigma X)(\Sigma Y)}{\sqrt{[n(\Sigma X^2) - (\Sigma X)^2][n(\Sigma Y^2) - (\Sigma Y)^2]}}$$

where

n　The number of paired observations.

ΣX　The X variable is summed.

ΣY　The Y variable is summed.

ΣX^2　The X variable is squared and then summed.

$(\Sigma X)^2$　The X variable is summed and then squared.

ΣY^2　The Y variable is squared and then summed.

$(\Sigma Y)^2$　The Y variable is summed and then squared.

Problem

The data on high-school grade point averages and college grade point averages of five students cited in Table 13-1 are repeated.

Student	High-School GPA X	College GPA Y
Janet Artz	3	6
Lee Jackson	5	9
James Coble	2	5
Sue Brown	3	7
Marita Estrada	4	8

Determine the coefficient of correlation.

Table 13-2

Calculations Needed to Determine the Coefficient
of Correlation

Student	High-School GPA X	College GPA Y	XY	X²	Y²
Janet Artz	3	6	18	9	36
Lee Jackson	5	9	45	25	81
James Coble	2	5	10	4	25
Sue Brown	3	7	21	9	49
Marita Estrada	4	8	32	16	64
	17	35	126	63	255

Solution

The totals and the sums of squares needed are in Table 13-2. Computing r:

$$r = \frac{n(\Sigma XY) - (\Sigma X)(\Sigma Y)}{\sqrt{[n(\Sigma X^2) - (\Sigma X)^2][n(\Sigma Y^2) - (\Sigma Y)^2]}}$$

$$= \frac{5(126) - (17)(35)}{\sqrt{[5(63) - (17)^2][5(255) - (35)^2]}}$$

$$= \frac{35}{\sqrt{[26][50]}}$$

$$= \frac{35}{36.056}$$

$$= 0.97$$

Since 0.97 is very close to the perfect correlation value of 1.00, it is concluded that there is a very strong relationship between high-school performance (high-school GPA), and college performance (college GPA).

Self-Review 13-2

Is there a relationship between the number of votes received by candidates for public office and the amount spent on their campaigns? The following sample information was gathered for a recent election:

Candidate	Amount Spent on Campaign (000) X	Votes Received (000) Y
Weber	$3	14
Taite	4	7
Spencer	2	5
Lopez	5	12

a. Draw a scatter diagram.
b. Compute the Pearson coefficient of correlation.
c. Interpret the correlation coefficient.

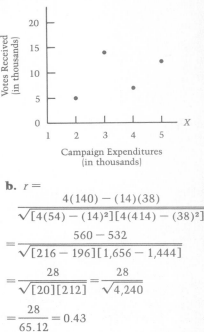

a.

b. $r =$

$$\frac{4(140) - (14)(38)}{\sqrt{[4(54) - (14)^2][4(414) - (38)^2]}}$$

$$= \frac{560 - 532}{\sqrt{[216 - 196][1,656 - 1,444]}}$$

$$= \frac{28}{\sqrt{[20][212]}} = \frac{28}{\sqrt{4,240}}$$

$$= \frac{28}{65.12} = 0.43$$

c. There is a moderate to weak relationship between the two variables.

Chapter Exercises

5. In an earlier exercise, an agronomist was studying the relationship between the average temperature (C) and the yield in bushels per acre for a certain fall crop. The data collected are repeated below.

Region	Temperature (C) X	Yield (in bushels per acre) Y
1	4	1
2	8	9
3	10	7
4	9	11
5	11	13
6	6	7

Compute the coefficient of correlation.

6. In an earlier exercise, a personnel trainee was studying the relationship between the number of years of employment with the company and the number of days absent from work last year. The following information was obtained from company records:

Employee	Length of Employment (in years) X	Absences Last Year (in days) Y
Phuong	1	8
Fadale	5	1
Sasser	2	7
Dilone	4	3
Thelin	4	2
Tortelli	3	4

Compute the coefficient of correlation.

7. Age is related to the length of stay of surgical patients in a hospital. The following sample data were obtained in a recent study:

Age	Days in Hospital
40	11
36	9
30	10
27	5
24	12
22	4
20	7

  a. Draw a scatter diagram. X is age, and Y is days in hospital.
  b. Compute the coefficient of correlation.
  c. Interpret the results.

8. Is there a relationship between the number of golf courses in the United States and the divorce rate? The following sample information was collected for several regions:

Golf Courses (number) X	Divorce Rate per 1,000 Population Y
28	2.2
38	2.5
47	3.5
54	4.1
62	4.8
66	5.0

  a. Draw a scatter diagram.
  b. Compute the coefficient of correlation.
  c. Interpret the results.

The Coefficient of Determination

The coefficient of correlation allowed us to make statements such as "the relationship between the two variables is very strong." But how can one measure "very strong"? A measure of correlation that does have a more precise meaning is known as the **coefficient of determination.**

Coefficient of Determination   The proportion of the total variation in one variable that is explained by the other variable.

This technique results in a proportion, or percent, that makes it relatively easy to arrive at a precise interpretation. It is computed by squaring the coefficient of correlation. The coefficient of determination may vary from 0 to 1.00 or, converted to a percent, from 0 to 100%. It is usually represented by r^2. To illustrate its computation and meaning, let us return to the coefficient of correlation r for the high-school and college grade-point-average data, computed to be 0.97. The coefficient of determination r^2 is 0.94, found by $(0.97)^2$. This means that 94% of the total variation in college grade point averages is explained, or accounted for, by variation in high-school grade point averages.

The **coefficient of nondetermination** is the proportion of the total variation *not* explained. It is $1 - r^2$. For the high-school and college GPA problem it is 0.06, found by $1 - (0.97)^2 = 1 - 0.94 = 0.06$. Interpreting, 6% of the total variation in college GPA is not explained (accounted for) by high-school GPA.

The magnitude of the coefficient of determination is smaller than that of the coefficient of correlation. For this reason, it is preferred by many statisticians. That is, it is a more conservative measure of the relationship between two variables. To put it another way, the coefficient of correlation tends to overstate the association between two variables. A correlation coefficient of 0.70, for example, would suggest that there is a fairly strong relationship between two variables. Squaring r, however, gives a coefficient of determination of 0.49, which is a somewhat smaller value.

Self-Review 13-3

In Self-Review 13-2 you computed a coefficient of correlation of 0.43 between campaign expenses and number of votes received.

a. What is the coefficient of determination?
b. What is the coefficient of nondetermination?
c. Interpret the two measures.

a. 0.18, found by $(0.43)^2$.
b. 0.82, found by $1 - (0.43)^2$.
c. Only 18% of the variation in the number of votes is accounted for by the level of campaign expenses; 82% is not.

Chapter Exercises

9. A study is made of the relationship between the number of passengers on an aircraft (X) and the total weight in pounds of luggage stored in the aircraft's baggage compartment (Y). The coefficient of correlation is computed to be 0.94.
   a. What is the coefficient of determination?
   b. What is the coefficient of nondetermination?
   c. Interpret these two measures.

10. The coefficient of correlation between the number of people on the beach at 4 P.M. and the high temperature for that day at Sunner's Creek is computed to be 0.961.
   a. What is the coefficient of determination?
   b. Find the coefficient of nondetermination.
   c. Interpret the two measures.

11. In a government study of the relationship between the tar content and nicotine content (in mg) of a cigarette, the coefficient of correlation is 0.432.
   a. Find the coefficient of determination.
   b. Compute the coefficient of nondetermination.
   c. Explain the meaning of these two numbers.

12. At Middletown High, a coefficient of correlation of -0.683 was found between the number of high-school activities offered and the number of suspensions for drug-related reasons.
   a. Determine the coefficient of determination.
   b. What is the coefficient of nondetermination?
   c. Interpret the two measures.

Testing the Significance of the Coefficient of Correlation

Only five students were selected for the study of the relationship of high-school GPA and college GPA. The coefficient of correlation was calculated to be $+0.97$, indicating a very strong, positive association between the two variables. The question arises, however, whether it is possible—due to the small size of the sample—that the correlation in the population is really 0 and the apparent relationship is due to chance. The "population" in this case might be *all* students entering a university.

In order to decide formally whether the correlation in the population could be 0, the hypothesis-testing procedure used in Chapters 9 - 12 is applied. The statement that the coefficient of correlation in the population is, in fact, 0 becomes our null hypothesis. The alternate hypothesis states that the coefficient of correlation in the population is not 0. The population correlation coefficient is usually represented by the Greek letter rho (ρ). Symbolically, then, the null hypothesis and the alternate hypothesis are

Null hypothesis, H_0: $\rho = 0$

Alternate hypothesis, H_a: $\rho \neq 0$

The appropriate test statistic follows the t distribution with $n - 2$ degrees of freedom and has this formula:

$$t = \frac{r\sqrt{n - 2}}{\sqrt{1 - r^2}}$$

where

r is the sample coefficient of correlation.

n is the number of items in the sample.

If the computed t value is in the rejection region, we reject the null hypothesis that there is no correlation between the variables. A t value in the rejection region indicates that there is a significant correlation in the population between the two variables. Shown in a diagram:

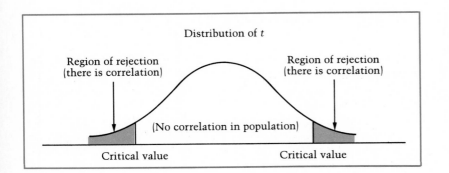

Distribution of t

Region of rejection (there is correlation)

Region of rejection (there is correlation)

(No correlation in population)

Critical value   Critical value

Problem

Recall that for the high-school and college GPA data the coefficient of correlation for the sample of five was computed to be 0.97. Is it possible that due to the small size of the sample the coefficient of correlation in the population is 0? We will now test this hypothesis at the 0.05 level of significance.

Solution

The usual five-step hypothesis-testing procedure will be followed.

Step 1. The null and alternate hypotheses are stated. They are

$$H_0: \rho = 0$$
$$H_a: \rho \neq 0$$

Step 2. The level of significance is selected. The null hypothesis is tested at the 0.05 level. Remember that the level of significance is the same as the probability of a Type I error. A Type I error is the probability of rejecting a null hypothesis when it is actually true.

Step 3. The appropriate test statistic is selected. It is

$$t = \frac{r\sqrt{n-2}}{\sqrt{1-r^2}}$$ with $n-2$ degrees of freedom

As noted, $n - 2$ degrees of freedom are connected with this test statistic. There are five paired observations in the high school and college GPA problem. Thus, n is 5 and $n - 2 = 5 - 2 = 3$ degrees of freedom.

Step 4. A decision rule is formulated. The alternate hypothesis states that ρ (the coefficient of correlation in the population) is not 0. Since it does not specify a direction, a two-tailed test is applied. The critical values of t are given in Appendix D. Go down the left column to 3 degrees of freedom. Then move horizontally to the critical value in the column for the 0.05 level of significance, two-tailed test. The value is 3.182. The decision rule, therefore, is: fail to reject the null hypothesis that ρ is 0 if the computed value of t is between -3.182 and 3.182. Otherwise, reject the null hypothesis and accept the alternate hypothesis that ρ is not 0.

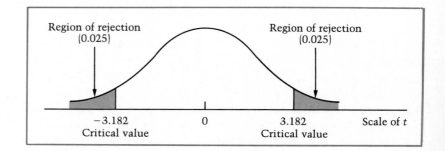

Step 5. A decision is made. In testing the hypothesis that $\rho = 0$ the final step is to compute t and to reach a decision based on the decision rule in the previous step. Computing t:

$$t = \frac{r\sqrt{n-2}}{\sqrt{1-r^2}}$$

$$= \frac{0.97\sqrt{5-2}}{\sqrt{1-(0.97)^2}}$$

$$= 6.91$$

The computed t value of 6.91 is in the rejection region. Therefore, H_0 that the population correlation coefficient is 0 is rejected at the 0.05 level of significance. H_a is accepted. The conclusion is that the correlation between high-school and college GPAs in the population is not 0.

Returning to the problem on the correlation of number of votes received and campaign expenses, r was computed to be 0.43.

a. Using a 0.01 significance level, test the hypothesis that correlation in the population is 0. The alternate hypothesis is that ρ is not 0.

b. Explain your results.

Self-Review 13-4

a. $t = \dfrac{0.43\sqrt{4-2}}{\sqrt{1-(0.43)^2}}$

$= \dfrac{0.6081}{0.9028}$

$= 0.67$

With 2 degrees of freedom for the 0.01 significance level and a two-tailed test statistic, the critical values are -9.925 and $+9.925$. The computed value of t is in the acceptance region.

b. Correlation in the population could be 0.

Chapter Exercises

13. A major airline selected a random sample of 25 flights and found that the correlation between the number of passengers and the total weight in pounds of luggage stored in the luggage compartment is 0.94. Using a 0.05 significance level, test the hypothesis that the correlation in the population is 0.

14. A sociologist claims that the success of students in college (measured by their GPA) is unrelated to their families' incomes. For a random sample of 20 students the coefficient of correlation was computed as 0.40. Using the 0.01 significance level, test the hypothesis that the coefficient of correlation in the entire population is 0.

15. An Environmental Protection Agency study of 12 cars revealed a correlation of 0.47 between the engine size and performance as measured in gasoline mileage. Use the 0.01 significance level

to test the hypothesis that the population correlation coefficient is 0.

16. A study of college soccer games revealed the correlation between shots attempted and goals scored to be 0.21 in a sample of 20 games. Use the 0.05 significance level to test the hypothesis that the population correlation coefficient is 0.

Rank-Order Correlation

In the previous section, the scale of measurement was interval. However, measures of correlation are also available for ordinal (ranked) data. The most widely used one was introduced in 1904 by Charles Edward Spearman (1863 - 1945), a British statistician. Spearman's rank-order correlation coefficient, also known as Spearman's rho, is similar to Pearson's r and is designated by the symbol r_s. It is determined by

$$r_s = 1 - \frac{6\Sigma d^2}{n(n^2 - 1)}$$

where

d is the difference between the ranks for each pair.

n is the number of paired observations.

Spearman's rank-order correlation coefficient ranges (as does Pearson's r) from -1.00 to $+1.00$ with a $+1.00$ indicating that the ranks are in perfect agreement. A -1.00 indicates that the variables are inversely related but in perfect agreement. An r_s of 0 reveals that there is no relationship between the ranks. The scatter diagram in Figure 13-4 portrays perfect agreement.

Candidate	Ranking by Party Bosses	Ranking by Sample of Voters
Lon Moore	3	3
Peter Flick	1	1
Sue Ganz	2	2
Paul Murray	5	5
Norm Bolt	4	4

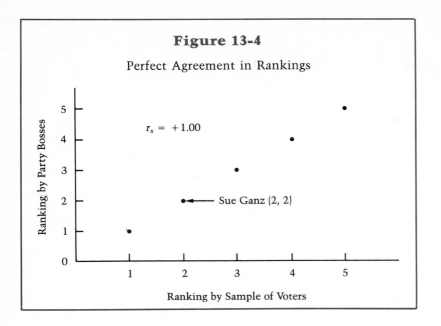

Figure 13-4

Perfect Agreement in Rankings

$r_s = +1.00$

Sue Ganz (2, 2)

Ranking by Party Bosses

Ranking by Sample of Voters

Figure 13-5 on page 386 portrays a case of perfect disagreement.

Weekly Television Program	Ranking by Teenagers	Rankings by Senior Citizens
Mother Knows Best	6	1
Wash Out	5	2
Ms. Detective	4	3
Tales of the Golden West	3	4
Loner Kate	2	5
Chopper Clem	1	6

Problem

The director of nursing asked two staff nurses to rank ten patients according to the difficulty of care required. A "difficult" assignment would include giving the patient a complete bath and monitoring the intravenous equipment. A less difficult assignment would be to give medication to a patient only at bedtime. Table 13-3 summarizes the ranking of each patient by the two nurses. What is Spearman's rank-order correlation coefficient for the data?

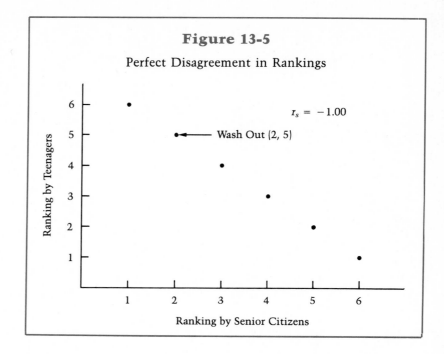

Figure 13-5

Perfect Disagreement in Rankings

$r_s = -1.00$

Wash Out (2, 5)

Ranking by Teenagers

Ranking by Senior Citizens

Table 13-3

Two Nurses' Rankings of Difficulty of Care
for Ten Patients

Patient	Nurse Scott Rank	Nurse Palmer Rank
Ms. Garcia	1	2
Ms. Hibner	2	1
Ms. Schaefer	3	5
Ms. Gerko	4	3
Ms. Owens	5	7
Mr. Bunt	6	6
Mr. Kolby	7	4
Mr. Cassow	8	10
Mr. Bianchi	9	9
Mr. Loftop	10	8

Solution

Table 13-4 repeats the rankings and extends them to include the calculations needed to compute the rank-order correlation coefficient.

Table 13-4

Two Nurses' Rankings and Calculations Needed
for the Rank-Order Correlation Coefficient

Patient	Nurse Scott Rank	Nurse Palmer Rank	Difference Between Ranks d	Difference Squared d^2
Ms. Garcia	1	2	−1	1
Ms. Hibner	2	1	1	1
Ms. Schaefer	3	5	−2	4
Ms. Gerko	4	3	1	1
Ms. Owens	5	7	−2	4
Mr. Bunt	6	6	0	0
Mr. Kolby	7	4	3	9
Mr. Cassow	8	10	−2	4
Mr. Bianchi	9	9	0	0
Mr. Loftop	10	8	2	4
				$\Sigma d^2 = 28$

Determining Spearman's rank-order correlation coefficient:

$$r_s = 1 - \frac{6\Sigma d^2}{n(n^2 - 1)}$$

$$= 1 - \frac{6(28)}{10(99)}$$

$$= 1 - 0.170$$

$$= 0.83$$

The coefficient of 0.83 indicates a very strong agreement between the two nurses with respect to the rankings given the patients.

In the previous problem, each nurse *ranked* a patient with respect to difficulty of care. In some problems, it may be necessary to convert data to ranks before computing the rank-order correlation coefficient. For example, two executives rated the performance of six subordinate executives on a scale of 0 to 100, with 0 representing very undesirable performance and 100 representing outstanding performance (see the first two columns in Table 13-5). Because the

Table 13-5

Ratings of Subordinates by Two Executives,
with Conversion to Rankings

Subordinate Executive	Ratings Harn's	Kander's	Rankings Harn's	Kander's
Glasser	42	55	4	3
Kingman	81	40	1	4
Cardellini	27	10	6	5
O'Hara	60	90	3	1
Compton	36	8	5	6
Wong	76	86	2	2

data are not known to be of interval scale, Pearson's *r* is not appropriate. However, if we convert each of the executive ratings to rankings, the rank-order correlation coefficient may be used to describe the agreement between the two executives' ratings.

Referring to the first column, note that Harn rated Kingman highest (81). Therefore, Kingman is ranked 1. Wong was rated next-highest (76) and is ranked 2, and so on. Kander's ratings are treated the same way—that is, O'Hara with a 90 rating is ranked 1, Wong with an 86 rating is ranked 2, and so on. (Note that the ratings for each executive were ranked from high to low. However, had each set of ratings been ranked from low to high, the rank-order correlation coefficient would be the same.)

Suppose, now, that Harn had rated Kingman's and Wong's performance as equal, assigning 83 points to each. How would their respective ranks be determined? Both would be ranked 1.5, which is the average of ranks 1 and 2. Had Kander rated O'Hara, Wong, and Glasser highest, with 91 points each, how would their ranks be determined? These three would tie for ranks 1, 2, and 3. The average of the three tied ranks would be 2, found by (1 + 2 + 3)/3. Therefore, each of the three would be given the rank 2.

Self-Review 13-5

A professional person and a blue-collar worker were asked to rank 12 occupations according to the social status they attached to each. A ranking of 1 was assigned to the occupation with the highest status, a 2 to the occupation with the next-highest status, and so on. Their rankings are:

Occupation	Rank of Professional Person	Rank of Blue-Collar Worker
Physician	1	1
Dentist	4	2
Attorney	2	4
Pharmacist	6	5
Optometrist	12	9
School teacher	8	12
Veterinarian	10	6
College professor	3	3
Engineer	5	7
Accountant	7	8
Health care administrator	9	11
Government administrator	11	10

a. Compute the rank-order coefficient of correlation.
b. Interpret the coefficient.

a.

Rank	Rank	d	d^2
1	1	0	0
4	2	2	4
2	4	−2	4
6	5	1	1
12	9	3	9
8	12	−4	16
10	6	4	16
3	3	0	0
5	7	−2	4
7	8	−1	1
9	11	−2	4
11	10	1	1
		0	60

$$r_s = 1 - \frac{6(60)}{12[(12)^2 - 1]}$$

$$= 1 - \frac{360}{1716}$$

$$= 0.79$$

b. Quite strong agreement between the two rankings.

Chapter Exercises

17. There is an opening for police sergeant. Applicants for promotion to sergeant are rated separately by both the police board and the county commissioners. On a scale of 1 to 50, the ratings of six applicants are

Applicant	Police Board's Rating	County Commissioners' Rating
Arbuckle	42	40
Cantor	36	40
Silverman	16	21
Trepinski	42	40
Lopez	49	47
Condon	8	9

a. Rank the ratings of the applicants. (Watch the ties.)
b. Compute Spearman's rank-order correlation coefficient.

18. A sales manager and a personnel manager were asked to rank sales trainees who had completed the company's training program in terms of the likelihood of their success on the job. Determine the coefficient of rank correlation.

Trainee	Sales Manager	Personnel Manager
A	3	2
B	1	5
C	8.5	9
D	2	1
E	4	6
F	8.5	10
G	5	3
H	7	4
I	10	8
J	6	7

Caution: Reasoning About Cause and Effect. If a strong correlation, such as 0.85, is found between two variables, it is tempting to assume that a change in one variable *causes* the other variable to change. This may not be true. For example, it has been discovered that the birthrate goes up when beer consumption goes up. An increase in the consumption of beer, however, does not cause the birth rate to increase. This only indicates that there is a relationship between the two variables. Results of this nature are called nonsense, or spurious, correlation.

Summary

Several techniques are available to describe the relationship between two variables. One of the simplest is the scatter diagram. It allows us to visualize the potential association, or correlation, between two variables. If the data are interval scaled, the Pearson's coefficient of correlation can be computed to measure the strength of the relationship. Perfect agreement between two variables will result in a coefficient of either -1.00 or $+1.00$. Coefficients of, say, -0.92 or $+0.92$ would reveal that the correlation is very strong, while -0.21 or $+0.21$ would show it to be very weak.

Other measures of the strength of the association between two interval-scaled variables are the coefficient of determination and the coefficient of nondetermination. They can assume any values from 0 to $+1.00$ or 0 to 100%. The coefficient of determination is computed by r^2 and the coefficient of nondetermination by $1 - r^2$. The coefficient of determination is the proportion of the total variation in one variable explained, or accounted for, by the variation in the other variable. The coefficient of nondetermination is the proportion not accounted for. Instead of the correlation coefficient, many researchers prefer to use the coefficient of determination to measure the relationship because it is a smaller value and thus does not overstate the association between the two variables.

A small sample selected from the population often invites the question, "Based on the sample coefficient of correlation, is it possible that the correlation in the population is 0?'' This question can be answered by applying the five-step hypothesis-testing procedure examined in earlier chapters. In brief, the five steps are: (1) The null hypothesis tested is: ρ (the population coefficient of correlation) is 0. The alternate hypothesis is: ρ is not 0. (2) A level of significance is selected. Usually, it is 0.05 or 0.01. (3) An appropriate test statistic is selected. In this case it is t. (4) A decision rule is formulated, and (5) Based on sample data, the decision is made to either reject or fail to reject the null hypothesis.

The strength of the relationship between two sets of ordinal-scaled data is measured by Spearman's rank-order correlation coefficient. This measure also ranges from -1.00 to $+1.00$. As implied by its name, the rank-order coefficient, is obtained from data organized in ranked order.

Correlation Analysis

I. Interval Level of Measurement

A. Objective. To measure the strength of the relationship between two variables of interval scale.

B. Procedure

1. Draw a scatter diagram, using the independent variable (X) and dependent variable (Y) as axes and coordinates.

2. Compute Pearson's coefficient of correlation, r:

$$r = \frac{n(\Sigma XY) - (\Sigma X)(\Sigma Y)}{\sqrt{[n(\Sigma X^2) - (\Sigma X)^2][n(\Sigma Y^2) - (\Sigma Y)^2]}}$$

3. Compute the coefficient of determination r^2. It is defined as the proportion of the total variation in one variable explained by the other variable.

4. Compute the coefficient of nondetermination $1 - r^2$. It is the proportion of the total variation not explained.

5. Test the significance of the coefficient of correlation. This is done using the five-step hypothesis-testing approach:

   (1) State H_0 and H_a.

   (2) Decide on a level of significance.

   (3) Choose an appropriate test statistic.

   (4) Give the decision rule.

   (5) Take a sample, compute t, and based on the decision rule either reject or fail to reject the null hypothesis.

II. Ordinal Level of Measurement.

A. Objective. To measure the strength of the agreement between two sets of data of ordinal scale.

B. Procedure
    1. Draw a scatter diagram, placing one ranking on the
       X-axis, the second ranking on the Y-axis.
    2. Compute Spearman's rank-order correlation coefficient:

$$r_s = 1 - \frac{6\Sigma d^2}{n(n^2 - 1)}$$

Chapter Exercises

19. The following table shows the percentage of the vote actually
received by the candidates for mayor in five large cities and
the corresponding percentages predicted by a national polling
organization:

Predicted %	Actual %
59	62
47	51
43	42
55	56
57	57

a. Plot the data on a scatter diagram.
b. Compute the coefficients of correlation, determination, and
    nondetermination.
c. Test the hypothesis that the coefficient of correlation in the
    population is 0. Use a 0.05 significance level.
d. Interpret your findings.

20. Suppose you were interested in determining the strength of the
relationship between the amount of time spent studying for
mid-term examinations and the scores received. A sample of
eight students you selected at random reveals the following
paired data:

Minutes Studied X	Score Y
60	70
70	65
80	75
100	86
120	90
150	80
200	92
210	90

a. Plot the data on a scatter diagram.
b. Compute the coefficients of correlation, determination, and nondetermination.
c. Test the hypothesis that the coefficient of correlation in the population is 0. Use a 0.05 significance level.
d. Interpret your findings.

21. A researcher studied the relationship between the years of education beyond high school and monthly salary of a sample of 27 welfare workers. The coefficient of correlation was computed to be 0.25. Test the hypothesis that there is no correlation in the population, using the 0.01 significance level.

22. The crime rate in a large city has been increasing. A special task force studying the problem has suggested that more full-time law enforcement personnel be employed. The police commissioner, however, is not convinced that this is the best strategy; he doubts that there is a relationship between the number of full-time law enforcement personnel and the annual number of crimes. To investigate, 11 large cities are randomly selected and both the total number of reported crimes and the total number of full-time enforcement officers are obtained. The results are

City	Number of Offenses X (000)	Full-Time Law Enforcement Personnel Y (000)
Atlanta	49.5	1.9
Buffalo	29.9	1.4
Chicago	214.1	14.8
Cleveland	53.1	2.3
Denver	52.9	1.7
Houston	106.3	3.0
Louisville	23.5	1.0
Milwaukee	37.0	2.3
Pittsburgh	32.0	1.4
Seattle	40.0	1.4
Tampa	27.7	0.8

Source: U.S. Department of Justice, *Uniform Crime Report, Crime in the U.S. in 1976*, Sept., 1977.

a. Compute the coefficients of correlation, determination, and nondetermination.
b. Test the hypothesis that the coefficient of correlation is 0; use the 0.05 significance level.
c. Interpret the findings.

23. The design department of a large automobile manufacturer developed ten different mock-up models of an intermediate-size automobile. The design group is interested in determining the opinions of teenagers and senior citizens with respect to the models. A group of teenagers was asked to rank each model. Based on their rankings, a composite rank for each model was determined. For example, the teenagers ranked mock-up model E the most appealing, and model J the next most appealing. The same procedure was followed for the senior citizens. The rankings of the two groups are

Model	Rankings by Teenagers	Rankings by Senior Citizens
A	6	8
B	4	1
C	8	10
D	3	3
E	1	2
F	10	7
G	7	9
H	5	5
I	9	6
J	2	4

a. Draw a scatter diagram. Place the rankings of the teenagers on the X-axis and the rankings of the senior citizens on the Y-axis.
b. Compute Spearman's rank-order correlation coefficient.
c. Interpret your findings.

24. Professor P. Dant believes that students who complete her examinations in the shortest time receive the highest grades and that those who take the longest to complete them receive the lowest grades. In order to verify her suspicion, she numbers the midterm examination paper as each student completes it.

Name	Order of Completion	Grade (50 possible)
Gorney	1	48
Gonzales	2	48
McDonald	3	43
Sadowski	4	49
Jackson	5	50
Smythe	6	47
Carlson	7	39
Archer	8	30
Namath	9	37
MacFearson	10	35

a. Rank the grades. (Watch the ties.)
b. Draw a scatter diagram. Put order of completion on the X-axis and the rankings of the grades on the Y-axis.
c. Compute the rank-order correlation coefficient.
d. Interpret your findings.

25. Two panels, one composed of all men and the other of all women, were asked by a consumer-testing agency to rank eight colas according to taste. A rank of 1 was given to the best-tasting cola and a rank of 8 to the worst. The purpose of the test was to determine the similarity in men's and women's tastes for cola.

Brand	Panel of Men	Panel of Women
Red Ribbon Cola	8	7
Bluebell Cola	6	5
Steel City Cola	2	4
Glatz Cola	1	2
Pearl Cola	3	1
Deer Run Cola	4	3
Boca Cola	5	6
Krolla Cola	7	8

Compute the rank-order correlation coefficient and interpret your findings.

First answer all the questions. Then check your answers against those given in the Answer section of the book.

Chapter Achievement Test

I. Multiple-Choice Questions. Select the response that best answers each of the questions that follow (7 points each).

Questions 1 through 5 are based on the following graphs:

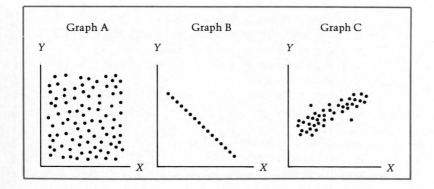

1. These graphs are called
   a. tree diagrams
   b. shotgun blasts
   c. scatter diagrams
   d. frequency polygons
   e. none of these are correct
2. Graph A. The coefficient of correlation, if computed, would be approximately
   a. 100
   b. −1.00
   c. +1.00
   d. 0
   e. none of these are correct
3. Graph B. The coefficient of nondetermination, if computed, would be approximately
   a. −1.00
   b. 0.50
   c. 0
   d. +1.00
   e. none of these are correct
4. Graph B. The linear relationship between the two variables is
   a. very weak
   b. moderate
   c. perfect
   d. cannot be determined from this graph
   e. none of these are correct
5. Graph C. The coefficient of determination, if computed, would be
   a. about 60%
   b. negative
   c. about 0
   d. near infinity
6. Pearson's product-moment coefficient of correlation for a certain set of paired observations was computed to be 0.70.
   a. the coefficient of nondetermination is 0.30
   b. the coefficient of determination is 0.49
   c. the correlation in the population is 0
   d. the correlation in the population is 0.49
   e. none of these are correct
7. Pearson's r was computed to be −0.55. Which of the following values of r represents a stronger relationship than −0.55?
   a. 0
   b. +0.50

   c. −0.70

   d. 20.9

   e. none of these are correct

8. Which of the following statements is false?

   a. Spearman's rank-order correlation coefficient is used when the data are of ordinal level of measurement.

   b. Pearson's product-moment coefficient of correlation requires interval level of measurement

   c. Spearman's rank-order coefficient may be between −1.00 and +1.00

   d. Pearson's r may assume any value between −1.00 and +1.00

   e. Spearman's rank-order coefficient cannot be negative.

II. Computation Problems (Problem 9, 30 points; Problem 10, 14 points).

9. The ages and corresponding prices of five bottles of wine selected at random and sold at a wholesale auction are shown in the list that follows. The relationship between age and price is to be explored.

Age (in years) X	Price (in dollars) Y
2	5
6	16
10	18
5	9
8	15

   a. Portray the paired data in a graph.

   b. Compute Pearson's coefficient of correlation.

   c. Compute both the coefficient of determination and of nondetermination.

   d. Test the hypothesis that the coefficient of correlation in the population is 0. Use a 0.05 significance level.

   e. Interpret your findings.

10. Early in the basketball season ten teams appeared to be outstanding. A panel of sportswriters and a panel of college basketball coaches were asked to rank the ten teams. Their composite rankings were:

Team	Coaches	Sportswriters
Notre Dame	1	1
Indiana	2	5
Duke	3	4
North Carolina	4	6
UCLA	5	2
Louisville	6	3
Ohio State	7	10
Syracuse	8	8
Georgetown	9	7
LSU	10	9

a. Compute the appropriate coefficient of correlation to evaluate the agreement between the sportswriters and coaches with respect to the rankings.

b. Interpret your findings.

14
Regression Analysis

OBJECTIVES

When you have completed this chapter, you will be able to
- Describe a linear relationship between two variables.
- Determine the linear equation using the method of least squares.
- Develop a measure of the error around the regression equation.
- Establish confidence intervals for predictions.

Introduction

Chapter 13 dealt with measures of association between two variables—Pearson's product-moment correlation coefficient and Spearman's rank-order correlation coefficient. In this chapter we will continue the study of two variables. We will develop a mathematical equation that allows us to "predict" one variable based on another variable. For example, if a student's high-school grade point average is known, it may be possible to "predict" his or her college grade point average. Similarly, the sales of a product may be predicted by the level of advertising expenditures. The technique used to make these predictions is called **regression analysis.**

Linear Regression

The word "regression" was first used by Sir Francis Galton in the late nineteenth century. A study by Galton revealed that the height of the children of tall parents was above average, but seemed to move toward, or regress, toward the mean height of the population. Thus, the general process of predicting one variable (such as a child's height) based on another variable (the parents' heights) became known as **regression.**

The usual first step in studying the relationship between two variables is to plot the paired data in the form of a scatter diagram.

$XY \quad X^2 \quad Y^2$

$\overline{\Sigma XY} \quad \overline{\Sigma X^2} \quad \overline{\Sigma Y^2}$

Table 14-1

High-School and College Grade Point
Averages

Student	High-School GPA X	College GPA Y
Janet Artz	3	6
Lee Jackson	5	9
James Coble	2	5
Sue Brown	3	7
Marita Estrada	4	8

In order to illustrate the concept of regression, we will reexamine the high-school and college GPA problem from Chapter 13. The data are repeated in Table 14-1 and plotted in the form of a scatter diagram in Figure 14-1. This time we are interested in *predicting* a college grade point average based on a student's high-school average. Recall that the variable being predicted (college GPA) is the *dependent* variable—designated Y—and the variable used to make

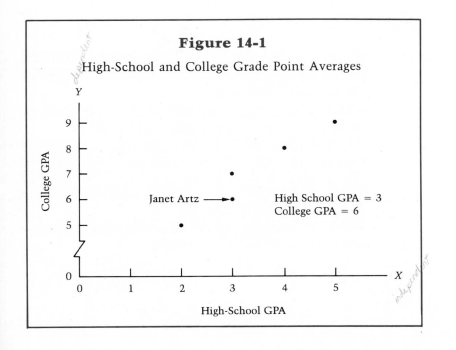

Figure 14-1

High-School and College Grade Point Averages

the prediction (high-school GPA) is the *independent* variable—designated *X*.

A straight line is perhaps the simplest of all mathematical relationships. Whenever possible, therefore, we try to find a straight line which—when translated into equation form—can describe the relationship between the two variables. An important use of the equation for the straight line is to approximate a value of the dependent variable based on a selected value of the independent variable. For example, in the grade-point-average problem we may want to approximate Terri Mancuso's college grade point average based on his high-school GPA of 2.1. The equation for that line is called the **regression equation.**

> **Regression Equation** A mathematical equation that defines the relationship between two variables.

The linear equation that expresses the relationship between two variables is

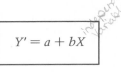

$$Y' = a + bX$$

where

Y'   (read "*Y* prime") is the predicted value of *Y* for a selected *X* value. In the example, it is the predicted college GPA for a given high-school GPA.

a   is a constant: the value where the straight line intersects the *Y*-axis. It is also the value of *Y* when *X* = 0.

b   is also a constant and is the *slope* of the straight line. It is the change in *Y'* for each change of one (either increase or decrease) in *X*. In this problem, *b* is the increase in college grade point average for an increase of one point in high-school grade point average.

X   is any value of *X* that is selected. In this problem, *X* is any high-school GPA that is chosen.

To further illustrate the meaning of *a* and *b*, suppose that the regression equation in the form of $Y' = a + bX$ for the data in Figure 14-2 had been computed to be $Y' = 20 + 10X$ (in thousands

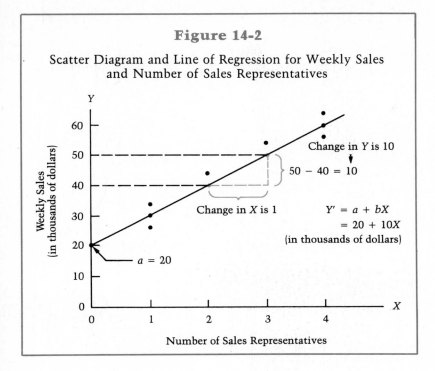

Figure 14-2

Scatter Diagram and Line of Regression for Weekly Sales and Number of Sales Representatives

of dollars). Note that when $X = 0$ the straight line intercepts the Y-axis at \$20,000. This is a, sometimes called the Y-intercept. Figure 14-2 also shows that weekly sales increase from \$40,000 a week when two sales representatives call on clients, to \$50,000 when three representatives do. Thus, with each increase of one sales representative, weekly sales increase by \$10,000. This is b.

The regression equation $Y' = 20 + 10X$ (in thousands of dollars) can be used to estimate the average weekly sales if four sales representatives are employed, $X = 4$. Then $Y' = 20 + 10(4) = 60$, or \$60,000.

A reminder: cover the answers in the margin.

Self-Review 14-1

Some data dealing with the education and weekly incomes of residents of Precinct 13 have been collected. The purpose of the study is to predict incomes based on education.

a. Don has <u>five years</u> of education beyond high school; his weekly income is $450.

b. $a = 250$

c. $b = 100$. One way to compute: when $X = 3$, $Y = 550$ and when $X = 4$, $Y = 650$; then, $650 - 550 = 100$. It indicates that for each one additional year beyond high school, weekly income increases $100.

d. $Y' = 250 + 100X$ (in dollars).

e. $650, found by $Y' = 250 + 100(4)$.

$$Y' = 250 + 100 \times (4) = 650$$

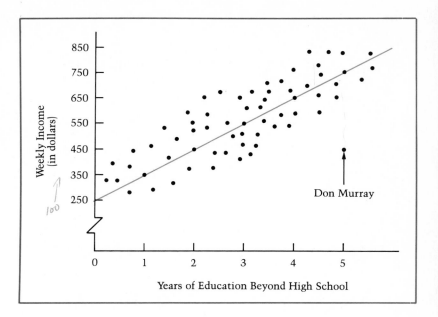

a. Interpret the meaning of the point for Don Murray.
b. What is the value of a in the regression equation $Y' = a + bX$?
c. What is the value of b in the regression equation? What does it indicate?
d. What is the regression equation?
e. Based on the regression equation, estimate the average weekly income for a person with four years of education beyond high school.

Chapter Exercises

1. The length of confinement in days (X) and the dollar cost (Y) for six patients at St. Matthew's Hospital were tabulated and then plotted on a scatter diagram:

Length of Confinement X	Dollar Cost Y
5	$1,110
2	490
12	2,500
4	880
7	1,530
1	270

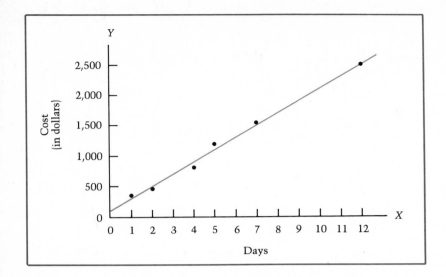

a. What is the general form of the regression equation?
b. Based on the scatter diagram, what is *a?*
c. What is *b* (the slope) of this regression line?
d. If a patient stayed six days, what would you predict his or her bill to be?
e. Interpret in words the values of *a* and *b.*

2. A study attempting to relate the number of minority employees (*Y*) to the number of nonminority employees (*X*) in eight firms revealed the following:

Firm	Number of Nonminority Employees *X*	Number of Minority Employees *Y*
London Mfg.	4,628	182
ABC Chemicals	4,272	299
Cork Floors Inc.	1,663	173
Sanson Industries	691	44
Martin Trucking	5,355	1,069
Cable Lead	5,618	777
Parson Farms	2,934	802
Tayo Electric	3,650	790

a. What is the general form of the regression equation?
b. Based on the scatter diagram on page 406, what is the *Y*-intercept?
c. What is *b,* the slope of this regression line?

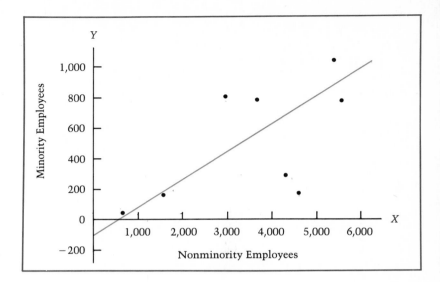

d. If a firm employs 3,000 nonminority persons, predict the number of minority employees.
e. Interpret the values of *a* and *b*.

3. An equation that relates the cost to heat a single-family residence (Y) to its size in square feet of living area (X) is $Y' = -200 + 0.3X$.
   a. Predict the cost of heating a 3,500-square-foot home.
   b. What is the slope of this line?
   c. What is the Y-intercept?
   d. Interpret in words the meaning of *a* and *b*.

4. The equation $Y' = 39 - 0.002X$ relates the birth rate per thousand population (Y) to the median annual income (X).
   a. Predict the birth rate for a country whose median annual income is $7,000.
   b. What is the Y-intercept of the line of regression?
   c. What is the slope of this line?
   d. Interpret the meaning of slope *b*.

Fitting a Line to Data

Freehand Method

How should the line through the plots on a scatter diagram be drawn? It could be drawn through the points simply by placing a ruler on the scatter diagram and drawing a line through the points. If this job of drawing the best-fitting line "right through the middle of

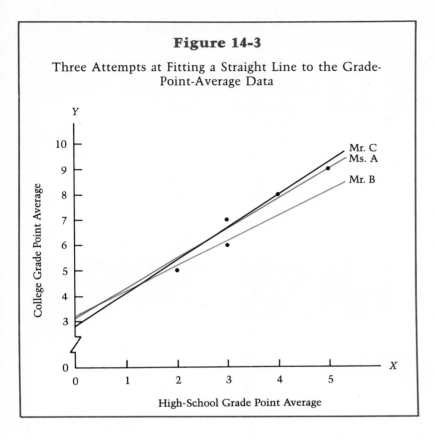

Figure 14-3

Three Attempts at Fitting a Straight Line to the Grade-Point-Average Data

the points'' were given to three persons, however, their lines would probably all be different. Consider the example of predicting college GPA on the basis of high-school GPA. (The grade point averages for the five students in the sample were shown in Table 14-1.) Figure 14-3 shows how three different people might draw a straight line through the data. Each is different. Therefore, if, in order to predict a college GPA from a high-school GPA, ruler-drawn lines were used, each prediction would be different. What is needed is a precise method of arriving at the best-fitting straight line. Then the same line will be obtained every time, regardless of who does the calculation.

Least-Squares Principle

Reliance on personal judgment as a way to find the exact location of the straight line can be avoided by employing a method called the **least-squares principle.** It provides us with the ''best-fitting'' straight line. How does it work? Note that in Figure 14-2 the plotted points were not exactly on the straight line. The difference between an actual value (a plotted point) and a predicted value (a corresponding value on the straight line) can be thought of as the "error" we make when using the regression equation to predict a specific value of the dependent variable. To say it another way, this difference, or error, can be considered to be the deviation of the actual value from the predicted value.

By using this principle we obtain a *unique* regression equation that is "best" in the sense that the squared errors around the regression line are smaller than around any other line. That is, the sum of the squares of the deviations between the actual values (*Y*) and the predicted values (*Y'*) is minimized. The regression line is unique—meaning that there is only one least-squares line and this line can be determined by anyone doing the calculations.

> **Least-Squares Principle** A method used to determine the regression equation by minimizing the sum of the squares of the distances between the actual *Y* values and the predicted values of *Y*.

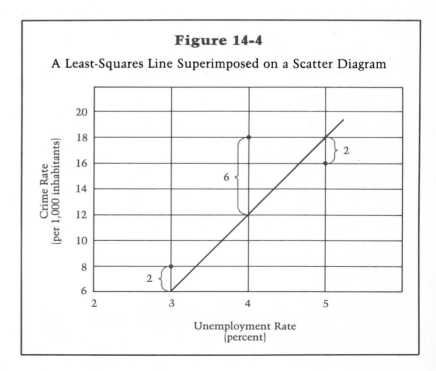

Figure 14-4

A Least-Squares Line Superimposed on a Scatter Diagram

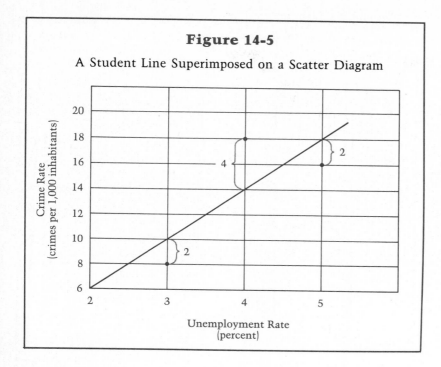

Figure 14-5

A Student Line Superimposed on a Scatter Diagram

Figure 14-4 on page 408 illustrates this concept. The straight line was determined by the method of least squares. The first plot ($X = 3$, $Y = 8$) deviates vertically by 2 units from the predicted value of 10 on the straight line. The squared deviation is 4, found by 2^2. The squared deviation for $X = 4$, $Y = 18$ is 16, found by $(18 - 14)^2$. The squared deviation for $X = 5$, $Y = 16$ is 4. The sum of the squared deviations is 24, found by $4 + 16 + 4$.

The same crime and unemployment data in Figure 14-4 also appear in Figure 14-5. The straight line through the data in Figure 14-5, however, is not determined using the least-squares method. A beginning student thought the line should be positioned as shown. The sum of the vertical squared deviations from the freehand line is 44, found by $2^2 + 6^2 + 2^2$. The 44 is greater than 24, the sum of the squared deviations around the least-squares line.

The same crime and unemployment data used in Figures 14-4 and 14-5 appear again in the following scatter diagram. This line was also drawn by a student.

Self-Review 14-2

a. 132, found by

X	Deviation	Squared Deviation
3	8	64
4	2	4
5	8	64
		132

b. 132 is greater than 24. The sum of squared deviations from the least-squares line is always smaller.

c. The crime rate (it is on the Y-axis).

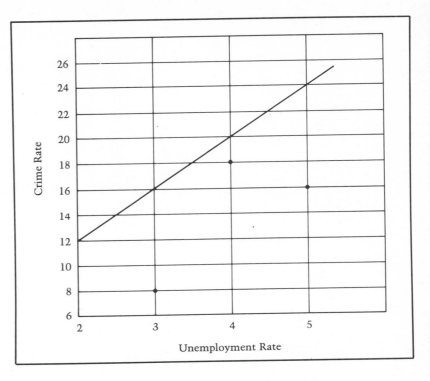

a. Determine the sum of the squared vertical deviations from the line.

b. Is the sum greater or smaller than the sum from Figure 14-4? Explain.

c. What is the dependent variable in this diagram?

The Method of Least Squares

In the previous section we explained why the "best" equation is the one that minimizes the sum of the squared deviations. Here you will learn how this line is determined and whether it is unique. The determination is made by the mathematical process of minimization. That is, among all possible lines, the one where $\Sigma(Y - Y')^2$ is the smallest is obtained, and it is unique. The following formulas, based on the least-squares principle, are used to compute the unique values of b and a.

$$b = \frac{n(\Sigma XY) - (\Sigma X)(\Sigma Y)}{n(\Sigma X^2) - (\Sigma X)^2}$$

$$a = \frac{\Sigma Y - b(\Sigma X)}{n}$$

b then A then Y

b = slope of the line

A = Y intercept

where

X is a value of the independent variable.

Y is a value of the dependent variable.

n is the number of items in the sample.

Problem

The high-school and college GPAs for five students are repeated:

Student	High-School GPA X	College GPA Y
Janet Artz	3	6
Lee Jackson	5	9
James Coble	2	5
Sue Brown	3	7
Marita Estrada	4	8

Determine the linear regression equation that could be used to predict these students' college GPA based on their high-school GPA by the method of least squares. Predict the college GPA for a student who had a high-school GPA of 2.75.

Solution

The products, squares, and totals needed for the method of least squares are shown in Table 14-2.

Solving for *b* and *a*:

$$b = \frac{n(\Sigma XY) - (\Sigma X)(\Sigma Y)}{n(\Sigma X^2) - (\Sigma X)^2}$$

Table 14-2

Calculations Needed for Least-Squares
Regression Equation

Student	High-School GPA X	College GPA Y	XY	X^2	Y^2
Janet Artz	3	6	18	9	36
Lee Jackson	5	9	45	25	81
James Coble	2	5	10	4	25
Sue Brown	3	7	21	9	49
Marita Estrada	4	8	32	16	64
	17	35	126	63	255

$$= \frac{5(126) - (17)(35)}{5(63) - (17)^2}$$

$$= \frac{630 - 595}{315 - 289}$$

$$= \frac{35}{26}$$

$$= 1.35$$

$$a = \frac{\Sigma Y - b(\Sigma X)}{n}$$

$$= \frac{35 - 1.35(17)}{5}$$

$$= 2.41$$

Therefore, the regression equation is $Y' = 2.41 + 1.35X$. The predicted college GPA for a student who had a high-school GPA of 2.75 is 6.12, found by $Y' = a + bX = 2.41 + 1.35X = 2.41 + 1.35(2.75) = 2.41 + 3.71 = 6.12$.

Drawing the Line of Regression

Now that we have the regression equation, the least-squares line of regression can be plotted on the scatter diagram. The first step is to select an X value, say 0. When $X = 0$, $Y' = 2.41$, found by

$Y' = 2.41 + 1.35(0)$. The computations for some other points on the line follow.

When X Is	Y' Is	Found by
0	2.41	$Y' = 2.41 + 1.35(0)$
1	3.76	$Y' = 2.41 + 1.35(1)$
2	5.11	$Y' = 2.41 + 1.35(2)$
3	6.46	$Y' = 2.41 + 1.35(3)$

Figure 14-6 shows the original scatter diagram with the line of regression superimposed on it.

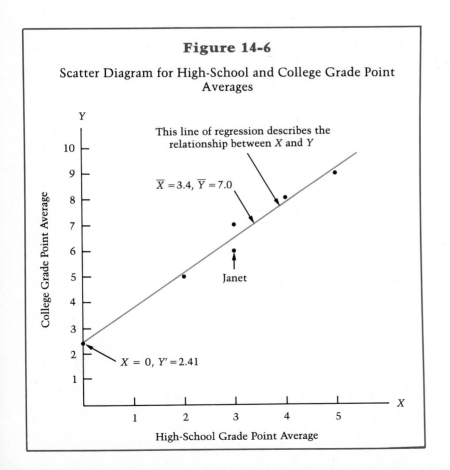

Figure 14-6

Scatter Diagram for High-School and College Grade Point Averages

Columns

Self-Review 14-3

There is interest in determining if the number of votes received in an election can be predicted, based on the amount of money spent during the campaign. A tabulation of available data follows.

a.

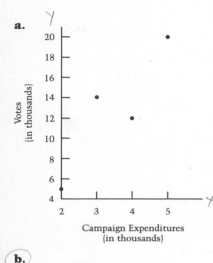

Votes (in thousands) vs Campaign Expenditures (in thousands)

Candidate	Amount Spent on Campaigns (000) X	Votes Received (000) Y
Weber	$3	14
Taite	4	12
Spence	2	5
Henry	5	20

a. Draw a scatter diagram.
b. Determine the regression equation using the method of least squares.
c. How many additional votes could a candidate expect for each additional $1,000 spent?
(This problem will be continued; save your calculations.)

b.

$$b = \frac{4(200) - (14)(51)}{4(54) - (14)^2}$$

$$= \frac{86}{20} = 4.3$$

$$a = \frac{51 - 4.3(14)}{4}$$

$$= -2.3$$

$Y' = -2.3 + 4.3X$ (in thousands of votes)

c. For each additional $1,000, the candidate could expect 4,300 votes.

Chapter Exercises

5. (This set of problems will be continued; save your calculations.) Listed below are the populations of five counties (X) and the number of physicians (Y) who practice there:

County	Population (000)	Number of Physicians
A	38	21
B	76	35
C	36	13
D	62	24
E	32	18

a. Draw a scatter diagram of the data.
b. Find the least-squares regression equation.
c. Interpret the values of a and b.
d. Predict the number of physicians for a county with a population of 45,000.

6. The number of adult-movie theaters (X) and the annual number of sex-related crimes (Y) reported for five cities are:

Number of Movie Theaters	Number of Sex-related Crimes
15	175
20	220
10	120
12	152
16	181

Handwritten annotations:

X X^2 Y $X \cdot Y$

225	2625
400	4400
100	1200
144	1824
256	2896
73 1125	848 12945

$b = \dfrac{5(12945) - (73)(848)}{5(1125) - (73)^2}$

$64725 - 61904 = 2821$
$5625 - 5329 = 296$

$\dfrac{2821}{296} = 9.53$

$A = \dfrac{848 - 9.53(73)}{5}$

$\dfrac{848 - 695.69 = 152.31}{5}$

30.46

$y' = 30.46 + 9.53X$

$9.53 \times (18) = 171.54$

202.00

a. Draw a scatter diagram of the data.
b. Find the least-squares regression equation.
c. Interpret the values of *a* and *b*.
d. Predict the number of sex-related crimes for a city with 18 movie theaters.

7. The mean nighttime low temperature (*X*) and an index of collective aggression during the month of July (*Y*) are recorded for six urban areas:

Mean Nighttime Low (in degrees Fahrenheit)	Index of Collective Aggression
73	51
82	63
90	67
60	35
51	23
40	20

a. Draw a scatter diagram of the data.
b. Find the least-squares regression equation.
c. Interpret the values of *a* and *b*.
d. Predict the value of the index when the mean nighttime low is 80.

8. Five patients of various ages (*X*) undergoing the same medical treatment suffer asthma attacks of different durations (*Y*):

Age (in yrs.)	Duration of Attack (in mins.)
30	15
25	28
65	30
50	22
40	24

Handwritten annotations:

in class

X X^2 Y $X \cdot Y$

900	450
625	700
4225	1950
2500	1100
1600	960
210 9850	119 5160

$n = 5$

a)

$a = \dfrac{\Sigma Y - b(\Sigma X)}{n}$

$a = 119 - .16(210) = 85.4$

$\dfrac{85.4}{5} = 17.08$

a. Draw a scatter diagram of the data.
b. Find the least-squares regression equation.

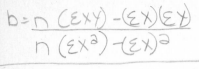

$$b = \frac{n(\Sigma XY) - (\Sigma X)(\Sigma Y)}{n(\Sigma X^2) - (\Sigma X)^2}$$

$$b = \frac{5(5160) - (210)(119)}{5(9850) - (44100)}$$

$25800 - 24990 = 810$
$49250 - 44100 = 5150$

$\frac{810}{5150} = .16$

c) $Y' = a + bX$

$Y = 17.08 + 16X$

$Y = 17\underline{.08} + .16(42)^{6.72}$

$Y = 23.8$

$b = \frac{\Delta Y}{\Delta X} = \frac{.16}{1},$

3 c. Interpret the values of a and b.
4 d. Predict the duration of an attack if the patient is 42 years old. 42 ×

The Standard Error of Estimate

If all the data points fall precisely on the line of regression, then our prediction was exact. In the previous chapter, this situation was described as perfect correlation. The coefficient of correlation would be either −1.00 or +1.00. The coefficient of correlation, then, is a possible measure of how well the line of regression fits the paired data. Figures 14-7 and 14-8 show the cases of perfect correlation and perfect prediction.

Perfect predictions of Y based solely on X, as illustrated in the two figures, are practically nonexistent in real-world situations. Instead, there is usually some scatter around the regression line. The **standard error of estimate** is another measure of the scatter of the observed Y values around the Y' values on the line of regression.

> **Standard Error of Estimate** A measure of the variability of the observed values around the regression line.

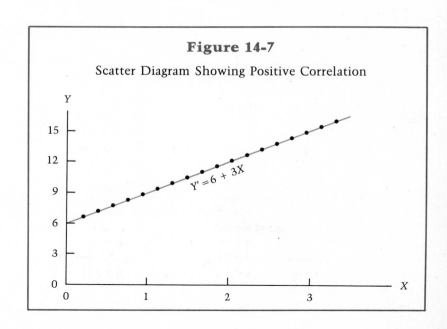

Figure 14-7

Scatter Diagram Showing Positive Correlation

$Y' = 6 + 3X$

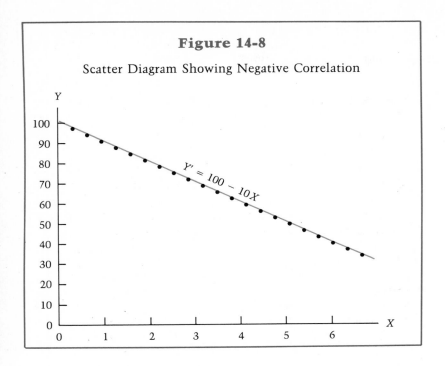

Figure 14-8

Scatter Diagram Showing Negative Correlation

$Y' = 100 - 10X$

The standard error, as it is usually called, is an indicator of how well the predicted Y' values compare to the actual Y values for a particular X. It is designated by $s_{Y \cdot X}$ and has the following formula:

$$s_{Y \cdot X} = \sqrt{\frac{\Sigma(Y - Y')^2}{n - 2}}$$

where the $\Sigma(Y - Y')^2$ component refers to the sum of the squared differences between the actual Y value and the value predicted for Y, called Y'. The subscript $Y \cdot X$ reminds us that we are predicting Y based on values of X. n, of course, refers to the size of the sample. The standard error measures variation in a manner similar to the standard deviation, except that, instead of summing the squared deviations from the mean as the standard deviation does $[\Sigma(Y - \bar{Y})^2]$, it employs the squared deviation between the actual values and the predicted values of $Y'[\Sigma(Y - Y')^2]$. In fact, it may be thought of as a conditional standard deviation of the dependent variable Y for a given value of the independent variable X.

The previous formula for the standard error requires that the difference between each Y observation and its corresponding Y'

value be determined and then squared. These are very time-consuming and error-prone calculations, especially for large samples. An equivalent form is

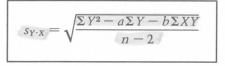

$$s_{Y\cdot X} = \sqrt{\frac{\Sigma Y^2 - a\Sigma Y - b\Sigma XY}{n-2}}$$

Problem

The high-school and college grade point averages discussed earlier and the needed calculations for the standard error of estimate are shown in Table 14-3. Recall that the regression equation has been found to be $Y' = 2.41 + 1.35X$. What is the standard error of estimate?

Solution

Inserting the appropriate values in the formula for the standard error of estimate:

$$s_{Y\cdot X} = \sqrt{\frac{\Sigma Y^2 - a\Sigma Y - b\Sigma XY}{n-2}}$$

$$= \sqrt{\frac{255 - 2.41(35) - 1.35(126)}{5-2}}$$

$$= \sqrt{0.1833}$$

$$= 0.43$$

Further interpretation of this value will follow shortly.

Table 14-3

Data Needed to Calculate the Standard Error of Estimate

Student	High-School GPA X	College GPA Y	XY	X^2	Y^2
Janet Artz	3	6	18	9	36
Lee Jackson	5	9	45	25	81
James Coble	2	5	10	4	25
Sue Brown	3	7	21	9	49
Marita Estrada	4	8	32	16	64
	17	35	126	63	255

The problem begun in Self-Review 14-3 is continued here. The data are

Candidate	Amount Spent on Campaign (000) X	Votes Received (000) Y	XY	X²	Y²
Weber	$ 3	14	42	9	196
Taite	4	12	48	16	144
Spence	2	5	10	4	25
Henry	5	20	100	25	400
	14	51	200	54	765

$$s_{Y \cdot X} = \sqrt{\frac{\Sigma Y^2 - a\Sigma Y - b\Sigma XY}{n - 2}}$$

$$= \sqrt{\frac{765 - (-2.3)(51) - 4.3(200)}{4 - 2}}$$

$$= \sqrt{\frac{765 + 117.3 - 860}{2}}$$

$$= 3.3$$

The regression equation is $Y' = -2.3 + 4.3X$. Compute the standard error of estimate.

Chapter Exercises

9. Refer to Exercise 5. The relationship between the number of physicians (Y) and the population (X) is

$$Y' = 3.259 + 0.388X, \ n = 5, \ \Sigma Y = 111,$$
$$\Sigma XY = 5{,}990, \text{ and } \Sigma Y^2 = 2{,}735$$

Compute the standard error of estimate.

10. Refer to Exercise 7 on the index of aggression and its relationship to the nighttime low temperature in July.

$$Y' = -25.7 + 1.04X, \ n = 6, \ \Sigma Y = 259,$$
$$\Sigma XY = 18{,}992, \text{ and } \Sigma Y^2 = 13{,}213$$

What is the standard error of estimate?

11. Sample data on the unemployment rate (X) and the crime rate (Y) are as follows:

X	Y
3	8
4	18
5	16

a. Calculate the regression equation.

b. Find the standard error of estimate.

12. Refer to Exercise 1. The data showing length of confinement in days (X) and hospital cost (Y) at St. Matthew's Hospital are repeated:

X	Y
5	$1,110
2	490
12	2,500
4	880
7	1,530
1	270

a. Calculate the regression equation.

b. Find the standard error of estimate.

Assumptions About Regression

Further analysis about the line of regression is based on certain assumptions:

1. For a given value of X, the Y observations are normally distributed around the line of regression.

2. The standard deviation of each of these normal distributions is the same. This common standard deviation is estimated by the standard error of estimate $(s_{Y \cdot X})$.

3. The deviations from the regression line are independent. This means that if there is a large deviation from the line of regression for an X value, it does not necessarily indicate that a large deviation must appear for other X values.

These assumptions are graphically summarized in Figure 14-9, wherein each of the distributions depicted (a) is normal; and (b) has the same standard deviation, estimated by $s_{Y \cdot X}$.

Assuming the observed Y values are normally distributed around the line of regression, we can say

$Y' \pm 1 s_{Y \cdot X}$ encompasses about 68% of the points;

$Y' \pm 2 s_{Y \cdot X}$ encompasses about 95.5% of the points;

$Y' \pm 3 s_{Y \cdot X}$ encompasses about 99.7% of the points.

What does the standard error of estimate in the high-school and college GPA problem tell us? It is a measure of the accuracy

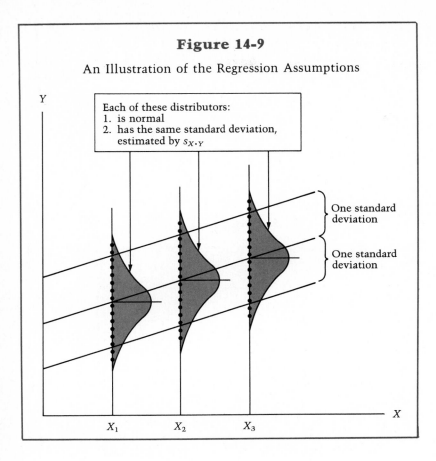

Figure 14-9

An Illustration of the Regression Assumptions

Each of these distributors:
1. is normal
2. has the same standard deviation, estimated by $s_{X \cdot Y}$

One standard deviation

One standard deviation

X_1 X_2 X_3

of our prediction. Suppose we were to repeat this experiment with a large number of students and tried to predict their college GPA. Then, the difference between our prediction and their actual college GPA would be less than one standard error of the estimate (0.43) in 68% of the cases. Ninety-five percent of our predictions would be "off" by no more than 2(0.43). And virtually all of our predictions will be "off" by no more than 3(0.43).

Computing Confidence-Interval Estimates

Suppose we wanted to calculate the actual confidence interval for our grade-point-average problem. We would have to consider two factors. One is the size of the sample. The other is that errors of prediction will be larger as we move away from the mean value of the independent variable (X).

We will consider two applications. In the first case, the *mean* value of Y is estimated for a given value of X. In the other case, an *individual* value of Y is estimated for a given value of X.

The formula of the confidence interval for the *mean value of Y* is

$$Y' \pm t(s_{Y \cdot X}) \sqrt{\frac{1}{n} + \frac{(X - \overline{X})^2}{\Sigma X^2 - \dfrac{(\Sigma X)^2}{n}}}$$

The formula of the confidence interval for an *individual value of Y* is

$$Y' \pm t(s_{Y \cdot X}) \sqrt{1 + \frac{1}{n} + \frac{(X - \overline{X})^2}{\Sigma X^2 - \dfrac{(\Sigma X)^2}{n}}}$$

where

- t is the value of t from Student's t Distribution in Appendix D. For the two-variable linear case, two degrees of freedom are lost because two least-squares coefficients, a and b, are being estimated.
- X is any selected value of the independent variable.
- $\overline{X}$ is the mean of the independent variable.
- n is the number of paired observations.

Note that there is only a slight difference in the two formulas. There is a 1 in the formula for finding the confidence interval for an individual value. Logically, the inclusion of 1 under the radical widens the prediction for the confidence interval for the individual. This is reasonable, because individuals are more unpredictable than group averages.

Problem

Returning again to the data concerning the high-school and college grade point averages:

Student	High-School GPA X	College GPA Y	XY	X^2	Y^2
Janet Artz	3	6	18	9	36
Lee Jackson	5	9	45	25	81
James Coble	2	5	10	4	25
Sue Brown	3	7	21	9	49
Marita Estrada	4	8	32	16	64
	17	35	126	63	255

What is the 95% confidence interval for the *mean* college GPA for all students with a 3.0 high-school grade point average?

Solution

The predicted *average* GPA in college, using the regression equation computed previously, is 6.46, found by

$$Y' = 2.41 + 1.35X$$
$$= 2.41 + 1.35(3.0)$$
$$= 6.46$$

The value of t is obtained by referring to Student's t Distribution in Appendix D. Read down the left column until you locate $n - 2$, or $5 - 2 = 3$ degrees of freedom. Then move horizontally to the t value under the 0.05 level of significance for a two-tailed test ($t = 3.182$).

Other values needed to set the confidence-interval estimate were computed previously: $s_{Y \cdot X} = 0.43$, and from Table 14-2, $n = 5$, $\Sigma X = 17$, $\Sigma X^2 = 63$ and $\bar{X} = 17/5 = 3.4$. Inserting these values:

$$Y' \pm t(s_{Y \cdot X}) \sqrt{\frac{1}{n} + \frac{(X - \bar{X})^2}{\Sigma X^2 - \frac{(\Sigma X)^2}{n}}}$$

$$= 6.46 \pm (3.182)(0.43) \sqrt{\frac{1}{5} + \frac{(3 - 3.4)^2}{63 - \frac{(17)^2}{5}}}$$

$$= 6.46 \pm 1.36826\sqrt{0.2307692}$$

$$= 5.80 \text{ and } 7.12$$

Interpreting: the expected college performance (GPA) for a group of high-school students with a GPA of 3.0 is 6.46. (This corresponds to the point on the regression line for $X = 3.0$, $Y' = 6.46$.) Further, the probability is 0.95 that the college grade point average for high-school students with a GPA of 3.0 will be in the interval between 5.80 and 7.12.

Problem

Estimate the college GPA for an individual (Armando Lopez) whose high-school grade point average is 3.0. This differs from the previous illustration in that we are concerned about a *specific person* with a GPA of 3.0 rather than the *average* of all those with a GPA of 3.0.

Solution

The formula for a confidence interval for an individual value is

$$Y' \pm t(s_{Y \cdot X}) \sqrt{1 + \frac{1}{n} + \frac{(X - \bar{X})^2}{\Sigma X^2 - \frac{(\Sigma X)^2}{n}}}$$

For Armando Lopez's 3.0 high-school GPA:

$$6.46 \pm (3.182)(0.43) \sqrt{1 + \frac{1}{5} + \frac{(3.0 - 3.4)^2}{63 - \frac{(17)^2}{5}}}$$

$$6.46 \pm 1.36826\sqrt{1.23077}$$

$$4.94 \text{ and } 7.98$$

Interpreting: for a specific high-school student with a GPA of 3.0, the 95% confidence interval is between 4.94 and 7.98.

Self-Review 14-5

In Self-Reviews 14-3 and 14-4, which deal with campaign expenditures and the number of votes received, we computed these values:

$$Y' = -2.3 + 4.3X, \; s_{Y \cdot X} = 3.3, \; n = 4, \; \Sigma X^2 = 54, \; \Sigma X = 14, \text{ and}$$
$$\bar{X} = 3.5$$

a. $Y' = -2.3 + 4.3(3.8) = 1.40$

Then:

$$14.0 \pm 2.920(3.3) \sqrt{\frac{1}{4} + \frac{(3.8 - 3.5)^2}{54 - \frac{(14)^2}{4}}}$$

$$= 9.0 \text{ and } 19.0$$

a. Determine the 90% confidence interval for the *mean* number of votes gained by all candidates who spent \$38,000 on their campaigns.

b. Interpret your answer to **a.**

c. Compute the 90% confidence interval for an *individual* whose campaign cost $38,000.

d. Interpret your answer to **c.**

e. Compare your responses to **b** and **d.**

Chapter Exercises

13. Referring back to the data for the high-school and college GPAs problem:

   a. What is the 90% confidence interval for the mean college GPA of students with a 3.0 high-school GPA?

   b. What is the 90% confidence interval for the college GPA of an individual student with a 3.0 high-school GPA?

14. Exercise 7 dealt with data relating an aggression index (Y) to the mean nighttime low temperature (X). The equation was $Y' = -25.7 + 1.04X$. The predicted value of the index was 57.8 when the temperature was 80°F. In Exercise 10, the standard error of estimate was found to be 3.6, $n = 6$, $\Sigma X = 396$, and $\Sigma X^2 = 27,954$.

   a. Construct a 95% confidence interval for the *mean* value of the aggression index when the temperature stands at 80°F.

   b. On a *particular* night the temperature is 80°F. Find the 95% confidence interval for the aggression index on this evening.

15. Exercise 5 presented some data relating the population of a county (X) to the number of physicians who practice in that county (Y). Exercise 9 asked you to compute the standard error of estimate.

   a. Based on these data, construct a 95% confidence interval for the mean number of physicians practicing in counties with a population of 45,000.

   b. Based on these data, construct a 95% confidence interval for the number of physicians practicing in a particular county with a population of 45,000.

16. Refer to the data from St. Matthew's Hospital (see Exercises 1 and 12).

   a. Construct a 90% confidence interval for the *mean* cost of a six-day stay.

   b. Construct a 90% confidence interval for the cost of an individual whose length of stay is six days.

b. The probability is 0.90 (that is, the odds are 9 to 1) that a candidate who spends $38,000 on his or her campaign will receive between 9,000 and 19,000 votes.

c. $14.0 \pm 2.920(3.3) \cdot$

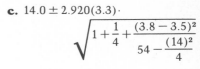

$$= 14.0 \pm 10.85$$
$$= 3.15 \text{ and } 24.85$$

d. The probability is 0.90 that if individual candidates spend $38,000 on their campaigns they will receive between 3,150 and 24,850 votes.

e. The wider interval is the result of comparing individuals and comparing means.

A Comprehensive Illustration

The problem that follows summarizes linear regression as well as correlation, as discussed in the previous chapter. The objectives of this study are (1) to predict income of the chief executive officer of large firms based on the age of the executive, and (2) to measure the association between the two variables. However, to keep calculations at a minimum, paired data (income and age) for only ten executives will be considered (see Table 14-4).

Problem

In order to review the techniques studied:

a. Draw a scatter diagram.

b. Compute Pearson's product-moment correlation coefficient.

c. Compute the coefficients of determination and nondetermination.

d. Determine the regression equation using the method of least squares.

e. Plot the line of regression on the scatter diagram.

f. Predict the annual income of the average 50-year-old executive.

g. Find the standard error of estimate.

h. Set a 95% confidence-interval estimate for the average 50-year-old executive.

Table 14-4

Total Annual Income and Age of the Chief
Executive Officer for Selected Firms, 1977

Company	Chief Executive Officer	Total Income	Age
Southland Corp.	John P. Thompson	$325,000	51
Boise Cascade	John B. Fery	444,000	47
McDonald's	Fred L. Turner	268,000	44
Gulf & Western Ind.	Charles G. Bluhdorn	605,000	50
White Consolidated	Roy H. Holdt	569,000	56
Marriott	J. Willard Marriott, Jr.	190,000	45
Amer. Broadcasting	Leonard H. Goldenson	946,000	71
Data General	Edson D. deCastro	75,000	38
Food Fair	Jack M. Friedland	100,000	52
Texaco	Maurice F. Granville	661,000	61

Source: *Forbes*, May 15, 1977, p. 244.

Solution

a. *The scatter diagram.* The variable to be predicted (income) is the dependent variable and is plotted on the Y-axis. The independent variable (age) is plotted on the X-axis.

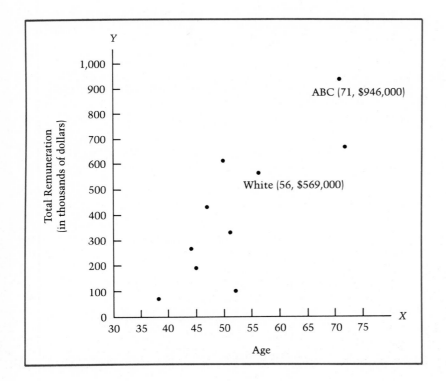

b. The *correlation coefficient* is 0.85. The essential sums and squares are in Table 14-5.

$$r = \frac{n(\Sigma XY) - (\Sigma X)(\Sigma Y)}{\sqrt{[n(\Sigma X^2) - (\Sigma X)^2][n(\Sigma Y^2) - (\Sigma Y)^2]}}$$

$$= \frac{10(235,437) - (515)(4,183)}{\sqrt{[10(27,317) - (515)^2][10(2,447,933) - (4,183)^2]}}$$

$$= 0.85$$

c. The *coefficient of determination* is $r^2 = (0.85)^2 = 0.72$. The *coefficient of nondetermination* is $1 - r^2 = 1 - 0.72 = 0.28$.

d. The *regression equation* is $Y' = a + bX = -878.934 + 25.189X$ (in thousands of dollars), found by:

$$b = \frac{n(\Sigma XY) - (\Sigma X)(\Sigma Y)}{n(\Sigma X^2) - (\Sigma X)^2}$$

$$= \frac{10(235,437) - (515)(4,183)}{10(27,317) - (515)^2}$$

$$= 25.19$$

$$a = \frac{\Sigma Y - b(\Sigma X)}{n}$$

$$= \frac{4,183 - 25.189(515)}{10}$$

$$= -878.9$$

e. *Drawing the line of regression on the scatter diagram.* Three points on the line of regression are

Age X	Income Y′	Found by
40	$128,700	Y′ = −878.9 + 25.19(40) (in thousands of dollars)
52.5	$443.575	Y′ = −878.9 + 25.19(52.5) (in thousands of dollars)
68	$834,020	Y′ = −878.9 + 25.19(68) (in thousands of dollars)

Table 14-5

Sums and Squares Needed to Calculate the Coefficient
of Correlation

Chief Executive Officer	Total Income (000) Y	Age X	XY	X²	Y²
John P. Thompson	$ 325	51	16,575	2,601	105,625
John B. Fery	444	47	20,869	2,209	197,136
Fred L. Turner	268	44	11,792	1,936	71,824
Charles G. Bluhdorn	605	50	30,250	2,500	366,025
Roy H. Holdt	569	56	31,864	3,136	323,761
J. Willard Marriott, Jr.	190	45	8,550	2,025	36,100
Leonard H. Goldenson	946	71	67,166	5,041	894,916
Edson D. deCastro	75	38	2,850	1,444	5,625
Jack M. Friedland	100	52	5,200	2,704	10,000
Maurice F. Granville	661	61	40,321	3,721	436,921
	$4,183	515	235,437	27,317	2,447,933

Plotting these points and drawing the straight line:

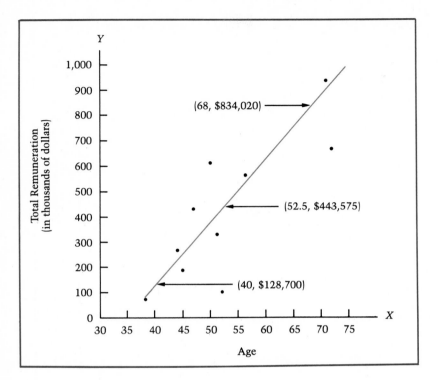

f. *Predicting the annual income of 50-year-old executives.* The best estimate is $380,566, found by $-878.934 + 25.189(50)$ (in thousands of dollars).

g. *Determining the standard error of estimate:*

$$s_{Y \cdot x} = \sqrt{\frac{\Sigma Y^2 - a(\Sigma Y) - b(\Sigma XY)}{n-2}}$$

Inserting the appropriate sums,

$$s_{Y \cdot x} = \sqrt{\frac{2,447,933 - (-878.9)(4183) - 25.189(235,437)}{10-2}}$$

$$= \sqrt{24,214.209}$$

$$= 155.6 \text{ (in thousands of dollars)}$$

h. *Setting the 95% confidence-interval estimate for the average 50-year-old executive:*

$$Y' \pm t(s_{Y \cdot x}) \sqrt{\frac{1}{n} + \frac{(X - \overline{X})^2}{\Sigma X^2 - \dfrac{(\Sigma X)^2}{n}}}$$

$$= 380.566 \pm 2.306(155.6) \sqrt{\frac{1}{10} + \frac{(50 - 51.5)^2}{27,317 - \dfrac{(515)^2}{10}}}$$

$$= 380.566 \pm 115.2$$

$$= 265.366 \text{ and } 495.766 \text{ (in thousands of dollars)}$$

Interpreting the 95% confidence-interval estimate, the probability is 0.95 (that is, the odds are 95 to 5) that for *all* 50-year-old executives the mean annual income is in the interval between $265,366 and $495,766.

A Computer Application

The calculations for the regression equation and other measures become quite tedious as the number of paired observations increases. Instead of doing them by hand, computer packages such as the Statistical Package for the Social Sciences (SPSS) can be employed. To illustrate, the income data for the ten executives given in Table 14-5 were fed into a computer:

```
THOMPSON, JOHN P.            325    51
FERY, JOHN B.                444    47
TURNER, FRED L.              268    44
BLUHDORN, CHARLES            605    50
HOLDT, ROY H.                569    56
MARRIOTT, J WILLARD          190    45
GOLDENSON, LEONARD H.        946    71
DECASTRO, EDSON D.            75    38
FRIEDLAND, JACK M.           100    52
GRANVILLE, MAURICE F         661    61
```

The mean and the standard deviation for both variables were calculated:

VARIABLE	MEAN	STANDARD DEV	CASES
SALARY	418.3000	278.5247	10
AGE	51.5000	9.3956	10

The coefficient of correlation, the coefficient of determination, and the standard error of estimate were calculated:

```
SALARY ANALYSIS

DEPENDENT VARIABLE..      SALARY

VARIABLE(S) ENTERED ON STEP NUMBER  1..      AGE


MULTIPLE R              0.84966
R SQUARE                0.72193
ADJUSTED R SQUARE       0.68717
STANDARD ERROR        155.78176
```

The regression equation is

```
         VARIABLES IN THE EQUATION

         VARIABLE              B

         AGE              25.18754
         (CONSTANT)     -878.8583
```

Note that, except for small differences due to rounding and perhaps to slightly different computing formulas, the results determined by the computer method are the same as those calculated by hand. The computer can offer substantial savings in time when large problems are encountered.

Summary

This chapter was concerned with regression analysis. The intent of regression analysis is to express the relationship between two variables through a mathematical equation called the regression equation. It has the form $Y' = a + bX$. The letter a is the Y-intercept, and b is the change in Y for each change of one unit in X. The equation can be used to predict one variable, called the dependent variable, based on a value of the independent variable. In plotting a scatter diagram, the dependent variable is scaled on the Y-axis and the independent variable on the X-axis.

A precise regression equation can be developed using the least-squares technique. This technique minimizes the sum of the squared deviations between the actual point Y and the predicted point Y' [written $\Sigma(Y - Y')^2 = $ minimum]. A measure of the accuracy of the prediction equation is obtained by determining the standard error of estimate, which can also be used to provide confidence-interval estimates. That is, statements can be made regarding the likelihood that a predicted (dependent) variable will be in a particular interval, given a value of the predictor (independent) variable. Confidence intervals can be obtained for either the mean of the Y values for any given X, or for a specific X value.

Chapter Outline

Regression Analysis

I. Objective. To develop an equation that expresses the relationship between a dependent variable Y and an independent variable X.

II. Procedure
   A. Draw a scatter diagram.

   Example:

City	Crime Rate Per 1,000 Population Y	Unemployment Rate (%) X
Chicago	52	8
New York	80	10
Denver	40	6
Dallas	60	8

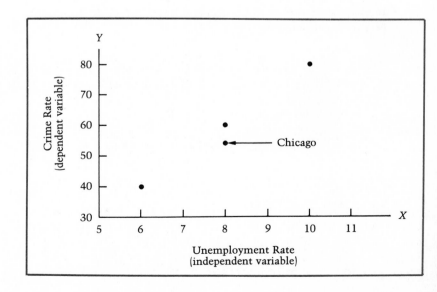

B. Determine the regression equation in the form of $Y' = a + bX$, where

    Y' is the average value of the Y variable for a selected X value.

    a is the Y-intercept. It is the value of Y when $X = 0$.

    b is the average change in Y' for each change of one (either increase or decrease) in X.

    X is any value of X that is selected.

C. Computations

$$b = \frac{n(\Sigma XY) - (\Sigma X)(\Sigma Y)}{n(\Sigma X^2) - (\Sigma X)^2}$$

$$a = \frac{\Sigma Y - b(\Sigma X)}{n}$$

D. Predictions. Suppose the equation for the previous unemployment and crime-rate problem had been computed to be $Y' = -25 + 10X$. For cities with an unemployment rate of 9.5%, the predicted crime rate per 1,000 population would be 70, found by $Y' = -25 + 10(9.5)$.

E. Measuring the error in the prediction. Calculate the standard error of estimate, $s_{Y \cdot X}$:

$$s_{Y \cdot X} = \sqrt{\frac{\Sigma(Y - Y')^2}{n - 2}}$$

or, using a more convenient formula,

$$s_{Y \cdot X} = \sqrt{\frac{\Sigma Y - a(\Sigma Y) - b(\Sigma XY)}{n - 2}}$$

F. Setting confidence-interval estimates. For the mean of the Y values for any given X value:

$$Y' \pm t(s_{Y \cdot X}) \sqrt{\frac{1}{n} + \frac{(X - \bar{X})^2}{\Sigma X^2 - \frac{(\Sigma X)^2}{n}}}$$

For a specific X value:

$$Y' \pm t(s_{Y \cdot X}) \sqrt{1 + \frac{1}{n} + \frac{(X - \bar{X})^2}{\Sigma X^2 - \dfrac{(\Sigma X)^2}{n}}}$$

Chapter Exercises

17. The number of requests for information (X) and actual enrollments (Y) for the past six years at a small university are

Requests for Information	Actual Enrollments
3,000	3,300
3,500	4,100
4,200	5,600
4,800	5,200
5,000	5,900
5,100	5,500

a. Draw a scatter diagram.
b. Determine the least-squares regression equation.
c. Calculate the standard error of estimate.
d. Compute a 90% confidence interval for the mean enrollment when the requests for information number 4,500.

18. A student of criminology, interested in predicting the age at incarceration (Y) using the age at first police contact (X), collected the following data:

Age at First Contact X	Age at Incarceration Y
11	21
17	20
13	20
12	19
15	18
10	23
12	20

a. Draw a scatter diagram.
b. Find the regression equation by the least-squares method.
c. Compute the standard error of estimate.
d. Predict the age at incarceration for an individual who at age

14 had his first contact with police. Develop a 95% confidence interval for your estimate.

19. The length of time children are exposed to a common virus (X) and the percentage of the group (Y) who contract the disease are

Weeks X	Percentage Y
2	0.1
3	0.3
3	0.5
4	0.8
4	1.2
6	1.8
7	2.5
8	3.4

a. Draw a scatter diagram of the data.
b. Compute the least-squares line of regression.
c. Find the standard error of estimate.
d. Make a prediction, with confidence coefficient 99%, about the percentage of a group who will contract the virus after a five-week exposure.

20. The length of various magazine articles (X) and an individual's respective reading times (Y) were recorded:

Length of Article (in pages)	Reading Time (in minutes)
5	13
6	15
6	15
7	18
4	10
3	8
7	18
12	25

a. Draw a scatter diagram.
b. Calculate the regression line.
c. What is the standard error of estimate?
d. Predict the average reading time for a ten-page article, and construct a 95% confidence interval about your prediction.

21. A statistics instructor wishes to study the relationship between the number of cuts (absences) students take and their final course grade. The data obtained are:

Number of Absences	Grade
1	98
2	90
4	83
3	88
5	71
2	85
4	76
3	81
6	71

a. Draw a scatter diagram and determine the regression equation.

b. Compute the standard error of estimate and develop a 90% confidence interval for all students with three absences.

c. If a particular student has three absences, what is the 90% confidence interval for his or her score?

22. The marriage-license bureau records the ages of men and women when they apply for a marriage license. You suspect that the age of the prospective wife can be predicted from her prospective husband's age. You gather the following data to verify your hypothesis:

Husband's Name	Husband's Age	Prospective Wife's Age
Cain	25	25
Behner	22	21
Freeman	26	23
Rops	37	24
Sarantou	30	25
Crosby	38	36
Gasser	24	20
Labash	46	35
Bailey	19	18
Saelzler	31	34
Smith	41	22
Baden	21	17
Snyder	21	20
Quilter	23	23
Cole	24	21
Schwamberger	20	18
Mominee	34	28
Sweney	21	20

a. Draw a scatter diagram and determine the regression equation.

b. Compute the standard error of estimate.

c. Predict the prospective wife's age for the average 40-year-old man and construct 90% confidence limits for your prediction.

23. The life expectancy and the per-capita incomes (in U.S. dollars) for selected countries follow:

Country	Life Expectancy Y	Per-Capita Income X
Algeria	50.7	$ 430
Ecuador	52.3	360
Indonesia	47.5	110
Iran	50.0	1,280
Iraq	52.1	3,010
Libya	36.9	180
Saudi Arabia	42.3	1,530
Venezuela	66.4	1,240

a. Draw a scatter diagram.
b. Compute the regression equation.
c. Compute the standard error of estimate.

24. A study was made of the relationship between the height and weight of a sample of 14 college men.

Height (in inches) X	Weight (in pounds) Y
64.7	165
67.2	116
71.6	158
65.0	153
72.0	149
71.8	181
73.0	173
64.5	120
68.3	125
72.4	163
66.3	125
72.5	173
68.0	146
67.0	139

a. Compute the regression equation.
b. Compute the standard error of the mean.
c. Determine a 95% confidence interval for the weight of all men who are six feet tall.
d. John Kuk is six feet tall. What is the 95% confidence interval on his weight?

Chapter Achievement Test

First answer all the questions. Then check your answers against those given in the Answer section of the book.

I. Multiple-Choice Questions. Select the response that best answers each of the questions that follow (5 points each).

1. A variable about which predictions or estimates are to be made is called
   a. the dependent variable
   b. the discrete variable
   c. the independent variable
   d. the correlation variable
   e. none of these are correct

2. In the regression equation, the value that gives the amount by which Y changes for every unit in X is called
   a. the coefficient of correlation
   b. the coefficient of determination
   c. the slope
   d. none of these are correct

3. In the equation $Y' = a + bX$, the letter a stands for
   a. the coefficient of correlation
   b. the coefficient of determination
   c. the slope of the regression line
   d. the Y-intercept of the regression line
   e. none of these are correct

4. What kind of linear relationship exists if Y decreases as X increases?
   a. inverse
   b. direct
   c. significant
   d. no relationship
   e. none of these are correct

Questions 5 through 7 are based on the following information: the number of chin-ups a child is able to perform, based on age, is given by the equation $Y' = -1.0 + 1.5X$.

5. A 12-year-old child could be expected to do how many chin-ups?
   a. 2.5
   b. 1.5
   c. 17.0
   d. 19.0
   e. none of these are correct

6. As age increases, the number of chin-ups is expected to
   a. increase
   b. decrease

   c. age has no effect

   d. cannot be determined

7. For each additional one year of age, the child can be expected to do how many more chin-ups?

   a. 1.0

   b. 1.5

   c. 3.0

   d. −1.0

   e. none of these are correct

8. The standard error of estimate is a measure of

   a. total variation

   b. the explained variation

   c. the unexplained variation

   d. all of the above

   e. none of these are correct

9. The width of the confidence-interval estimate for the predicted value of Y' is dependent on

   a. the standard error of estimate

   b. the size of the sample

   c. the value of X for which the prediction is being made

   d. the desired level of confidence

   e. all of the above

   f. none of these are correct

II. Computation Problem (55 points).

10. A commercial grower wants to correlate the mean height of a new variety of dahlias and the number of days since emergence above ground of a sample of seven plants.

Days Above Ground X	Height (in centimeters) Y
6	10
22	19
34	31
42	39
45	47
48	58
47	66

   a. Draw a scatter diagram.

   b. Determine the least-squares regression equation.

   c. Determine the standard error of estimate.

   d. Compute a 90% confidence interval for the height of all dahlias that have been above ground 25 days.

15

Multiple Regression and Correlation Analysis

OBJECTIVES

When you have completed this chapter, you will be able to
- Describe the relationship between one dependent variable and two or more independent variables.
- Compute a multiple standard error of estimate.
- Compute measures of multiple correlation.
- Interpret the SPSS stepwise procedure, given access to this computer package.

Introduction

Our study of correlation and regression began in Chapter 13. The relationship between two sets of interval-scaled measurements is provided by Pearson's product-moment coefficient of correlation. A coefficient of, say, 0.91 indicates that the relationship between the two variables is strong. A coefficient near 0, say 0.17, would indicate a very weak relationship. If the relationship is linear, the equation $Y' = a + bX$ is used to predict the dependent variable Y based on the independent variable X.

In this chapter we continue our study of correlation and regression. However, instead of using one independent variable to predict the dependent variable, *two or more* independent variables will be considered. The purpose of using several independent variables is to obtain a more accurate prediction or to explain more of the variation in the dependent variable. When more than one independent variable is used, we refer to the procedure as **multiple regression and correlation analysis.**

Assumptions About Multiple Regression and Correlation

Multiple regression and correlation analysis is based on four assumptions:

1. The independent variables and the dependent variable have a linear relationship.
2. The dependent variable must be continuous and at least of interval scale.
3. The variation around the regression line must be the same for all values of X. This means that Y varies the same amount when X is a low value as when it is a high value. Statisticians call this assumption **homoscedasticity.**
4. Successive observations of the dependent variable must be uncorrelated. Violation of this assumption is called **autocorrelation.** Autocorrelation often occurs when data are collected over a period of time.

Multiple regression and correlation techniques seem to work well even where one or more of these assumptions are violated, which often happens in practice. The results, however, are still adequate. The regression equation and measures of correlation provide the researcher with predictions that are better than any that could otherwise be made.

The Multiple Regression Equation

Recall from Chapter 14 that for *one* independent variable the linear regression equation has the form $Y' = a + bX$. The multiple regression case extends this equation to include other independent variables. An independent variable is the variable used to predict the dependent variable Y. For two independent variables the regression equation is

$$Y' = a + b_1X_1 + b_2X_2$$

where

> X_1 and X_2 are the two independent variables being considered.
>
> a is the point of intercept with the Y-axis.
>
> b_1 is the rate of change in Y for each unit change in X_1, with X_2 held constant. It is called a **regression coefficient.**

b_2 is the rate of change in Y for each unit change in X_2, with X_1 held constant. It is also called a regression coefficient.

For any number of independent variables (k) the general multiple regression equation is

$$Y' = a + b_1X_1 + b_2X_2 + b_3X_3 + \cdots + b_kX_k$$

Problem

The director of admissions at Mizlo University uses high-school grade point average (X_1) and IQ score (X_2) to predict first-quarter grade point averages (GPA) of entering students. The director has determined the multiple regression equation to be

$$Y' = 0.60 + 0.75X_1 + 0.001X_2$$

Predict the first-quarter GPA for a student whose high-school GPA is 3.0 and whose IQ is 100. Also, predict the first-quarter GPA for an entering student with a high-school GPA of 4.0 and an IQ of 100. What does 0.60 and the regression coefficient 0.75 in the equation indicate?

Solution

The predicted GPA for a student with a 3.0 GPA in high school and with an IQ of 100 is 2.95, found by

$$\begin{aligned} Y' &= 0.60 + 0.75X_1 + 0.001X_2 \\ &= 0.60 + (0.75)(3.0) + (0.001)(100) \\ &= 2.95 \end{aligned}$$

The predicted GPA for a student with a 4.0 GPA in high school and with an IQ of 100 is 3.70, found by $Y' = 0.60 + 0.75(4.0) + 0.001(100)$.

The 0.60 indicates that the regression equation crosses through the Y-axis at 0.60. The regression coefficient of 0.75 indicates that for each increase of 1 in the high-school grade point average, the first-quarter GPA at Mizlo will increase 0.75 *regardless of the student's IQ.*

A reminder: cover the answers in the margin.

An economist is studying a sample of households to determine the weekly amount saved by each. Three independent variables seem to hold promise as predictors of the amount saved, namely, weekly income, weekly amount spent on food, and weekly amount spent on entertainment. This equation was computed:

$$Y' = 20.0 + 0.50X_1 - 1.20X_2 - 1.05X_3$$

where

Y' is the weekly amount saved.

X_1 is the weekly income of the household.

X_2 is the weekly amount spent on food.

X_3 is the weekly amount spent on entertainment.

a. If a household had an income of $300 per week, spent $70 on housing, and $50 on entertainment, how much would be saved per week?

b. Interpret the results.

Self-Review 15-1

a. $33.50 a week.
$$Y' = 20.0 + 0.50X_1 - 1.20X_2$$
$$- 1.05X_3$$
$$= 20.0 + 0.50(300) - 1.20(70)$$
$$- 1.05(50)$$
$$= 33.5 \text{ (in dollars)}$$

b. Households with a weekly income of $300 who spend $70 on food and $50 on entertainment are expected to save $33.50 per week.

Chapter Exercises

1. A medical researcher is studying the systolic blood pressure of executives. Two independent variables are being used as predictors: income and age. The following regression equation has been computed:

$$Y' = 130 + 0.32X_1 + 0.22X_2$$

where

Y' is the systolic pressure.

X_1 is the income (in thousands of dollars).

X_2 is the age.

a. What is the expected systolic pressure reading of a 60-year-old executive earning $80,000 a year?

b. Regardless of age, how much does the systolic pressure increase for each $10,000 additional annual income?

2. A sample of women were polled to determine the degree of satisfaction with their marriages. The women were all between the ages of 35 and 50 and all had at least one child. An index of satisfaction was developed for the study. Three factors were used as predictor variables: number of children (X_1), number of years since the birth of the last child (X_2), and the mother's yearly earnings outside the home (X_3). The regression equation is

$$Y' = 30 + 1.05X_1 + 6.5X_2 + 0.005X_3$$

a. Predict the index of marriage satisfaction for a woman with three children who earns $12,000 outside the home and had her last child eight years ago.
b. Which would contribute more to marriage satisfaction, another year without a child, or an additional $5,000 in income?

3. A real estate agent is studying the selling price of homes in a certain district of the city. The following equation has been developed:

$$Y' = 24.2 + 9.80X_1 + 3.50X_2$$

where

Y' is the selling price in thousands of dollars.
X_1 is the number of bedrooms.
X_2 is the number of bathrooms.

a. What do you estimate the selling price to be for a three-bedroom, two-bath home in the area?
b. Does the addition of a bedroom add more to the selling price than the addition of a bathroom?

4. From a study of the number of points scored by teams in the National Football League, the following multiple regression equation was developed:

$$Y' = 6.0 + 0.06X_1 + 0.05X_2 - 2.75X_3$$

where

Y' is the number of points scored.
X_1 is the number of yards gained running the ball.
X_2 is the number of yards gained passing the ball.
X_3 is the number of turnovers (the number of fumbles plus the number of passes intercepted).

a. If the Cleveland Browns gained 150 yards running the ball and 310 yards passing and had three turnovers, how many points would you expect them to score?
b. How many points are lost for each turnover?

The Multiple Standard Error of Estimate

If the multiple regression equation fits the data perfectly, there is no error in the predicted value of Y. Of course, this seldom occurs in research situations. Therefore, we need a measure that can provide the size of the error in our prediction. The **multiple standard error of estimate** is just such a measure and is similar to the standard error of estimate discussed in Chapter 14.

$$s_{Y \cdot 12} = \sqrt{\frac{\Sigma(Y - Y')^2}{n - k - 1}}$$

where

$s_{Y \cdot 12}$ is the multiple standard error of estimate. The subscript $Y \cdot 12$ indicates that Y is the dependent variable. The 12 indicates that there are two independent variables labelled X_1 and X_2.

n is the number of observations in the sample.

k is the number of independent variables.

Problem

The administrator of Valley General Hospital predicted the length of hospital confinement based on two variables, the patient's age (X_1) and sex (X_2). The administrator wants to examine the discrepancy (error) between actual lengths of confinement for the sample group of patients and predicted length of their confinement. Compute the multiple standard error of estimate.

Solution

The multiple regression equation (computations not shown) is $Y' = 3.3040 + 0.1566X_1 + 2.3513X_2$, where X_1 is the patient's age and X_2 his or her sex (male $= 0$, female $= 1$). Here is the computation for the estimated length of stay Y' of the first patient in the sample, Ms. Kulla:

$$Y' = 3.3040 + 0.1566(42) + 2.3513(1)$$
$$= 12.2325 \text{ days}$$

Table 15-1 shows that Ms. Kulla's actual confinement was 12 days. Thus, the prediction error is −0.2325 days, found by $Y - Y' = 12.0 - 12.2325$. The squared error $(Y - Y')^2$ is 0.0541. The complete set of calculations needed to determine the standard error of estimate are also in Table 15-1. Computing the multiple standard error of estimate:

$$s_{Y \cdot 12} = \sqrt{\frac{\Sigma(Y - Y')^2}{n - k - 1}}$$

$$= \sqrt{\frac{11.5336}{9 - 2 - 1}}$$

$$= 1.3865 \text{ days}$$

Recall that in the section on assumptions, point 3 stated that the variation around the regression line must be the same for all values of X. The multiple standard error of estimate is a measure of this common variation. The use of the multiple standard error

Table 15-1

Length of Confinement, Age, and Sex for a Sample of Nine Patients and Calculations Necessary for the Standard Error of Estimate

Patient	Age	Sex (male = 0, female = 1) X_2	Length of Stay Y	Estimated Length of Stay Y'	$Y - Y'$	$(Y - Y')^2$
Kulla	42	1	12	12.2325	−0.2325	0.0541
Miller	36	0	10	8.9416	1.0584	1.1202
Quinn	32	1	11	10.6665	0.3335	0.1112
Reckner	29	0	6	7.8454	−1.8454	3.4055
Russell	26	0	9	7.3756	1.6244	2.6387
Siravo	24	0	6	7.0624	−1.0624	1.1287
Bancer	22	1	9	9.1005	−0.1005	0.0101
Waterfield	18	0	5	6.1228	−1.1228	1.2607
Zorn	15	0	7	5.6530	1.3470	1.8144
					−0.0003*	11.5336

* Due to rounding.

of estimate is similar to that of the standard error of estimate described in Chapter 14. It can be used to construct confidence intervals, although that feature will not be shown here.

Coefficients of Multiple Correlation and Determination

In Chapter 13 Pearson's coefficient of correlation described the strength of the relationship between two variables. Recall that its range is from -1.00 to $+1.00$. Similarly, the **coefficient of multiple correlation** measures the relationship between a dependent variable and *two or more independent* variables. It ranges from 0 to $+1.00$. Zero implies no correlation and 1.00 perfect correlation.

The **coefficient of multiple determination,** R^2, converts the variation in a variable explained by the regression equation to a percent (proportion). An $R^2 = 0.75$, for example, indicates that 75% of the variation in Y can be explained by the independent variable under consideration. Both these coefficients are elements in the computer application that follows.

A Computer Application Using the Stepwise Multiple Regression Procedure

Calculating the multiple regression equation, multiple standard error of estimate, coefficient of multiple determination, and other multiple regression and correlation measures can be tedious. Therefore, such calculations are usually done on a computer. Many college centers make available to students statistical packages such as BMDP, SPSS, SAS, or MINITAB. Check with your computer center to determine what package or packages are available to you. The illustration that follows is part of SPSS.

Applied research often deals with a large number of independent variables. Frequently, however, just a few of them will account for most of the variation in the dependent variable. How can the researcher determine which independent variables are most important in the predicting equation? One very useful technique is called the **stepwise** multiple regression procedure. First, the stepwise procedure identifies the independent variable most highly correlated with the dependent variable. Next, variables are added one step at a time. The following example will serve to illustrate.

A study has been undertaken to assess the degree of racial prejudice in a randomly selected sample of 21 people. Four independent variables have been isolated:

VAR1   is a person's mobility, measured in number of months since his or her last move.

VAR2   is the person's age in years.

VAR3   is the person's "index of social and economic status." The norm for this index is 100. Someone very poor and with practically no social status might have an index score of, say, 10. A person with a high income and high social status might have a score of 192.

VAR4   is the person's political orientation. Based on an interview, each person sampled is assigned a score from 1 (very conservative) to 10 (very liberal).

The data collected are shown in Table 15-2, which is a listing of the data by the SPSS program.

Table 15-2
Degree of Racial Prejudice (Based on Mobility, Age, Socio-economic Status and Political Orientation)

STUDY OF PREJUDICE

FILE   LIND

CASE-N	DEPVAR	VAR1	VAR2	VAR3	VAR4
1	80.	38.	23.	135.	2.
2	86.	44.	30.	170.	6.
3	48.	28.	38.	82.	4.
4	40.	35.	30.	77.	5.
5	66.	26.	19.	72.	3.
6	64.	38.	23.	122.	2.
7	94.	33.	17.	159.	3.
8	32.	42.	70.	68.	7.
9	48.	14.	51.	129.	3.
10	32.	23.	31.	67.	1.
11	88.	38.	21.	129.	6.
12	76.	24.	22.	126.	4.
13	52.	33.	33.	53.	8.
14	40.	25.	31.	75.	6.
15	76.	43.	22.	123.	4.
16	14.	22.	65.	43.	2.
17	62.	18.	26.	153.	2.
18	40.	49.	38.	79.	8.
19	72.	24.	22.	116.	3.
20	60.	34.	29.	91.	3.
21	76.	24.	52.	133.	5.

Using these sample data, how will the stepwise SPSS output appear? How is the output interpreted?

Correlation Matrix

The first output is usually a **correlation matrix.** A correlation matrix is simply a table that lists all possible correlations. These are the Pearson's product-moment correlation coefficients r computed in Chapter 13. An analysis of the following output shows, for example, that the coefficient of correlation between the dependent variable (degree of prejudice, labelled DEPVAR) and the socioeconomic index (labelled VAR3) is 0.82463. Further, Pearson's coefficient of correlation between the two independent variables (political orientation, labelled VAR4; and age, labelled VAR2) is 0.22351.

	DEPVAR	VAR1	VAR2	VAR3	VAR4
DEPVAR	1.00000	0.20747	-0.66203	0.82463	-0.05212
VAR1	0.20747	1.00000	-0.12727	0.05192	0.52251
VAR2	-0.66203	-0.12727	1.00000	-0.42308	0.22351
VAR3	0.82463	0.05192	-0.42308	1.00000	-0.19762
VAR4	-0.05212	0.52251	0.22351	-0.19762	1.00000

Step 1. The stepwise procedure begins with the selection of the independent variable that has the highest correlation with the dependent variable. It is designated STEP NUMBER 1. Note that socioeconomic status was selected first because it is most highly correlated with the dependent variable. Its value is 0.82463. Referring to the following output for STEP NUMBER 1:

1. R SQUARE, the coefficient of determination, is 0.68001—meaning that about 68% of the variation in degree of prejudice is explained by the variation in socioeconomic status.

2. The regression equation for this step is $Y' = 8.821154 + 0.4817238X_3$, where X_3 is the designation for VAR3, socioeconomic status.

3. For purposes of prediction, the standard error of estimate is 12.45475.

It should be noted that the output for this step and the following ones includes additional statistical measures that are not considered here.

* * * * * * M U L T I P L E   R E G R E S S I O N * * * * * *

DEPENDENT VARIABLE..   DEPVAR   DEGREE OF PREJUDICE

VARIABLE(S) ENTERED ON STEP NUMBER 1..   VAR3   SOCIO-ECONOMIC

```
MULTIPLE R              0.82463
R SQUARE                0.68001
ADJUSTED R SQUARE       0.66317
STANDARD ERROR         12.45475.
```

---------------- VARIABLES IN THE EQUATION ------------------

VARIABLE	B	BETA	STD ERROR B	F
VAR3	0.4817238	0.82463	0.07581	40.377
(CONSTANT)	8.821154			

* *

------------ VARIABLES NOT IN THE EQUATION --------------

VARIABLE	BETA IN	PARTIAL	TOLERANCE	F
VAR1	0.16510	0.29147	0.99730	1.671
VAR2	-0.38142	-0.61096	0.82101	10.721
VAR4	0.11535	0.19989	0.96095	0.749

* *

How does the computer decide which independent variable to bring into the equation next? When more than two independent variables are being considered, **partial correlations** are computed. A partial correlation is an index that measures the amount of the variation left *unexplained*.

At the bottom of the previous printout there are three independent variables not in the equation computed for STEP NUMBER 1. They are age (VAR2), mobility (VAR1), and political orientation (VAR4). The column headed PARTIAL lists the partial correlations for each of these three variables. The variable with the strongest partial correlation is VAR2, age. That partial correlation is −0.61096. Therefore, *age* is the independent variable that will be added next.

Step 2. Notice in the following exhibit that when age, VAR2, is entered in the stepwise procedure as an additional explanatory variable, the coefficient of multiple correlation (MULTIPLE R) is increased from 0.82463 in STEP NUMBER 1 to 0.89412 in STEP NUMBER 2. Further, by including a second independent variable,

79.946% of the variation in prejudice is explained. (Only 68.001% is explained using just one variable.) After the two variables have been entered, the line of best fit is $Y' = 36.96286 + 0.3874553X_3 - 0.5532419X_2$, where X_3 is VAR3, socioeconomic status, and X_2 is VAR2, age.

```
VARIABLE(S) ENTERED ON STEP NUMBER  2..      VAR2         AGE


MULTIPLE R                0.89412
R SQUARE                  0.79946
ADJUSTED R SQUARE         0.77717
STANDARD ERROR           10.13011


-------------------- VARIABLES IN THE EQUATION --------------------

VARIABLE             B           BETA      STD ERROR B         F

VAR3          0.3874553       0.66326       0.06805        32.417
VAR2         -0.5532419      -0.38142       0.16897        10.721
(CONSTANT)   36.96286


------------- VARIABLES NOT IN THE EQUATION --------------

VARIABLE      BETA IN      PARTIAL     TOLERANCE          F

VAR1          0.12654      0.28027      0.98380          1.449
VAR4          0.17522      0.37878      0.93711          2.848
```

Step 3. Study the printout that follows, then answer the questions in Self-Review 15-2.

```
* * * * * * * * * * * * * * * * * * * * * * * * *

DEPENDENT VARIABLE..     DEPVAR      DEGREE OF PREJUDICE

VARIABLE(S) ENTERED ON STEP NUMBER  3..      VAR4   POLITICAL-ORIENT


MULTIPLE R                0.91007
R SQUARE                  0.82823
ADJUSTED R SQUARE         0.79792
STANDARD ERROR            9.64710


-------------------- VARIABLES IN THE EQUATION --------------------

VARIABLE             B           BETA      STD ERROR B         F

VAR3          0.4003045       0.68525       0.06525        37.635
VAR2         -0.5965496      -0.41128       0.16294        13.403
VAR4          1.828634        0.17522       1.08365         2.848
(CONSTANT)   29.46891


* * * * * * * * * * * * * * * * * * * * * * * * * * * *
```

Self-Review 15-2

a. VAR1 and VAR4.

b. VAR4 will enter the equation next, because the partial of 0.37878 for VAR4 is larger than 0.28027 for VAR1.

c. 0.91007

d. 0.82823. About 82.8% of the degree of prejudice is explained by VAR2, VAR3, and VAR4.

e. $Y' = 29.46891 - 0.5965496\,X_2 + 0.4003045\,X_3 + 1.828634\,X_4$

f. It decreased from 10.13011 to 9.64710.

Referring to the printout for STEP NUMBER 2, note that two variables were not in the equation.

a. Name these two independent variables.
b. Which of the two will enter the equation in STEP NUMBER 3? Why?
c. What is the coefficient of multiple correlation?
d. What is the unadjusted coefficient of determination? Explain what it means.
e. What is the multiple regression equation?
f. What happened to the standard error of estimate between steps 2 and 3?

Summary Table.   This stepwise procedure could continue for all independent variables. Then a SUMMARY TABLE would be printed. For this problem it would look like this:

DEPENDENT VARIABLE..   DEPVAR   DEGREE OF PREJUDICE

SUMMARY TABLE

VARIABLE		MULTIPLE R	R SQUARE	RSQ CHANGE
VAR3	SOCIO-ECONOMIC	0.82463	0.68001	0.68001
VAR2	AGE	0.89412	0.79946	0.11944
VAR4	POLITICAL-ORIENT	0.91007	0.82823	0.02877
VAR1	MOBILITY	0.91072	0.82942	0.00119
(CONSTANT)				

Notice the column headed by R SQUARE, the coefficient of multiple determination. The first variable entered (socioeconomic status) explained about 68% of the variation in the degree of prejudice. The second variable entered (age) explained an additional 12% (actually, 11.944%, as shown in the column RSQ CHANGE). The independent variable, political orientation, was added in Step 3. It added less than 3% to the explained variation (actually, 2.877%). Finally, had the variable mobility been entered, it would only contribute about $\frac{1}{10}$ of 1% to R^2. Since mobility contributes very little, most researchers would probably omit it and use only the three independent variables—socioeconomic status, age, and political orientation—in making predictions about the degree of a person's racial prejudice.

Using the information in Self-Review 15-2, predict the degree of prejudice of a person with a socioeconomic index of 120, who is 35-years old and has a political-orientation score of 3.

63.2, found by
$$Y' = 29.46891 - 0.5965496(35)$$
$$+ 0.4003045(120)$$
$$+ 1.828634(3)$$

Summary

The discussion in Chapters 13 and 14 was limited to the relationship between a dependent variable and one independent variable. The techniques used in those two chapters have been extended in this chapter to include the relationship between a dependent variable and two or more independent variables.

The multiple regression equation for two independent variables has the form

$$Y' = a + b_1X_1 + b_2X_2$$

where a is the point of intercept with the Y-axis, and b_1 and b_2 are the regression coefficients associated with X_1 and X_2. This equation can be expanded to include any number of independent variables. For four independent variables it would be $Y' = a + b_1X_1 + b_2X_2 + b_3X_3 + b_4X_4$.

If the multiple regression equation fits the data perfectly, there is no error in predicting Y, the dependent variable. This rarely happens in real-world research. The multiple standard error of estimate is available to measure the error in a prediction.

The strength of the relationship between the dependent and independent variables is measured by the coefficient of multiple correlation, R^2. It ranges from 0 to 1.00, with 0 indicating no relationship and 1.00 perfect relationship. The coefficient of multiple determination R^2 is a measure of the proportion of the variation in the dependent variable Y, which is explained by the multiple regression equation.

Four assumptions should be met in order to employ these techniques, namely, (1) there should be a linear relationship between the dependent and independent variables, (2) the dependent variable must be continuous and at least of interval scale, (3) the variation around the regression line must be the same for all values of X, and (4) successive observations should not be correlated. However, most research projects cannot fully satisfy all of these assumptions. Despite such limitations, the use of these techniques is encouraged—

but caution should be used when drawing conclusions about the results of regression and correlation analysis.

Many computer programs and packages are available that can quickly handle a large number of independent variables. One of the most commonly used is the stepwise procedure in the Statistical Package for the Social Sciences (SPSS). The independent variables that are of greatest importance for prediction are brought into the multiple regression equation one at a time. The multiple regression equation, the coefficient of multiple correlation, the coefficient of multiple determination, and other statistics are printed out at each step. Using the stepwise procedure, a researcher with, say, 20 possible independent variables can determine which ones are most important and eliminate those that add little or nothing to the prediction.

Chapter Outline

Multiple Regression and Correlation Analysis

I. Multiple Regression
A. Objective. To develop a regression equation that can predict the dependent variable Y based on two (or more than two) independent variables and that can measure the error in the prediction.
B. The equation for two independent variables is

$$Y' = a + b_1 X_1 + b_2 X_2$$

II. Multiple Correlation
A. Objective. To measure the accuracy of a multiple regression equation.
B. *Coefficient of multiple correlation* measures the strength of the relationship between the dependent variable and the independent variables. It may assume a value between 0 and 1.00, with 0 indicating no relationship and 1.00 perfect relationship.
C. *Coefficient of multiple determination* measures the proportion of the variation in the dependent variable attributed to the independent variables in the equation.
D. The stepwise procedure is often used to build the multiple regression equation. From the independent variables being considered, it selects the ones that will increase R^2 most rapidly.

Chapter Exercises

5. In the course of a study of the yearly amount spent on food, a sociologist found that annual income and number of persons in the family explained 86.3% of the variation in the yearly amount spent. The regression equation was computed to be

$$Y' = 500 + 0.023X_1 + 252X_2$$

where

X_1 is the annual family income.

X_2 is the number of family members.

a. How much does the food expenditure increase with each additional family member?
b. If a family's income increases by $1,000, how much would you expect the expenditure on food to increase?
c. Predict the amount to be spent on food by a family of five with an annual income of $20,000.

6. An insurance company is analyzing the profile of its individual policyholders as it relates to the amount of life insurance coverage they carry. Three independent variables are being considered: policyholder's annual income (X_1), policyholder's number of children under 21 (X_2), and policyholder's age (X_3).

$$Y' = 2.43 + 0.32X_1 + 8.89X_2 + 0.14X_3$$

where Y' and X_1 are measured in thousands of dollars.
a. How much life insurance would a 40-year-old man with an income of $50,000 and three children be expected to carry?
b. How much does the amount of coverage increase with each child?
c. If two policyholders were both 40 years old and each had five children, but one made $40,000 per year and the other $50,000, how much difference would you expect there to be in their respective insurance coverages?

7. A study is being conducted to isolate factors that can be used to predict marital happiness for males over 50. Five factors have been considered—years of education (VAR1), number of leisure hours per week (VAR2), net worth (VAR3), age (VAR4), and number of visits to family per month (VAR5). The sample consisted of 20 males. The data on the first few cases are shown in the tabulation that follows.

Name	DEPVAR Marital Happiness	VAR1 Years of Education	VAR2 Leisure Hours	VAR3 Net Worth (000)	VAR4 Age	VAR5 Visits to Family
Greene	170	17	43	$250	62	26
Jackson	82	19	8	125	59	24
Hernandez	103	18	21	162	50	27
⋮	⋮	⋮	⋮	⋮	⋮	⋮

a. The SPSS output gave the following correlation matrix. Explain the meaning of 0.92117 in the top row.

```
           DEPVAR      VAR1       VAR2       VAR3       VAR4       VAR5

DEPVAR    1.00000    0.30824    0.92117   -0.18741    0.08057   -0.29346
VAR1      0.30824    1.00000    0.38910   -0.31947   -0.23576    0.15553
VAR2      0.92117    0.38910    1.00000   -0.19213   -0.03037   -0.22962
VAR3     -0.18741   -0.31947   -0.19213    1.00000    0.09647    0.14782
VAR4      0.08057   -0.23576   -0.03037    0.09647    1.00000   -0.24110
VAR5     -0.29346    0.15553   -0.22962    0.14782   -0.24110    1.00000
```

b. Which of the independent variables will be entered first?
c. STEP NUMBER 1 follows. What is the regression equation? What variable will enter the equation next as STEP NUMBER 2?

```
DEPENDENT VARIABLE..     DEPVAR     MARITAL HAPPINESS

VARIABLE(S) ENTERED ON STEP NUMBER  1..     VAR2       LEISURE HOURS


MULTIPLE R            0.92117
R SQUARE             0.84856
ADJUSTED R SQUARE    0.84015
STANDARD ERROR      16.87593


----------------- VARIABLES IN THE EQUATION ------------------

VARIABLE          B          BETA       STD ERROR B        F

VAR2          2.681022     0.92117       0.26696        100.858
(CONSTANT)   60.74774


------------- VARIABLES NOT IN THE EQUATION --------------

VARIABLE        BETA IN      PARTIAL     TOLERANCE          F

VAR1          -0.05914     -0.13999      0.84860         0.340
VAR3          -0.01082     -0.02728      0.96308         0.013
VAR4           0.10865      0.27907      0.99908         1.436
VAR5          -0.08651     -0.21636      0.94728         0.835
```

d. STEP NUMBER 3 follows. What variables are in the equation? What is the multiple regression equation? What is the strength of the correlation?

```
DEPENDENT VARIABLE..     DEPVAR      MARITAL HAPPINESS

VARIABLE(S) ENTERED ON STEP NUMBER 3..      VAR5  VISITS PER MONTH


MULTIPLE R              0.92939
R SQUARE                0.86377
ADJUSTED R SQUARE       0.83822
STANDARD ERROR         16.97706


--------------- VARIABLES IN THE EQUATION ---------------

VARIABLE          B           BETA      STD ERROR B        F

VAR2         2.647769       0.90975      0.27707        91.320
VAR4         0.3541514      0.09323      0.36266         0.954
VAR5        -0.2361119     -0.06209      0.37284         0.401
(CONSTANT)  44.00119


------------ VARIABLES NOT IN THE EQUATION --------------

VARIABLE      BETA IN      PARTIAL   TOLERANCE           F

VAR1        -0.01861     -0.04389     0.75796        0.029
VAR3        -0.01327     -0.03479     0.93688        0.018
```

---

e. The following is the SUMMARY TABLE assembled after all independent variables have been entered into the equation. Explain the meaning of an R SQUARE of 0.86437. What variable entered the equation last, and how much did it increase R SQUARE?

```
DEPENDENT VARIABLE..     DEPVAR      MARITAL HAPPINESS
                         SUMMARY TABLE
```

VARIABLE		MULTIPLE R	R SQUARE	RSQ CHANGE	SIMPLE R
VAR2	LEISURE HOURS	0.92117	0.84856	0.84856	0.92117
VAR4	AGE	0.92755	0.86035	0.01179	0.08057
VAR5	VISITS PER MONTH	0.92939	0.86377	0.00341	-0.29346
VAR1	YEARS OF EDUCATION	0.92953	0.86403	0.00026	0.30824
VAR3	NET WORTH	0.92972	0.86437	0.00034	-0.18741
(CONSTANT)					

If a computer package such as SPSS is available to you, you may attempt Problems 8, 9, and 10. First enter the data into the computer, then, for Problems 8 and 9, answer the questions that follow, in the sequence given.

a. Referring to the correlation matrix, determine which variable will be entered first as STEP NUMBER 1 and state your reason for answering as you have.

b. For STEP NUMBER 2, determine (1) the multiple regression equation, (2) the strength of the correlation, and (3) the coefficient of multiple determination.

8. The production supervisor in a factory that assembles electric toasters wants to develop a multiple regression equation that will predict the amount of the bonus employees can be expected to earn each week based on their productivity. He decides to use both the employees' score on a manual dexterity test and their years of experience with the company as independent variables. The following sample information is obtained:

Weekly Bonus Earned (in dollars)	Experience (in years)	Manual Dexterity Score
70	7	21
20	14	9
40	10	16
70	8	18
30	12	9
50	9	19
60	6	17
10	12	7
20	11	12
65	15	10
40	10	5
60	4	20
50	8	8
50	6	12

9. The director of evening sessions at Ludlow University is studying the enrollment in various noncredit courses. Specifically, she is interested in determining if there is a relationship between the enrollment in a particular course, the number of course information packets requested, and the number of radio advertisements about the course. A tabulation of the information gathered follows:

Enrollment	Information Packets	Radio Advertisements
53	85	3
44	80	2
30	60	1
63	150	4
60	84	2
57	108	2
60	210	3
65	200	4
54	100	4
30	40	1
35	60	2
50	90	3
50	90	2
42	85	2
36	80	1

10. Select a dependent variable which you think can be predicted by several independent variables. One suggested problem might involve predicting sales, another might be the tensile strength of a wire based on the outside diameter, and so on. Stock prices, the gross national product, employment, production, and grade point average in college are other possible dependent variables. Sports enthusiasts might be interested in predicting the American League baseball champions based on runs batted in, home runs, and so on, for each team as early as the All-Star break in July. Football, soccer, tennis, and so on are other possibilities, where past history is available.

a. Collect the data and prepare them for the computer.

b. Analyze the computer printout and write a summary of the findings. Include in your discussion such measures as the coefficient of multiple correlation, the multiple regression equation, and the standard error of estimate.*

First answer all the questions. Then check your answers against those given in the Answer section of the book.

Chapter Achievement Test

I. Multiple-Choice Questions. Select the response that best answers each of the questions that follow (5 points each).

1. A procedure for selecting the order in which independent variables enter the multiple regression equation.

a. correlation

b. hypothesis testing

* This question is adapted from *Statistical Techniques in Business and Economics*, fifth edition, by Robert D. Mason © 1982 by Richard D. Irwin, Inc. Used by permission.

   c. autocorrelation
   d. stepwise

2. The variation around the regression line must be the same
   for all values of X. This requirement is called
   a. stepwise
   b. autocorrelation
   c. homoscedasticity
   d. correlation

3. If successive observations of the dependent variable are
   themselves correlated, then this is called
   a. autocorrelation
   b. homoscedasticity
   c. regression
   d. stepwise

4. Once we have developed a regression equation, we can infer
   that by changing values of X (the independent variable)
   we *cause* values of Y to change.
   a. true
   b. false

5. To employ the multiple regression technique, the dependent
   variable must be at least of _____scale.
   a. interval
   b. ratio
   c. ordinal
   d. nominal

Questions 6 through 9 are based on the following information. A
study has been undertaken to predict annual income based on the
number of years on the job, age of employee, and years of education
beyond the eighth grade. This equation has been developed:
$Y' = 10 + 0.2X_1 + 0.1X_2 + 2.5X_3$ (in thousands of dollars).

6. The equation is called a
   a. coefficient of regression equation
   b. simple regression equation
   c. multiple regression equation
   d. dependent variable equation
   e. none of these are correct

7. How many dependent variables are there?
   a. 0
   b. 1
   c. 2
   d. 3
   e. none of these are correct

8. How many independent variables are there?
   a. 0
   b. 1
   c. 2
   d. 3
   e. none of these are correct

9. What are the numbers 0.2, 0.1, and 2.5 called?
   a. partial correlation coefficients
   b. regression coefficients
   c. coefficients of determination
   d. coefficients of standard errors
   e. none of these are correct

10. Variable 1 is number of years on the job, variable 2 is age, and variable 3 is the number of years of education beyond eighth grade. An employee has been on the job 20 years, is 50 years old, and has eight years of education beyond the eighth grade. What is the employee's predicted annual salary?
    a. $12,800
    b. $39,000
    c. $88,000
    d. $16,000
    e. none of these are correct

11. For each additional year of education beyond the eighth grade, annual income increases
    a. $2,500
    b. $1,000
    c. $1,280
    d. $39,000
    e. none of these are correct

II. Computation Problems. Questions 12 through 16 are based on the following simulated SPSS stepwise output. Give the correct answers (9 points each).

VARIABLE(S) ENTERED ON STEP NUMBER 2..   VAR4

MULTIPLE R	0.67152
R SQUARE	0.45094
ADJUSTED R SQUARE	0.43193
STANDARD ERROR	8.13885

----------------------VARIABLES IN THE EQUATION------------------

	B
VAR4	1.5634668
VAR3	−0.4690281
(CONSTANT)	12.49883

```
--------------- VARIABLES NOT IN THE EQUATION ----------------
                                          PARTIAL
VAR5                                      0.23671
VAR2                                     −0.38672
VAR1                                      0.18263
```

12. What is the multiple regression equation for this step?
13. What percent of the variation in the dependent variable is explained by the multiple regression equation?
14. Evaluate the strength of the correlation at this step.
15. Which variable will enter the equation in STEP NUMBER 3? Explain.
16. How many independent variables will be in the equation after STEP NUMBER 3?

HIGHLIGHTS
From Chapters 13, 14, and 15

In the last three chapters you were introduced to the fundamental concepts of correlation and regression analysis. We examined various measures used to describe the degree of relationship between a dependent variable and one or more independent variables. We also developed a mathematical equation that allows us to predict a dependent variable based on one or more independent variables.

Key Concepts

1. **Correlation analysis** is important because, based on various measures of association, it allows us to assess the strength of the relationship between two or more variables, and what proportion of the total variation is explained by the independent variable(s).

   a. **Pearson's product-moment coefficient of correlation (r).** This measure can answer the question, "How strong is the relationship between dependent and independent variables?" Its use assumes that the data are of an interval scale. For two variables, r can have any value from -1.00 to $+1.00$ inclusive. The *strength* of the relationship is not dependent on the *direction* of the relationship. For example, correlation coefficients of -0.09 and $+0.09$ are equal in strength—both very weak.

   For more than two variables, the coefficient of multiple correlation may assume a value from 0 to $+1.00$, inclusive.

   b. **Coefficient of determination.** This measure answers the question, "What proportion of the total variation is explained by the independent variable (or variables)?" For two variables, it is the proportion of the total variation in one variable explained by the other variable. Likewise, for more than two independent variables, it is the proportion of the total variation in the dependent variable explained by the independent variables. It varies from 0 to $+1.00$. A coefficient of 0.80, for example, indicates that 80% of the total variation in the dependent variable is explained by the independent variable or variables.

c. **Coefficient of nondetermination.** This measure answers the question, "What proportion of the total variation in the dependent variable is *not* explained by the independent variable (or variables)?" It is determined by subtracting the coefficient of determination from 1.00. Thus, it can assume any value from 0 to $+1.00$, inclusive. A coefficient of 0.20, for example, reveals that 20% of the total variation in the dependent variable is not explained by the independent variable or variables.

d. **Spearman's rank-order correlation coefficient.** This statistic is used if there is interest in determining the relationship between two sets of ordinal-level data, that is, data that you can rank from high to low, or vice versa. Spearman's coefficient can assume any value between -1.00 and $+1.00$, inclusive.

2. **Regression analysis** is important because, by using a mathematical equation, we can estimate the value of one variable based on another variable.

a. **Regression equation.** For one dependent and one independent variable, it has the form $Y' = a + bX$. For one dependent and three independent variables it is $Y' = a + b_1X_1 + b_2X_2 + b_3X_3$.

b. **Standard error of estimate** is a measure that allows us to assess the accuracy of the line of regression.

Key Terms

Scatter diagram
Pearson's product-moment correlation coefficient
Spearman's rank-order coefficient of correlation
Coefficient of determination
Dependent variable
Independent variable
Slope

Y-intercept
Least-squares method
Standard error of estimate
Confidence-interval estimates
Coefficient of partial correlation
Regression equation
Line of regression
Correlation analysis
Coefficient of nondetermination

Key Symbols

r The coefficient of correlation for two variables.

r^2 The coefficient of determination.

$1 - r^2$ The coefficient of nondetermination.

R^2 The coefficient of multiple determination.

$Y' = a + bX$ The regression equation.

$s_{Y \cdot X}$ The standard error of estimate.

r_s Spearman's rank-order correlation coefficient.

Review Problems

1. A manufacturer is interested in determining the relationship between number of years of service and number of sick leaves granted to staff members. If there is a strong relationship, the personnel director wants to predict the number of sick leaves that are likely to be granted during the next year. A sample of the personnel files reveals the following data:

Number of Years of Service X	Total Number of Sick Leaves Y
11	16
8	12
20	27
2	3
4	7
5	4
14	22

   a. Draw a scatter diagram.

   b. Calculate Pearson's coefficient of correlation, the coefficient of determination, and the coefficient of nondetermination.

   c. Determne the line of regression.

   d. Plot the line of regression on the scatter diagram.

   e. Find the number of sick leaves *per year* for a person who has 12 years of service.

   f. Interpret your findings.

2. You want to explore the relationship between the annual birth rate and suicide rate for United Nations countries. A sample of countries yields these figures:

Country	Birth Rate (per 1,000 population)	Suicide rate (per 1,000 population)
Italy	12.5	5.8
Poland	19.0	12.1
United States	15.3	12.7
Australia	15.7	11.1
Finland	13.5	25.1
East Germany	13.9	30.5
Mexico	35.3	2.1
Spain	17.2	4.1
Czechoslovakia	18.4	21.9
Singapore	17.0	11.3

a. Draw a scatter diagram using suicide rate as the dependent variable and birth rate as the independent variable.
b. Compute the regression equation.
c. For a country with a birth rate of 15.0 per 1,000 population, what is the predicted suicide rate?
d. Determine the coefficients of correlation, determination, and nondetermination.
e. Interpret your findings.

3. An educator is examining the relationship between number of hours of classroom study per day and students' ability to speak a foreign language. A large group of students are randomly assigned to one of five classrooms. The number of hours of study varies in each. After ten weeks, a proficiency examination is given and the score determined.

Hours of Study	Score
1	55
1	51
1	43
2	67
2	59
2	63
3	85
3	80
3	72
4	89
4	80
4	84
4	74

a. Compute the regression equation.
b. What score would you estimate for a student studying four hours per week?
c. Compute the coefficient of correlation.
d. Test to determine if there is correlation in the population.

4. Eight physicians scored an experimental drug on a scale of 1 to 20 with respect to its effectiveness and its aftereffect on the patient.

Physician	Effectiveness Score	Aftereffect Score
Otto	8	2
Bono	16	17
Smythe	3	9
Apple	20	10
Butz	3	2
Archer	9	16
Marchand	6	4
Sabino	13	7

Calculate Spearman's rank-order correlation coefficient to measure the strength of the relationship between the effectiveness of the experimental drug and its aftereffect.

5. Ten professional rodeo contestants have competed in last season's California Roundup and Wyoming Roundup. Their scores are

Contestant	California Roundup Scores	Wyoming Roundup Scores
Best	69	70
Asner	86	79
Jones	81	82
Belk	96	91
Sampson	72	67
Sawicki	69	62
Aey	70	70
Damon	92	98
Bardi	57	70
Aslo	84	80

A trainer is interested in evaluating the ranking of the contestants at the two roundups.
a. Rank the points scored in each rodeo (watch ties).
b. Calculate Spearman's rank-order correlation coefficient to evaluate the consistency of each contestant, and interpret your findings.

Computer Exercise

6. The salary structure of the Evergreen School System is being analyzed. Two variables are thought to relate to a teacher's salary—number of years with the system and whether or not the individual has obtained a Master's degree. A sample of 15 teachers revealed the following:

Salary (000)	Experience (in years)	Master's (yes = 1, no = 0)
$17.5	4	1
18.6	8	1
19.1	10	1
20.3	16	1
23.0	22	1
22.0	18	1
16.2	3	0
17.1	8	0
18.7	11	0
19.4	13	0
19.8	19	0
20.0	20	0

a. Compute the regression equation.
b. What percent of the variation in salary is explained by the two dependent variables?
c. Estimate the salary for a teacher with a Master's degree and ten years' experience.
d. Holding experience constant, how much does a Master's degree add to salary?

Case Analysis

The McCoy's Market Case

(Data for these two cases can be found in the first Chapter Highlights, pp. 126-130.)

You are interested in the relationship between the number of items purchased and the total amount spent by a customer.

1. Make a scatter diagram in which the total amount spent is the dependent variable and the number of items purchased is the independent variable.

2. Calculate Pearson's coefficient of correlation between these two variables. Test whether this correlation is significantly different from 0.

3. What is the numerical value of the coefficient of determination? What does it mean?

4. Determine the line of regression and plot it on the scatter diagram. Do any of the observations appear unusual?

The St. Mary's Emergency Room Case

1. Calculate Pearson's coefficient of correlation between the cost and the number of staff involved. In addition, find the correlation between cost

and patient's age. In determining the cost, which of the two appears to be the more significant factor?

2. Compute the regression equation of cost in terms of its dependence on both number of staff needed to service the patient and age of the patient. In other words, cost is the dependent variable and there are two independent variables: number of staff and patient's age. Estimate the cost for a patient who is 70 years old and requires the services of three staff members.

3. Interpret each of the numbers in your regression equation, and test whether each is significantly different from 0.

16

Analysis of Nominal-Level Data: The Chi-Square Distribution

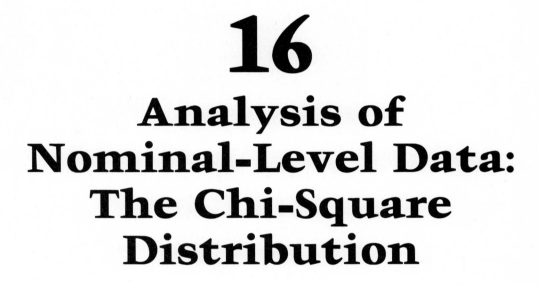

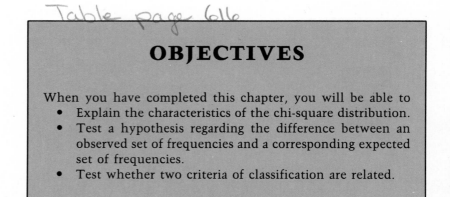

Table page 616

OBJECTIVES

When you have completed this chapter, you will be able to
- Explain the characteristics of the chi-square distribution.
- Test a hypothesis regarding the difference between an observed set of frequencies and a corresponding expected set of frequencies.
- Test whether two criteria of classification are related.

Introduction

The hypothesis tests we examined in Chapters 9, 11, and 12 dealt with interval level of measurement for problems in which the population was assumed to be normal. This chapter begins our study of statistical tests for *nominal* and *ordinal* scale of measurement or for cases in which no assumptions need be made about the shape of the parent population. Recall from Chapter 1 that the nominal level of data is the most "primitive," or the "lowest," type of measurement. Nominal level information, such as male or female, can only be classified into categories. Ordinal level of measurement assumes that one category is ranked higher than the next one. To illustrate, each staff therapist in a clinic might be rated as being superior, good, fair, or poor. It is assumed that a rating of "superior" is higher than a "good" rating, and a "good" rating is higher than a "fair" rating, and so on.

Tests involving nominal or ordinal levels of measurements are called **nonparametric** or **distribution-free** tests. In this chapter we will examine two tests that employ the **chi-square distribution** as the test statistic.

The Chi-Square Distribution

The chi-square distribution is appropriate for both nominal-level and ordinal-level data. It is designated χ^2, and because it involves squared observations *it is always positive.* As in the case of the *t* and F distributions discussed earlier, there are many chi-square distributions, each with a different shape that depends on the number of degrees of freedom. Figure 16-1 shows the shape of various distributions for selected degrees of freedom (df). Note that, as the number of degrees of freedom increases, the distribution approaches symmetry.

The number of degrees of freedom is determined by the number of categories minus one, that is, $k - 1$, and not by the size of the sample. For example, if a sample of 150 undergraduate students were classified as freshmen, sophomores, juniors, or seniors, there would be $k - 1 = 4 - 1 = 3$ degrees of freedom.

In summary, these are the three major properties of the chi-square distribution:

1. Chi-square is nonnegative—that is, it is either 0 or positive.

2. The chi-square distribution is not symmetrical. Its skewness is positive. However, as the number of degrees of freedom increases, chi-square approaches a symmetric distribution.

3. There is a family of chi-square distributions. There is a particular distribution for each degree of freedom.

Chi-Square χ^2

$$\chi^2 = \sum \frac{(f_o - f_e)^2}{f_o}$$

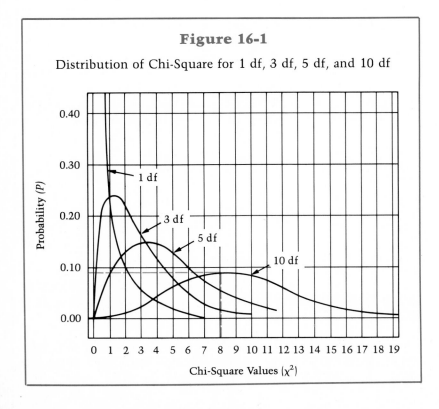

Figure 16-1

Distribution of Chi-Square for 1 df, 3 df, 5 df, and 10 df

Goodness-of-Fit Test: Equal Expected Frequencies

The chi-square distribution is used as a test statistic to determine how well an actual set of observations "fits" a theoretical, or expected, set of observations. To put it another way, the objective is to find out how well an *observed* set of frequencies abbreviated f_o, fit an *expected* set of frequencies, f_e. This test is called the **goodness-of-fit test.**

The general nature of the goodness-of-fit test can best be explained in terms of a specific application. The following problem illustrates the procedure to be followed when the expected frequencies f_e are all the same.

Problem

A psychologist is interested in determining whether mentally retarded children, given four choices of colors, prefer one over the other three. The researcher conjectures that color preference may have some effect on behavior. Eighty mentally retarded children are given a choice of a brown, orange, yellow, or green T-shirt. This is a tally of their selections:

Color	Number
Brown	25
Orange	18
Yellow	19
Green	18
Total	80

The question is: Do the children have a color preference?

Solution

The five-step hypothesis-testing procedure followed earlier is applicable.

Step 1. The null hypothesis and alternate hypothesis are stated. The null hypothesis H_0 is: there is no preference among the four colors. The alternate hypothesis H_a is: there is a preference among the four colors.

Step 2. A level of significance is selected. The 0.05 level has been chosen. Recall that this is the same as the Type I error. It means that the probability is 0.05 of incorrectly rejecting a true null hypothesis.

Step 3. An appropriate test statistic is chosen. It is chi-square and defined as

$$\chi^2 = \sum \frac{(f_o - f_e)^2}{f_e}$$

where

f_o is an observed frequency in a category (color in the problem).

f_e is an expected frequency in a particular category.

The number of degrees of freedom is $k - 1$. k is the number of categories.

Step 4. A decision rule is formulated. As in other hypothesis-testing situations, we look at the sampling distribution of the test statistic (chi-square, in this case) in order to arrive at a *critical value*. Recall that the critical value is the number that separates the region of acceptance from the region of rejection. The shape of the chi-square distribution depends on the number of degrees of freedom. In a goodness-of-fit test there are $k - 1$ degrees of freedom, where k stands for the number of categories. There are four categories (colors), therefore, $k - 1 = 4 - 1 = 3$ degress of freedom.

The critical value of chi-square is found in Appendix F. In that appendix, the number of degrees of freedom is shown in the left

Table 16-1

Critical Values of the Chi-Square Distribution

Degrees of Freedom df	Possible Values of χ^2 Right-Tail Area			
	0.10	0.05	0.02	0.01
1	2.706	3.841	5.412	6.635
2	4.605	5.991	7.824	9.210
3	6.251	7.815	9.837	11.345
4	7.779	9.488	11.668	13.277
5	9.236	11.070	13.388	15.086

margin. The various column headings, such as 0.10 and 0.05, represent the area, or probability, to the right of the particular χ^2 value. In this example, go down the left-hand column to 3 degrees of freedom (df). Then move across that row to the column headed 0.05, the level of significance selected for this problem. The critical value is 7.815. A portion of Appendix F is reproduced as Table 16-1 on page 477.

The decision rule is: do not reject the null hypothesis if the computed value of chi-square is 7.815 or less. Reject the null hypothesis and accept the alternate hypothesis if the computed value of chi-square is greater than 7.815. Shown graphically:

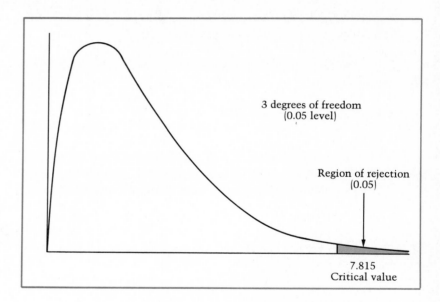

3 degrees of freedom
(0.05 level)

Region of rejection
(0.05)

7.815
Critical value

A large computed value of chi-square (over 7.815 in this problem) occurs when there is a *substantial* difference between observed and expected frequencies. If that happens, we reject the null hypothesis that there is no difference between the sets of observed and expected frequencies. If the differences between f_o and f_e are small, then the computed value of χ^2 is small (7.815 or less in this problem). Should this happen, it would indicate that the difference between the observed and expected frequencies *occurred by chance.*

Step 5.   Chi-square is computed and a decision made. Recall that out of the 80 children, 25 chose a brown T-shirt, 18 chose an orange one, 19 chose a yellow one, and 18 chose a green one. These are the *observed* frequencies denoted by f_o. Where are the *expected* frequencies? As noted in Step 1, the null hypothesis states there is no preference with respect to the four colors. If the null hypothesis is true, one would expect that one-fourth of the 80 children, or 20, would choose a brown shirt, one-fourth, or 20 would select orange, and so on. Thus, in this problem all the expected frequencies (f_e) are equal.

The observed and expected frequencies and the calculations needed for chi-square are in Table 16-2. The procedure is as follows:

1. Subtract each expected frequency from each observed frequency, that is, $f_o - f_e$.
2. Square the difference $(f_o - f_e)^2$.
3. Divide each squared difference by the expected frequency in that category.

$$\frac{(f_o - f_e)^2}{f_e}$$

4. Sum these quantities to obtain chi-square.

$$\chi^2 = \Sigma \frac{(f_o - f_e)^2}{f_e}$$

Note in the table that the computed value of χ^2 is 1.70.

The computed value of 1.70 is less than the critical value of 7.815. Therefore, we fail to reject the null hypothesis. The mentally retarded children have no significant preference with respect to the four colors.

Table 16-2

Calculations for Chi-Square

Color	f_o	f_e	$f_o - f_e$	$(f_o - f_e)^2$	$\dfrac{(f_o - f_e)^2}{f_e}$
Brown	25	20	5	25	25/20 = 1.25
Orange	18	20	−2	4	4/20 = 0.20
Yellow	19	20	−1	1	1/20 = 0.05
Green	18	20	−2	4	4/20 = 0.20
Total	80	80	0		1.70

must be equal   must be 0   computed χ^2

HW

Self-Review 16-1

a. H_0: there is no preference
H_a: there is a preference

b. Accept the null hypothesis if the computed χ^2 is equal to, or less than, 13.28. Otherwise, reject it. $(k - 1 = 5 - 1 = 4$ df.)

c.

$f_0 - f_e$	$(f_0 - f_e)^2$	$\dfrac{(f_0 - f_e)^2}{f_e}$	
$32 - 40$	$(8)^2$	$64/40 =$	1.6
$30 - 40$	$(10)^2$	$100/40 =$	2.5
$28 - 40$	$(12)^2$	$144/40 =$	3.6
$58 - 40$	$(18)^2$	$324/40 =$	8.1
$52 - 40$	$(12)^2$	$144/40 =$	$\underline{3.6}$
		$\chi^2 =$	19.40

d. Reject H_0 because 19.40 is in the region beyond 13.28, accept H_a. There is a difference with respect to taste preference.

A reminder: cover the answers in the margin.

A manufacturer of toothpaste hopes to market one or more toothpastes with these flavors: spearmint, root beer, lime, vanilla, or orange. The manufacturer wondered if consumers have a preference with respect to the flavors. Small tubes of each flavor were given to 200 consumers and each one was asked to state his or her preference. The 0.01 level of significance is to be used.

a. What are the null hypothesis and the alternate hypothesis?
b. What is the decision rule?

The preferences of the 200 in the sample are

Flavor	Number Preferring Flavor
Spearmint	32
Root beer	30
Lime	28
Vanilla	58
Orange	$\underline{52}$
	200

c. What is the calculated value of chi-square?
d. What is your decision with respect to the null hypothesis?

Chapter Exercises

1. A city has three television stations, each of which has its own evening news program from 6:00 P.M. to 6:30 P.M. The Acklin Survey Group is hired to determine if there is a preference among the viewing audience for any station. A random sample of 150 viewers revealed that 53 watched the evening news on WNAE-TV, 64 on WMWM-TV, and 33 on WRRN-TV. At the 0.05 level of significance, is there sufficient evidence to show that the three stations do not have equal shares of the evening news audience?

2. It is suspected that a particular six-sided die is "loaded," making it not a "true" die. As an experiment, this die is rolled 120 times. The results are:

Face of Die	Frequency	f_e	$(f_o - f_e)$	$(f_o - f_e)^2$	$(f_o - f_e)^2 / f_e$
⚀	15	20	−5	25	25/20 1.25
⚁	29	20	9	81	81/20 4.05
⚂	14	20	−6	36	36/20 1.8
⚃	17	20	−3	9	9/20 .45
⚄	28	20	8	64	64/20 3.2
⚅	17	20	−3	9	9/20 .45
	120	120	0		11.2

Applying the usual five steps and the 0.05 level, test the null hypothesis that there is no difference between the set of observed frequencies and the set of expected frequencies.

11.070

$K - 1 = 5$

Goodness-of-Fit Test: <u>Unequal Expected Frequencies</u>

The expected frequencies in the preceding problems were all equal. For the color-preference problem involving a sample of 80 mentally retarded children, we expected 20 to choose brown, 20 to choose orange, and so on. What if the expected frequencies were unequal? Chi-square can still be applied. The following problem will serve as an illustration.

Problem

Suppose that the following were the respective U.S. automobile manufacturers' shares of the domestic market:

Manufacturer	Number of Automobiles	Percent of Total
General Motors	22,500,000	45*
Ford	17,500,000	35
Chrysler	7,500,000	15
AMC	2,500,000	5
Total	50,000,000	100

* Computed by 22,500,000/50,000,000.

A sample of 2,000 owners of American-made automobiles in Texas revealed this ownership pattern: 858 General Motors cars, 687 Ford cars, 300 Chrysler cars, and 155 AMC cars. The question is: Does the ownership pattern in Texas differ from the national pattern? Use the 0.01 level of significance.

$K = 3$
11.345

Solution

The null and alternate hypotheses are

H_0: there is no difference in the ownership pattern in Texas and the rest of the country

H_a: the ownership pattern in Texas is different from that of the country as a whole

If the null hypothesis is true, we could expect 45% of the automobiles in the sample of 2,000 to be General Motors cars. Thus, 2,000(0.45) = 900. Likewise, 35% of the sample of 2,000 (or 700) could be expected to be Fords.

The frequencies observed and the frequencies expected are

Manufacturer	f_o	f_e	**Found by**
General Motors	858	900 ←	.45% × 2,000
Ford	687	700 ←	.35% × 2,000
Chrysler	300	300 ←	.15% × 2,000
AMC	155	100 ←	.05% × 2,000
	2,000	2,000	

must be equal

The computations for the test statistic chi-square are

$$\chi^2 = \sum \frac{(f_o - f_e)^2}{f_e}$$

$$= \frac{(858 - 900)^2}{900} + \frac{(687 - 700)^2}{700} + \frac{(300 - 300)^2}{300} + \frac{(155 - 100)^2}{100}$$

$$= 32.5$$

Note that there are four categories. Therefore, there are 3 degrees of freedom, found by $k - 1 = 4 - 1 = 3$. The critical value for the 0.01 level and 3 df is 11.345 (from Appendix F).

Since the calculated value of the test statistic χ^2 exceeds the critical value of 11.345, the null hypothesis is rejected at the 0.01 level. We conclude that the ownership pattern for American-made cars in Texas is different from that of the country as a whole.

Another illustration of the versatility of chi-square in problem solving follows.

Problem

A national study revealed that, within five years of their release from prison, 20% of criminals had not been arrested again, 38% had been arrested once, and so on. The complete observed distribution is shown in Table 16-3.

A social agency in a large city has developed what it considers a unique guidance program for former prisoners who settle there. Anxious to compare local results with the national figures in Table 16-3, the director of the social agency selected at random the files of 200 former prisoners who are or were in the guidance program. The distributions of the frequencies observed and the national experience are shown in Table 16-4.

How would the director of the social agency use chi-square to compare the local experience with the national experience? The 0.01 level of significance is to be used.

Solution

The *number* in each category resulting from local experience cannot be directly compared with the *percent* resulting from the national study. The national percentages can, however, be converted to

Table 16-3

Number of Arrests and Percent of the Total

Number of Arrests After Release from Prison	Percent of Total
0	20.0
1	38.0
2	18.0
3	10.5
4	4.5
5	3.5
6 or more	5.5
Total	100.0

Table 16-4

Comparison of Local and National Distributions

Number of Arrests After Release from Prison	Local Experience (number)	National Experience (percent of total)
0	58	20.0
1	62	38.0
2	28	18.0
3	16	10.5
4	14	4.5
5	10	3.5
6 or more	12	5.5
	200	100.0

expected frequencies (f_e). Logically, if there is no difference between the local experience and the national experience, 20% of the 200 sampled, or 40, would never be arrested after being released from prison. Likewise, 38% of the 200, or 76, would be arrested once, and so on. For a complete set of observed frequencies and expected frequencies see Table 16-5.

The null hypothesis (H_0) is: there is no difference between the local experience and the national experience. That is, any differences between the observed and the expected frequencies are due to chance (sampling).

Table 16-5

Comparison of Local and National Frequencies

Number of Arrests After Release from Prison	Local Experience f_o	National Experience f_e
0	58	40
1	62	76
2	28	36
3	16	21
4	14	9
5	10	7
6 or more	12	11
	200	200

The alternate hypothesis (H_a) is: there is a difference between the local experience and the national experience.

There are seven categories in Table 16-5. There are, therefore, $k - 1 = 7 - 1 = 6$ degrees of freedom. The critical value of χ^2 from Appendix F is 16.812. Shown graphically, the decision rule is:

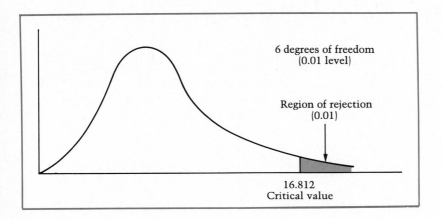

The computed value of chi-square is 17.801 (see Table 16-6).

The computed value of 17.801 is in the region beyond the critical value of 16.812. At the 0.01 level of significance, therefore, the

Table 16-6

Computation of Chi-Square

Number of Arrests After Release from Prison	Local Experience f_o	National Experience f_e	$f_o - f_e$	$(f_o - f_e)^2$	$\dfrac{(f_o - f_e)^2}{f_e}$
0	58	40	+18	324	324/40 = 8.100
1	62	76	−14	196	196/76 = 2.579
2	28	36	−8	64	64/36 = 1.778
3	16	21	−5	25	25/21 = 1.190
4	14	9	5	25	25/9 = 2.778
5	10	7	3	9	9/7 = 1.286
6 or more	12	11	1	1	1/11 = 0.090
	200	200	0		17.801

must be 0

χ^2

null hypothesis is rejected and the alternate hypothesis is accepted. The conclusion is that there is a difference between the local and national experiences. The director now has evidence that his program results in significantly fewer arrests.

Self-Review 16-2

a. Nominal.

b. H_0: there is no difference between local and national vacation destinations
H_a: there is a difference between local and national destinations

c.

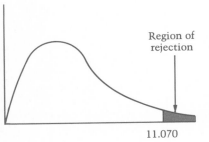

Region of rejection

11.070

d.

f_o	f_e	$\dfrac{(f_o - f_e)^2}{f_e}$
80	84	$16/84 = 0.190$
44	40	$16/40 = 0.400$
34	32	$4/32 = 0.125$
16	12	$16/12 = 1.333$
20	24	$16/24 = 0.667$
6	8	$4/8 = 0.500$
		$\chi^2 = 3.215$

Accept the null hypothesis because 3.215 is less than 11.070.

National figures revealed that 42% of the vacationers who travel outside the United States go to Europe, 20% to the Far East, 16% to South and Central America, 6% to the Middle East, 12% to the South Pacific, and 4% go elsewhere. A local travel agency wondered if its customers differ significantly from this breakdown with respect to their travel destination. A sample of the files of 200 of its customers revealed:

Destination	Number of Vacationers
Europe	80
Far East	44
South and Central America	34
Middle East	16
South Pacific	20
All others	6

a. What level of measurement is involved?
b. State both the null hypothesis and the alternate hypothesis.
c. Show the decision rule graphically. Use the 0.05 level.
d. Compute χ^2 and arrive at a decision.

Chapter Exercises

3. The occupation of parents is thought to influence the choice of occupation of their children. The distribution of career employment by categories of professional, technical, and service for a particular city is shown here:

Occupation	Percent
Professional	26%
Technical	64%
Service	10%

A sample of 360 children revealed the following about the occupations of their parents:

Occupation	Number
Professional	117
Technical	206
Service	37

Use the 0.01 level of significance to test the hypothesis that there has been no change in the distribution of occupations.

4. According to Mendel's theory of heredity, if plants with wrinkled, green seeds are crossbred with plants whose seeds are round and yellow, the offspring will follow this probability distribution:

Type of Skin	Probability
Round yellow	9/16
Wrinkled yellow	3/16
Round green	3/16
Wrinkled green	1/16

This indicates that out of every 16 seeds that germinate 9 will have round yellow skin, and so on. Use the 0.05 level of significance to check whether Mendel's theory is contradicted by an experiment that yields the results shown:

Type of Skin	Number
Round yellow	334
Wrinkled yellow	123
Round green	120
Wrinkled green	31
	608

Contingency-Table Analysis

The goodness-of-fit test discussed in the previous section was concerned only with a *single* trait, such as the color of a T-shirt. What testing procedure is followed if we are interested in the relationship between two characteristics, such as a person's adjustment to retirement and whether or not that person moved to a retirement community? By classifying adjustment as excellent, good, fair, or poor, the research data may be tallied into a table:

| | **Adjustment to Retirement** | | | |
Status	Excellent	Good	Fair	Poor
Moved to retirement community	卌 卌 卌 卌 卌 ///	卌 卌 卌 //	卌 卌 卌 卌	卌 卌 卌
Did not move to retirement community	卌 卌 卌 ///	卌 卌 卌 //	卌 卌 卌 /	卌 卌 卌 卌 卌 //

This table is called a **contingency table.** It is also popularly referred to as cross-tabulated data.

> **Contingency Table**   Frequency data from the simultaneous classification of *more than one* variable or trait of the observed item.

The chi-square statistic is employed again to determine whether two traits—home status and adjustment to retirement—are related. As before, we use observed frequencies (f_o) and expected frequencies (f_e). The observed frequencies are recorded in the preceding contingency table in the form of tallies. The expected frequencies must be computed.

Problem

A metropolitan law enforcement agency classifies crimes committed within its jurisdiction as either "violent" or "nonviolent." An investigation has been ordered to find out whether the type of crime (violent or nonviolent) depends on the age of the person who committed the crime. A sample of 100 crimes was selected at random from the police files. The results are cross-classified in the following contingency table (see Table 16-7).

Table 16-7

Classification of Crimes by Type and by Age of Person

Type of Crime	Age			Total
	Under 25	25 to 49	50 and Over	
Violent	15	30	10	55
Nonviolent	5	30	10	45
Total	20	60	20	100

Does it appear that there is any relationship between the age of a criminal and the nature of the crime? Use the 0.05 level of significance.

Solution

As usual, the initial step is to state the null hypothesis H_0 and the alternate hypothesis H_a:

H_0: there is no relationship between the type of crime committed and the age of the criminal

H_a: there is a relationship between the type of crime committed and the age of the criminal

The observed frequencies are shown in Table 16-7. In order to determine the expected frequencies, note first the number of crimes being studied (100). Referring to the marginal totals on the right side of the contingency table, observe that 55 of the 100 crimes, or 55%, were violent. If the null hypothesis is true (that there is no relationship between type of crime and age), logically it can be expected that 11 of the 20 criminals under age 25 (55%) committed a violent crime, found by 0.55×20.

Likewise, if the null hypothesis is true, one would expect 33 of the criminals in the 25 to 49 age bracket, or 55% to have committed a violent crime.

The same logic can be followed to find the expected frequencies for the nonviolent crimes. Again, from the right-hand marginal totals, 45/100, or 45%, were nonviolent crimes. Then:

$45/100 \times 20 = 9$ expected frequencies for the under 25 age group

$45/100 \times 60 = 27$ expected frequencies in the 25 to 49 age group

By now, no doubt you will have noticed that an expected frequency f_e can be computed by

$$\text{Expected frequency for a cell} = \frac{(\text{row total})(\text{column total})}{\text{grand total}}$$

For the nonviolent, 25 to 49 age cell—just computed to be 27 expected frequencies—the formula would be:

$$f_e = \frac{\text{(row total)(column total)}}{\text{grand total}}$$

$$= \frac{(45)(60)}{100}$$

$= 27$, the same answer as the one computed earlier

The observed frequencies and the corresponding expected frequencies are shown in the form of a contingency table (see Table 16-8).

A decision rule is formulated by first determining the number of degrees of freedom. For a contingency table, the number of degrees of freedom is found by multiplying number of rows minus one by number of columns minus one. The above contingency table has two rows and three columns. To find the degrees of freedom: (rows $-$ 1) (columns $-$ 1) = (2 $-$ 1) (3 $-$ 1) = 2 df. The critical value, which is the dividing point between the region of acceptance and the region of rejection, can now be located in Appendix F. Recall that the 0.05 level of significance had been selected. Go down the left margin in Appendix F to 2 df and read the critical value under the 0.05 column. It is 5.991. Thus, the decision rule is: do not reject the null hypothesis if the computed value of χ^2 is 5.991 or less. Otherwise, reject the null hypothesis.

As before, chi-square is computed by

$$\chi^2 = \Sigma \frac{(f_o - f_e)^2}{f_e}$$

Table 16-8

Observed Frequencies and Expected Frequencies

Type of Crime	Under 25 f_o	f_e	Age 25 to 49 f_o	f_e	50 and Over f_o	f_e	Total f_o	f_e
Violent	15	11	30	33	10	11	55	55
Nonviolent	5	9	30	27	10	9	45	45
Total	20	20	60	60	20	20	100	100

must be equal $\frac{(45)(60)}{100}$ must be equal

Inserting the observed frequencies f_o and the expected frequencies f_e from Table 16-8 in the formula for χ^2:

$$\chi^2 = \frac{(15-11)^2}{11} + \frac{(5-9)^2}{9} + \frac{(30-33)^2}{33} + \frac{(30-27)^2}{27}$$
$$+ \frac{(10-11)^2}{11} + \frac{(10-9)^2}{9}$$

$$= 1.45 + 1.78 + 0.27 + 0.33 + 0.09 + 0.11$$

$$= 4.03$$

The computed chi-square value of 4.03 is less than the critical value of 5.991. Therefore, we fail to reject the null hypothesis. The law-enforcement agency cannot conclude that there is any relationship between the age of the criminal and the degree of violence of the crime. To put it another way, the age of the criminal is independent of the degree of violence.

Caution. In the formula for χ^2, note that the expected frequencies f_e are in the denominator. If the expected frequency for any cell is quite small, the corresponding $(f_o - f_e)^2/f_e$ value for that cell may be disproportionately large. In turn, this one large value might be due to sampling errors resulting in a computed χ^2 greater than the critical value. The null hypothesis would, therefore, be rejected. Had the expected frequency for that one cell been larger, the null hypothesis would probably not be rejected. In order to avoid this problem, a useful rule of thumb is to look for an expected frequency of at least 5 in each cell. Whenever the number is smaller than that, you may want to consider combining several adjacent cells. The obvious alternative, of course, would be to increase the size of the sample.

In a city with a maximum-security prison, a poll of the residents has been conducted to determine if a relationship exists between marital status and a resident's stand on capital punishment.

a. What are the null hypothesis and the alternate hypothesis?

A random sample of 200 residents were asked for their opinion and for their marital status. The results were cross-classified into the following table:

Self-Review 16-3

a.
H_0: there is no relationship between a person's stand on capital punishment and marital status

H_a: there is a relationship between a person's stand on capital punishment and marital status.

b. A 2 × 2 contingency table.

c. 4, found by 2 × 2 = 4.

d. 1, found by $(2 - 1)(2 - 1)$.

e. 6.635, from Appendix F.

f. $\chi^2 = \dfrac{(100 - 90)^2}{90} + \dfrac{(20 - 30)^2}{30}$

$\qquad + \dfrac{(50 - 60)^2}{60} + \dfrac{(30 - 20)^2}{20}$

$\quad = 11.11$

g. Reject the null hypothesis. Conclusion: at the 0.01 level, there is a relationship between stand on capital punishment and marital status.

	Marital Status		
Stand on Capital Punishment	**Married**	**Not Married**	**Total**
Favor	100	20	120
Oppose	50	30	80
Total	150	50	200

b. What is this table called?

c. How many cells does the table contain?

d. How many degrees of freedom are there?

e. Using the 0.01 level of significance, what is the critical value of chi-square?

f. What is the computed value of chi-square?

g. What is your decision regarding the null hypothesis?

Chapter Exercises

5. Five hundred persons were divided into two groups to be sampled—one group consisting of people with church affiliation and the other of people without church affiliation. Each person was given a test to determine his or her degree of prejudice. The question to be explored is whether religious status and prejudice are related.

  a. State the null hypothesis and the alternate hypothesis. The results of the survey were cross-classified into the following contingency table:

	Degree of Prejudice		
Religious Status	**Highly Prejudiced**	**Somewhat Prejudiced**	**Not Prejudiced**
Church-affiliated	70	160	170
Not church-affiliated	20	50	30

  b. Using the 0.05 level, determine the critical value of chi-square.

  c. Compute χ^2 and arrive at a decision.

6. New employees hired for staff positions in a large social service agency immediately go into a six-week orientation and training program. At the end of the six weeks, each employee is rated as being either a below-average prospect, an average prospect, an above-average prospect, or an outstanding prospect. Each employee is then assigned to a supervisor. The supervisor rates the employee, after a period of two months, as being either poor, fair, good, or superior. The question to be explored is whether an employee's performance in the training program is correlated with the supervisor's ratings. A sample of the personnel files revealed:

- After two months, of the 30 employees who had been rated below-average prospects, 11 were rated poor by the supervisor, 8 were rated fair, 6 good, and 5 superior.
- Of the 40 employees who had been rated average, 9 were rated poor, 18 fair, 8 good, and 5 superior.
- Of the 60 employees who had been rated above average, 7 were rated poor, 11 fair, 28 good, and 14 superior.
- Of the 70 employees who had been rated outstanding, 5 were rated poor, 10 fair, 22 good, and 33 superior.

Using the 0.01 level of significance and the usual five-step hypothesis-testing procedure, organize the counts into a contingency table and reach a decision.

Summary

This chapter dealt with two types of hypothesis-testing problems involving the use of the chi-square distribution as the test statistic. If the objective is to determine whether a set of observed frequencies "fit" a set of expected or assumed frequencies, a goodness-of-fit test is applied. The usual hypothesis-testing procedure is followed: (1) Both the null hypothesis and the alternate hypothesis are stated. (2) A level of significance is chosen. (3) The appropriate test statistic is decided upon. It is chi-square in this type of problem. (4) A decision rule is formulated. The rule allows the researcher to reject or fail to reject the null hypothesis stated in step 1. (5) A sample is selected, the observed frequency f_o in each category is matched with the expected frequency f_e, and χ^2 is computed. Based on the decision rule and on the computed value of chi-square, a decision regarding the null hypothesis is made.

The second type of problem discussed in this chapter dealt with data cross-classified into a so-called contingency table. The null hypothesis tested is that the two criteria of classification are unrelated. Based on the marginal totals of the rows and the columns, an expected frequency f_e is computed for each cell. Chi-square is calculated and, given a predetermined level of significance, a decision is made regarding the null hypothesis.

The chi-square distribution has the following major characteristics:

1. Its value is always positive.
2. There is a family of chi-square distributions; for each number of degrees of freedom there is a particular distribution.
3. It is positively skewed but approaches a symmetric distribution as the number of degrees of freedom increases.

Chapter Outline

Analysis of Nominal-Level Data: The Chi-Square Distribution

I. Goodness-of-Fit Test

A. Objective. To determine if an observed set of frequencies differs from an expected set of frequencies.

B. Procedure

1. Set up a null hypothesis and an alternate hypothesis.
2. Select a level of significance.
3. Decide on the appropriate test statistic. It is chi-square, computed by

$$\chi^2 = \Sigma \frac{(f_o - f_e)^2}{f_e} \quad \text{with } k - 1 \text{ degrees of freedom}$$

4. Formulate a decision rule.
   Example: For five categories (cells) there are $k - 1 = 5 - 1 = 4$ degrees of freedom. Referring to Appendix F, for a 0.05 level of significance the critical value is 9.49. Shown graphically:

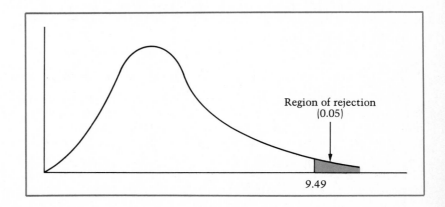

Region of rejection (0.05)

9.49

5. Take a sample, compute χ^2 and, based on the decision rule, arrive at a decision.

II. Contingency-Table Analysis

  A. Objective. To determine if two characteristics are related.

  B. Procedure. Use the five-step hypothesis-testing procedure outlined in I.

  C. A contingency table for age and first preference in music might appear as follows:

First Preference	Under 15	Age 15 - 20	22 - 30	Over 30	Total
Classical	7 ($f_e = 16.5$)	12	22	40	81
Jazz	10	82	64	8	164
Rock 'n' roll	81	45	16	9	151
Bluegrass	19	23	56	81	179
All others	7	8	10 ($f_e = 9.4$)	9	34
Total	124	170	168	147	609

$$\frac{(81)(124)}{609} \qquad \frac{(34)(168)}{609}$$

The frequency expected (f_e) for each cell is computed by:

$$\frac{(\text{row total})(\text{column total})}{\text{grand total}}$$

The degrees of freedom for the table are computed by: (rows $-$ 1) (column $-$ 1). For this table, (rows $-$ 1) (columns $-$ 1) = (5 $-$ 1) (4 $-$ 1) = 12 df.

Chapter Exercises

7. The number of military recruits from each of four regions of the United States is listed:

Region	Number
I	37
II	36
III	37
IV	50

Is there sufficient evidence at the 5% level of significance to demonstrate the existence of a difference in recruitment among the four regions?

8. A group of 385 mental patients have been classified according to parental social class, with the following results:

Social Class	Frequency
Upper	18
Upper-middle	31
Middle	46
Lower-middle	126
Lower	164

Test at the 0.05 level of significance to verify that the data are consistent with the assumption that all social classes are equally likely to be represented.

9. As part of a study of air traffic at the Express Airport, a record has been kept of the number of aircraft arrivals during any one half-hour period. The table that follows shows the number of half-hour periods in which there were 0, 1, 2, 3, or 4 or more arrivals. Test the hypothesis that the number of such arrivals follows the expected frequencies shown. Use the 0.05 significance level.

Number of Arrivals	Observed Frequency	Expected Frequency
0	19	14
1	23	27
2	28	27
3	18	18
4 or more	12	14
	100	

10. The Department of Education reports that the educational level of adults in the United States follows the distribution that follows. The Phoenix, Arizona, Chamber of Commerce wishes to compare a sample of 500 residents against this norm. What conclusion should be drawn from the results of this analysis? Use the 0.05 significance level.

Years of Education	Expected	Observed
1 - 8	100	77
9 - 12	150	198
12 - 16	200	178
over 16	50	47

11. Students claim not to like morning classes. As a test, a college statistics department offered sections of a basic statistics course at various times during the day. No limit was set on class size. The following table shows the number of students who selected each class. At the 0.01 level of significance, is there sufficient evidence to show that students have a time preference?

Time	Number of Students
Early A.M.	40
Late A.M.	62
Early P.M.	60
Late P.M.	35
Early evening	58
Late evening	45
Total	300

12. A group of executives were classified according to total income and their age. Test the hypothesis, at the 0.01 level, that age is not related to level of income.

Age	Less than $100,000	Income $100,000 to $399,999	$400,000 or more
Under 40	6	9	5
40 to 54	18	19	8
55 or older	11	12	17

13. A manufacturer of women's apparel is interested in determining if age is a factor in whether women believe they would buy a particular garment. Accordingly, the firm surveyed three age groups and asked each person to rate the garment as either excellent, average, or poor. The results follow. Test the hypothesis, at the 0.05 level, that rating is not related to age group.

	15 to 24	Age Group 25 to 39	40 to 55	Total
Excellent	40	47	46	133
Average	51	74	57	182
Poor	29	19	37	85
Total	120	140	140	400

14. Four coins are tossed 100 times; the number of heads for each trial is shown. At the 0.01 level, test the hypothesis that the coins are fair. (Hint: use the binomial distribution to determine the expected frequencies.)

Number of Heads	Observed Frequency
0	8
1	30
2	29
3	23
4	10
	100

Chapter Achievement Test

First answer all the questions. Then check your answers against those given in the Answer section of the book.

I. Multiple-Choice Questions. Select the response that best answers each of the questions that follow (4 points each).
1. For a study to determine if region of birth (four categories) is related to age (six categories), the number of degrees of freedom in the test statistic would be
   a. 10
   b. 24
   c. 15
   d. 2
   e. none of these are correct

2. In order to apply the chi-square test to data cross-classified in a contingency table:
   a. the expected frequencies cannot be determined
   b. the observed frequencies must total 100
   c. there must be at least ten expected frequencies in each cell
   d. none of these are correct

3. The chi-square distribution:
   a. is positively skewed
   b. is always greater than 0
   c. changes when the number of degrees of freedom changes
   d. all of the above are correct

4. A sample of 100 students consisted of 60 females and 40 males. The purpose of the study is to determine if the sex of the student is related to passing or failing the first test. Out of the 100 students in the sample, 70 students passed. The expected number of males passing is
   a. 70
   b. 40

c. 28

d. 12

e. none of these are correct

Questions 5 through 9 are based on the following problem. A supervisor wants to determine if unexcused absences in her department are distributed evenly throughout the workweek. A random sample of personnel files revealed:

Day	Frequency
Monday	20
Tuesday	16
Wednesday	17
Thursday	21
Friday	26
	100

5. How many degrees of freedom are there?

   a. 4

   b. 99

   c. 5

   d. 25

   e. none of these are correct

6. If the null hypothesis is true, what is the expected number of absences on Friday?

   a. 25

   b. 20

   c. 26

   d. 16

   e. none of these are correct

7. At the 0.01 significance level, the critical value of chi-square is

   a. 13.277

   b. 15.086

   c. 9.488

   d. 11.070

   e. none of these are correct

8. The computed value of χ^2 is

   a. 0

   b. 3.1

   c. 62

   d. none of these are correct

9. The decision is
   a. reject H_0
   b. do not reject H_0
   c. not enough information is available to make a determination

10. The correct statistical conclusion is
   a. the unexcused absences are evenly distributed throughout the workweek
   b. the unexcused absences are not evenly distributed throughout the workweek
   c. neither of these is a correct conclusion

II. Computation Problems (30 points each).
   11. Ninety congressmen were surveyed to determine if there is a relationship between party affiliation and their position on a proposed social-insurance-tax increase. Test the null hypothesis at the 0.05 significance level that there is no relationship. The survey results were

	Favor	Oppose	Undecided
Democrats	22	13	28
Republicans	8	7	12

   12. An analysis of last year's automotive sales revealed that for every one full-sized automobile purchased, two medium-sized, three compact, and four subcompact cars were purchased. A sample of purchases made during the last month shows car sales of 38 full-sized, 62 medium-sized, 41 compact, and 59 subcompact cars. Test at the 0.01 significance level to determine whether a change has occurred in the type of automobile purchased.

17

Nonparametric Methods: Analysis of Ranked Data

OBJECTIVES

When you have completed this chapter, you will be able to
- Calculate the sign test and describe its applications.
- Calculate the Wilcoxon signed-rank test and describe its applications.
- List and describe hypothesis tests using the Wilcoxon rank-sum test.
- Compute and describe the Kruskal-Wallis analysis of variance by ranks test.

Introduction

We began examining nonparametric hypothesis tests in Chapter 16. The chi-square goodness-of-fit test is one example of these distribution-free tests. For that particular nonparametric test, data of nominal level of measurement is the only requirement. We noted that such tests are especially useful if no assumptions—such as normality—can be made about the shape of the parent population.

All four nonparametric tests in this chapter, namely, the sign test, the Wilcoxon signed-rank test, the Wilcoxon rank-sum test, and the Kruskal-Wallis analysis of variance by ranks test, assume that the responses are at least of ordinal level of measurement—which is to say that the observations can be ordered (ranked) from lowest to highest. For example, if families were to be classified according to socioeconomic status (lower, lower-middle, middle, upper-middle, or upper class), the level of measurement would be ordinal scale.

First we will discuss the sign test and its applications.

The Sign Test

The sign test is one of the more widely used nonparametric tests. As the name implies, it is based on the sign of the difference in paired observations. For example, the pairs of measurement frequently represent "before" and "after" measurements taken on the same individual or item. Before the experiment, a person is given a rating, asked for an opinion, or asked to perform a specific task, the results of which are recorded. Then the experiment or treatment takes place, after which another observation is obtained and any changes are recorded again. An increase may be given a "+" sign and a decrease a "−" sign. The underlying assumption is that no change occurs (the null hypothesis). If the null hypothesis is true and the treatment had no effect, the distribution of positive differences will be binomial, with $p = 0.50$. Otherwise, if the treatment had some impact, the number of positive differences will be either significantly large or significantly small.

The test statistic is the number of plus signs; when the null hypothesis is true, it follows the binomial distribution, discussed in Chapter 6. Recall that the principal features of the binomial are: (1) the probability of a success remains constant from trial to trial, (2) successive trials are independent, and (3) in the binomial distribution we count the number of successes in a series of trials of fixed length. These traits are generally consistent with the sign test. The following example illustrates the details of a one-tailed test.

Problem

A special career center training program was devised with the objective of helping recently widowed homemakers reenter the job market by raising their self-confidence level. Thirteen widows were selected at random and, before the start of the training program, they were rated by a staff of psychiatrists as having either no self-confidence, some self-confidence, or great self-confidence. Then they enrolled in the training program. Upon completion, the same staff rated each widow's degree of self-confidence on the same scale as before training. Do the data demonstrate a significant increase in self-confidence? (See Table 17-1 on page 504.)

Solution

The self-confidence of each widow before and after the training program and the sign of the difference are shown in Table 17-1. A "+" sign indicates that the special training program was a *success*, that is, the widow had more confidence after the program. Note, for example, that widow 1 had no confidence in herself before the training program, but gained some confidence after the program. Therefore, the sign of the difference for her is "+." A "−" sign indicates a *failure*—that is, the widow was less self-confident about reentering the job market after the training program than before enrolling in it. If there was *no change* in self-confidence, that subject

Table 17-1
Self-Confidence of Widows Before and After
a Training Program

| | | Self-Confidence | | |
	Widow	Before	After	Sign of Difference
	1	none	some	+
	2	some	great	+
	3	none	great	+
	4	some	great	+
	5	some	none	−
	6	none	some	+
	7	some	great	+
	8	none	great	+
	9	none	some	+
	10	none	some	+
	11	great	none	−
	12	none	great	+
excluded from further analysis ←	13	none	none	0

will be omitted from further analysis. Widow 13 had no confidence in her chances in the job market either before the program or after it. Thus, she was excluded, and the size of the sample was reduced to 12.

It does seem logical that if the special training program was *not* effective in developing self-confidence among the widows, half of them would show increased self-confidence, the other half reduced self-confidence. That is, there would be an equal proportion of pluses and minuses. Thus, the *probability* of a success in the experiment would be $p = 1/2 = 0.50$. Therefore, the null hypothesis to be tested is $p \leq 0.50$. The null and alternate hypotheses are

H_0: there is no change in a widow's degree of self-confidence as a result of the training program ($p \leq 0.50$)

H_a: a widow's confidence is increased ($p > 0.50$)

The 0.10 level of significance is chosen. The appropriate test statistic is the binomial distribution, with p, the probability of a success, equal to 0.50, and n, the size of the sample, equal to 12.

The alternate hypothesis, stated earlier, is $p > 0.50$. Recall from previous hypothesis-testing problems that if a direction is predicted (such as that p is greater than 0.50), a one-tailed test is used. And, since only the tails of the sampling distribution are used in determining the critical region, only the right tail of the binominal distribution will be needed in this problem (because it is stated that p is *greater* than 0.50).

The region of rejection can be located by cumulating the probabilities starting with 12 successes, then 11 successes, and so on, until *we come as close as possible to the level of significance of 0.10 without exceeding it.* In this case, the probability of 12 successes is 0.000, the probability of 11 successes is 0.003, and so on, as shown in Table 17-2. (The probabilities are found in Appendix A for a p of 0.50 and an n of 12.)

Refer to the cumulative probabilities in the right-hand column of Table 17-2. A decision rule can now be formulated. The cumulative probability of 0.073 is as close as possible without exceeding the significance level of 0.10. Thus, if nine or more pluses (increases in self-confidence among the widows) appear in the sample of 12 usable pairs of ratings, the null hypothesis will be rejected. The rejection region is shown in Figure 17-1. The probabilities shown on the Y-axis are from Appendix A.

A count of the pluses and minuses in Table 17-1 shows that there are ten pluses (successes) and two minuses (failures). The ten successes are in the region of rejection. The null hypothesis that $p \leq 0.50$ is rejected; the alternate hypothesis that $p > 0.50$ is accepted. From a practical standpoint, the special training program was a success. Ten out of 12 widows had more self-confidence after the program than before it. In rejecting the null hypothesis, we

Table 17-2

Cumulative Probabilities for 12, 11, 10, 9, and 8 Successes

Number of Successes	Probability of Success		Cumulative Probability
.	.		.
.	.		.
.	.		.
.	.		.
8	0.121		0.194
9	0.054		0.073
10	0.016	add up	0.019
11	0.003		0.003
12	0.000		0.000

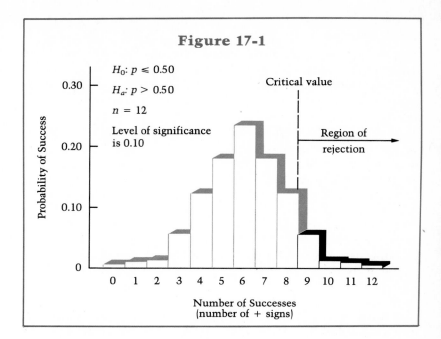

Figure 17-1

$H_0: p \leq 0.50$

$H_a: p > 0.50$

$n = 12$

Level of significance is 0.10

Critical value

Region of rejection

Probability of Success

Number of Successes
(number of + signs)

are saying that it is highly unlikely that such a large number of successes (ten successes out of 12 in the sample) *could be due to chance*. It is reasonable to assume, therefore, that this large number of successes is due to the effects of the special training program.

Self-Review 17-1

A reminder: cover the answers in the margin.

A physical-fitness instructor believes that his planned exercise program is effective in weight reduction. To investigate this claim, a random sample of seven persons enrolled in his program were selected. All were weighed before they started the program and again after completing it. These were the results:

a. H_0: there is no change as a result of the planned exercise program.

H_a: the program is successful in reducing weight

b. Refer to Appendix A for a p of 0.50 and an n of 7. The probability of seven successes is 0.008; of six successes 0.055. Adding gives 0.063, which is as close as possible to 0.10 without exceeding it. Therefore, reject the null hypothesis if there are six or seven pluses.

Student	Before	After
Dave	192	185
Rick	198	185
Jim	170	162
Kelly	130	118
Stacey	113	118
Rob	165	159
Andy	168	164

The sign test is to be applied to determine if there is evidence to demonstrate that the weight-reduction program is effective. The 0.10 significance level is used. (Note: in solving the problem, a *reduction* in a person's weight is considered a "success" and designated by a plus sign. However, an increase in a person's weight is considered a "failure" and designated by a minus sign.)

a. State the null hypothesis and the alternate hypothesis.
b. State the decision rule.
c. Should the null hypothesis be rejected? Explain.
d. Interpret your findings.

c. Yes; there are six pluses.
d. The planned exercise program is successful in reducing a person's weight.

The example that follows illustrates the use of the sign test in formulating a decision rule for a two-tailed hypothesis test.

Problem

A group of 20 male police officers are paired into ten teams that include one black and one white officer each. The "partners" are then observed and rated as to their degree of mutual trust. A high score indicates a great deal of trust. After this observation period, the ten pairs are subjected to a "wilderness experience" and then retested on their degree of mutual trust. Do the data in Table 17-3 suggest any significant changes attributable to the experience?

Table 17-3

Degree of Trust Before and After a Wilderness Experience

Partners	Rating Before Experience	Rating After Experience	Sign of Difference
A	68	32	−
B	55	55	0
C	49	55	+
D	40	75	+
E	20	50	+
F	18	25	+
G	30	23	−
H	70	49	−
I	52	62	+
J	50	41	−

Solution

Step 1. The null hypothesis is that there is no difference in the ratings before and after the wilderness experience. This will be tested against the alternate hypothesis that there is a difference. Note that this is a two-sided alternative. We are looking for *any* significant change.

Again, it seems logical that if the wilderness experience had *not* been effective in developing mutual trust among the police officers, half of the pairs would show increased trust, and the other half less trust. That is, there would be an equal proportion of pluses and minuses. Thus, if the wilderness experience is not effective, the probability of a success is $p = 0.50$. The null and alternate hypotheses are

$$H_0: p = 0.50$$
$$H_a: p \neq 0.50$$

Step 2. For this problem, the researcher selected a 20% level of significance.

Step 3. The test statistic is the binomial distribution.

Step 4. The decision rule is based on the binomial distribution. Notice that because the rating for partners B was neither an increase nor a decrease, their pair is removed from further calculations. Thus nine partner pairs remain. As in the previous example, the sampling distribution applicable for a true null hypothesis is constructed by referring to Appendix A, as shown in Table 17-4.

Step 5. Since this is a two-sided test with a 20% level of significance, each tail of the sampling distribution should contain a probability of 10%. It can be readily seen that the probability is 0.090 that there will be fewer than three "+" signs and, by symmetry, the probability is also 0.090 that there will be more than six "+" signs. Therefore, the decision rule for a significance level of approximately 20% is: reject the null hypothesis if there are fewer than three or more than six successes (plus signs). Otherwise, do not reject the null hypothesis. This decision rule is illustrated in Figure 17-2. The probabilities are based on Table 17-4.

Referring back to Table 17-3, note that there are five "+" signs. Five successes are not in the region of rejection. We conclude, therefore, that the wilderness experience did *not* change the degree of mutual trust between the partners. Essentially, we are saying that five successes could reasonably be attributed to chance.

Table 17-4

Binomial Probability Distribution
($n = 9$, $p = 0.50$)

Number of Successes	Probability of Success		Cumulative Probability
0	0.002		0.002
1	0.018	add down	0.020
2	0.070		0.090
			------- critical value
3	0.164		0.254
4	0.246		0.500
5	0.246		0.500
6	0.164		0.254
			------- critical value
7	0.070		0.090
8	0.018	add up	0.020
9	0.002		0.002

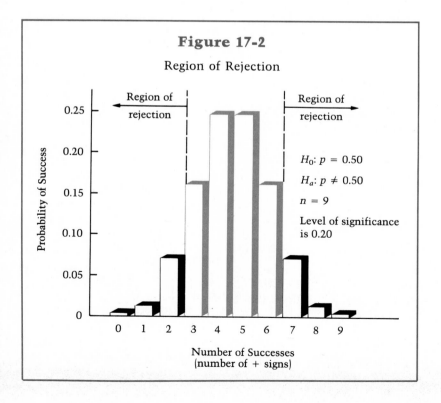

Figure 17-2

Region of Rejection

Region of rejection

Region of rejection

H_0: $p = 0.50$

H_a: $p \neq 0.50$

$n = 9$

Level of significance is 0.20

Probability of Success

Number of Successes
(number of + signs)

Self-Review 17-2

a. H_0: $p = 0.50$
H_a: $p \neq 0.50$

b. $n = 15$

c. The probability of 3 or fewer plus signs is 0.017. The probability of 12 or more plus signs is also 0.017; hence the probability of 3 or fewer, or 12 or more plus signs is 0.034. H_0 is accepted if the number of plus signs is between 4 and 11.

d. Since there are 5 plus signs, H_0 cannot be rejected.

Two groups, Calorie Counters and Fitness Fanatics, take different approaches to weight loss. Our interest is in determining if there is a difference in the percentage of weight loss experienced by the members of the two groups. A sample of 15 members from each of the two groups is selected and matched on the basis of age, current weight, and several other factors. The percentage of weight lost by each pair is recorded. Based on this information, would you conclude that the two weight loss methods differ in their success rates? Use the 0.05 significance level.

Pair	Calorie Counters	Fitness Fanatics	Sign
A	12	4	+
B	8	17	−
C	26	5	+
D	19	6	+
E	7	12	−
F	2	24	−
G	15	18	−
H	23	4	+
I	5	15	−
J	8	21	−
K	3	25	−
L	18	14	+
M	4	15	−
N	17	8	+
O	5	21	−

a. What is the null hypothesis? The alternate hypothesis?
b. What is the sample size?
c. What is the decision rule?
d. What is your decision?

Chapter Exercises

1. A random sample of eight college students and their parent of the same sex were administered a questionnaire that yields a score on their attitudes toward premarital sex. Seven of the parents were classified as more conservative than their offspring and one as less conservative. Use the sign test to determine if there is sufficient evidence at the 10% level of significance to claim that the attitudes of the two groups are significantly different.

2. The salaries paid six women and six men in similar positions are paired as shown. Use the sign test to see if there is a significant difference at the 5% level of significance.

Pair	Males	Females
1	$33,963	$28,673
2	37,271	28,697
3	26,130	18,180
4	33,016	38,265
5	60,718	57,653
6	17,957	13,214

The Sign Test with Large Samples

As discussed in Chapter 7, the binomial probability distribution closely approximates the normal probability distribution when both np and $n(1-p)$ exceed 5. Recall that when the normal probability distribution is used to approximate the binomial, the mean and standard deviation are computed by

Mean: $\mu = np$

Standard deviation: $\sigma = \sqrt{np(1-p)}$

and the standard normal deviate z is determined by

$$z = \frac{X - np}{\sqrt{np(1-p)}}$$

where

X is the number of plus signs.

p is the probability of a plus sign, which is 0.50.

n is the number of paired observations (disregarding ties).

z is the standard normal deviate.

Therefore, the normal approximation to the binomial may be used in applying the sign test when both np and $n(1-p)$ exceed 5.

Problem

Recall the earlier problem about the self-confidence level of widows before and after a training program (see Table 17-1). Did the training program increase the widows' self-confidence? As before, use the 0.10 significance level.

Solution

The calculations can be simplified if the normal approximation to the binomial can be used. Since $np = 12(0.5) = 6.0$ and $n(1 - p) = 12(0.5) = 6.0$, the qualifications for using the normal approximation to the binomial are met. The null and alternate hypotheses are as before:

$$H_0: p \leq 0.50$$
$$H_a: p > 0.50$$

The decision rule is determined by using the normal probability distribution table (Appendix C), with 0.10 in the upper tail of the curve. The decision rule is to reject H_0 if z is greater than 1.28, otherwise do not reject H_0.

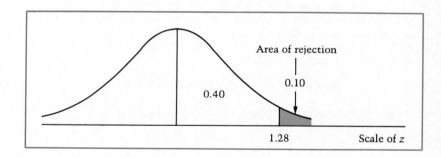

Reviewing the components in this problem:

$X = 10$ the number of widows exhibiting increased self-confidence after the training program.

$n = 12$ the size of the sample.

$p = 0.50$ the probability of a success, assuming the training program is not effective in developing confidence (H_0).

Substituting the values into the formula:

$$z = \frac{X - np}{\sqrt{np(1 - p)}}$$

$$= \frac{10 - 12(0.5)}{\sqrt{12(0.5)(1 - 0.5)}}$$

$$= 2.31$$

The computed value of 2.31 is the rejected region, therefore we reject H_0 and accept H_a. Again, we conclude that the training program does enhance the degree of the widows' self-confidence.

Refer to Self-Review 17-2, wherein 15 members from each of two health clubs were matched on the basis of similar characteristics. The percentage of weight loss for each person was determined. Six plus signs were noted and $p = 0.50$. Use the sign test, the normal approximation to the binomial, and the 0.05 significance level to determine if there is a difference in the effectiveness of the two weight-control programs.

Self-Review 17-3

Since $np = 15(0.5) = 7.5$ and $n(1 - p) = 15(0.5) = 7.5$, the normal approximation may be used.

$$H_0: p = 0.50$$
$$H_a: p \neq 0.50$$

The null hypothesis is rejected if z is less than -1.96 or greater than 1.96. Six plus signs were noted.

$$z = \frac{6 - 15(0.5)}{\sqrt{15(0.5)(0.5)}} = -0.77$$

H_0 is accepted. There is no difference in the two groups. Results are the same as those obtained earlier using the sign test.

Chapter Exercises

3. A gerontologist examined 20 elderly couples and rated both the husband and wife on a scale designed to measure various degrees of optimism. On this scale, 0 represents absolutely no optimism and 100 extreme optimism. The following data show the paired ratings. Use the sign test to determine if there is a significant difference between husband and wife with respect to optimism. Use the 0.05 significance level.

Husband	70	85	73	75	65	50	80	71	80	51	72	76	79	65	59	72	84	90	56	57
Wife	65	41	45	80	84	50	71	52	42	78	62	38	80	65	54	67	87	90	38	43
Difference	5	44	28	−5	−19	0	9	19	38	−27	10	38	−1	0	5	5	−3	0	18	14

4. Refer to Exercise 1, but assume the sample size to be 28, 20 of the parents to have been classified as being more conservative than their offspring, and 8 as less conservative. Use the 0.05 significance level and apply the sign test. Can the normal approximation be used? If so, solve the problem using the normal approximation.

The Wilcoxon Signed-Rank Test

The sign test, discussed in the previous section, considers only the sign of the difference between two paired observations. This appears to waste information because it ignores the value of the difference. The Wilcoxon signed-rank test, developed by Frank Wilcoxon in

1945, considers not only the sign of the difference, but also the magnitude (amount) of the difference between two paired observations. Thus it is considered more efficient than the sign test. The Wilcoxon signed-rank test is applied only to paired data. It is a replacement for the paired t test discussed in Chapter 11. If you suspect that the difference between the paired observations is not normally distributed, then the paired t test should be avoided and the Wilcoxon signed-rank test used in its place.

The Wilcoxon test can be calculated by following this procedure:

1. The differences between paired observations are found. As in the previous test, whenever the difference between the paired values is 0, the pair is not used.
2. These differences are ranked from highest to lowest, without regard to their signs. If ties occur, the ranks involved are averaged and each tied observation is awarded the average value.
3. The ranks with a positive difference are assigned to one column and those with a negative difference to another.
4. The sums of the positive (R^+) and negative (R^-) ranks are determined.
5. The smaller of the two sums (R^+ and R^-) is compared with the critical values found in Appendix G.

Problem

Consider again the earlier problem dealing with the wilderness experience of paired police officers. The before-and-after ratings are repeated in Table 17-5. The numerical differences between each paired rating, the ranks, and the ranks' signs have been added. Use the Wilcoxon signed-rank test to determine if there is a difference in the degree of trust before and after the police officers' wilderness experience.

Solution

The null and alternate hypotheses are

H_0: the trust ratings are the same before the wilderness experience and after it

H_a: the trust ratings are not the same before the wilderness experience and after it

If the null hypothesis is true, then it is to be expected that the total of the positive ranks is about equal to the total of the negative ranks. A large variation in the two sums is an indication that the

Table 17-5

Degree of Trust Before and After a Wilderness Experience

| | Rating | | | | | |
| | Before Experi- ence | After Experi- ence | Differ- ence | Rank | Posi- tive Ranks | Nega- tive Ranks |
Partners						
A	68	32	−36	9		9
B	55	55	0			
C	49	55	+6	1	1	
D	40	75	+35	8	8	
E	20	50	+30	7	7	
F	18	25	+7	2.5	2.5	
G	30	23	−7	2.5		2.5
H	70	49	−21	6		6
I	52	62	+10	5	5	
J	50	41	−9	4		4
					23.5	21.5

null hypothesis is false. The Wilcoxon test serves as a base for determining how large a difference between the positive and negative sums can reasonably be attributed to chance.

For a two-tailed alternative hypothesis, the smaller of R^+ and R^- is compared with the critical value. Appendix G is used to formulate the decision rule. Suppose the 0.05 significance level is to be used; to locate the critical value, find the column headed by 0.05. Next, move down that column to the row corresponding to the sample size. In this case, find the two-tailed probability 0.05 and move down to $n = 9$. (Recall that we started out with ten pairs, but because one pair—pair B—showed no difference, that pair was omitted, leaving a sample of only nine.) The critical value from Appendix G is 5. The decision rule is to reject H_0 if the smaller of R^+ or R^- is less than or equal to 5. Referring to Table 17-5, the one sum is 23.5, the other 21.5. The smaller of the two (21.5) is greater than 5, hence the null hypothesis cannot be rejected. Thus, we conclude the trust ratings are the same before the wilderness experience and after it.

The following problem illustrates a case where a one-tailed test is appropriate.

Problem

A study is to be conducted of the difference in ages between husbands and wives in first marriages. The question to be explored is:

Is there a statistically significant difference between the ages of husbands and wives?

Solution

Step 1.   The null hypothesis is that there is no significant difference between the two ages. To put it another way, the null hypothesis states that the sum of the positive ranks equals the sum of the negative ranks ($\Sigma R^+ = \Sigma R^-$). It is suspected that, in first marriages, the husbands are older than their wives; this is the *alternate hypothesis*. Because we are interested in determining if the husbands are *older* than their wives (only one direction) a one-tailed test is indicated.

Step 2.   A 10% level of significance has been chosen.

Step 3.   The Wilcoxon signed-rank test is an appropriate test to determine if there is a significant difference between two paired observations.

Step 4.   Eight married couples are randomly selected for inclusion in the study. The decision rule is formulated by referring to Appendix G. Go down the *n* column until the number in the sample (8) is located. Next, find the critical one-tail value in the column headed by 0.10; it is 8. Hence, the decision rule is: if the *smaller* of the sum of the positive or negative ranks is 8 or less, reject the null hypothesis and accept the alternate hypothesis. Otherwise, do not reject H_0.

Step 5.   The ages of both husbands and wives are listed in Table 17-6. The age differences are shown in column 4. These differences were *ranked* in column 5, disregarding the sign of the difference. There is a tie for first position: couple G had a difference of 1, couple H, −1 (see column 3). When the differences were ranked, this tie was resolved by giving each of the two couples a rank of 1.5, found by adding ranks 1 and 2, then dividing by 2. There is also a three-way tie—for third place (couples B, D, and E). This tie was resolved by adding ranks 3, 4, and 5 and dividing by 3. Each of these three couples was assigned a rank of 4.

Note that the sum of the *smaller* of the two columns of signed ranks is 5.5. This sum is less than the critical value of 8. Based on the decision rule in Step 4, the null hypothesis is rejected. The alternate hypothesis that, in first marriages, husbands are older than their wives is accepted. It is reasoned that such a large difference

Table 17-6

Comparison of Husband's and Wife's Ages

Couple	Husband's Age	Wife's Age	Difference	Rank	Signed Ranks R^+	R^-
A	24	21	3	6	6	
B	25	23	2	4	4	
C	27	19	8	8	8	
D	22	24	−2	4		4
E	20	18	2	4	4	
F	30	25	5	7	7	
G	18	17	1	1.5	1.5	
H	21	22	−1	1.5		1.5
					30.5	5.5

between the sum of the positive signed ranks (30.5) and the sum of the negative signed ranks (5.5) *could not be due to chance alone.* We conclude that, in first marriages, a preponderance of husbands are older than their wives.

Self-Review 17-4

The 1970 and 1982 burglary rates for a sample of ten major cities are shown in the following tabulation:

1970	1982
10.1	20.4
10.6	22.1
8.2	10.2
4.9	9.8
11.5	13.7
17.3	24.7
12.4	15.4
11.1	12.7
8.6	13.3
10.0	18.4

Is there sufficient evidence, at the 0.05 level of significance, to show that the rates have increased between 1970 and 1982?

a. State the null and alternate hypotheses.
b. State the decision rule.
c. State your conclusion and interpret.

a. H_0: the sum of the positive ranks (R^+) equals the sum of the negative ranks (R^-).

H_a: R^- is less than R^+.

b. Reject the null hypothesis if R^- is 10 or less.

c. $R^- = 0$, hence, we reject the null hypothesis. There has been a significant increase in the burglary rates during the period.

Chapter Exercises

5. Ten sets of identical twins were separated at the age of six. One twin in each set was educated at private schools while the other twin attended public schools. At the end of eight years, the twins' academic achievement was measured. Is there sufficient evidence, at the 0.05 level of significance, to claim that private schools generally produce a higher level of educational achievement?

Twin Pair	Public-School Educated	Private-School Educated
A	12	19
B	14	13
C	8	6
D	11	24
E	14	12
F	12	16
G	13	10
H	15	18
I	18	24
J	17	22

a. State the null and alternate hypotheses.
b. What is the decision rule, using the sign test?
c. What conclusion would you reach using the sign test?
d. If the Wilcoxon signed-rank test were used, what would the decision rule be?
e. What conclusion would be reached on the basis of the signed-rank test?

6. Apply the Wilcoxon signed-rank test to the data on weight reduction presented in Self-Review 17-2. Use the 0.05 significance level.

The Wilcoxon Rank-Sum Test

One test specifically designed to determine whether two independent samples come from the same population is the so-called **Wilcoxon rank-sum test.** It is an alternative to the two-sample t test described in Chapter 11. Recall that the t test required that the two populations be normally distributed and have equal variances. These conditions need not be met in order to apply the Wilcoxon rank-sum test.

The test is based on the sum of the ranks. The data are ranked as if the observations were part of a single sample. If the null hypothesis is true, then the ranks will be nearly evenly distributed between the two samples. That is, the low, medium, and high ranks should be about equally divided between the two samples. If the alternate hypothesis is true, one of the samples will have more of the lower ranks and thus its total will be smaller. The other sample will have the higher ranks and therefore its total will be larger. If both samples are eight, or larger, then the normal distribution may be used as the test statistic. The formula is

$$z = \frac{W - \dfrac{n_1(n_1 + n_2 + 1)}{2}}{\sqrt{\dfrac{n_1 \cdot n_2(n_1 + n_2 + 1)}{12}}}$$

where

n_1 is the number of samples from the first population.

n_2 is the number of samples from the second population.

W is the sum of ranks of the first population.

Problem

An insurance company separates its hospital claims into two categories: short-term (less than seven days), and long-term (seven days or more). It has been suggested that, among the long-term patients, male patients are confined longer than females. To investigate the validity of this suggestion, the insurance company randomly selects nine men and eight women from its files and records the length of their hospital confinement. That information is presented in Table 17-7 on page 520. Use the 0.05 significance level.

Solution

If long-term confinement is the same for men and women, then one can expect the totals of the ranks to be about the same.

Note there are two men and two women with confinements of 9 and 11 days. As before, when there is a tie, the ranks involved are averaged. That is, for the cases involving 9 days' confinement,

Table 17-7

Length of Hospital Confinement (in days) for a Sample
of Men and Women

Men	Women
13	11
15	14
9	10
18	8
11	16
20	9
24	17
22	21
25	

the ranks involved are 2 and 3. They are averaged and a rank of
2.5 assigned to each of the two cases. The confinements of 11 days
are handled similarly (see Table 17-8).

The insurance company believes that men are confined longer.
Therefore, a one-tailed test will be employed, with the region of
rejection in the upper tail. The two hypotheses are

H_0: the length of hospital confinement among extended-care patients is the same for men and women

H_a: the length of confinement is longer for men

Table 17-8

Ranked Length of Hospital Confinement

Men		Women	
Length	Rank	Length	Rank
13	7	11	5.5
15	9	14	8
9	2.5	10	4
18	12	8	1
11	5.5	16	10
20	13	9	2.5
24	16	17	11
22	25	21	14
25	17		—
	97		56

The test statistic follows the standard normal distribution. Using the 0.05 significance level, the critical value of z from Appendix C is 1.65. Therefore, the null hypothesis will be rejected if the computed value of z is greater than 1.65.

The alternate hypothesis concerns the claim that men are confined to the hospital longer. Therefore, the value of W is calculated for that group. W is 97, found by totaling the ranks of the men in the combined sample.

Computing z:

$$z = \frac{W - \dfrac{n_1(n_1 + n_2 + 1)}{2}}{\sqrt{\dfrac{n_1 \cdot n_2(n_1 + n_2 + 1)}{12}}}$$

$$z = \frac{97 - \dfrac{9(9 + 8 + 1)}{2}}{\sqrt{\dfrac{9 \cdot 8(9 + 8 + 1)}{12}}}$$

$$= 1.54$$

Since the value of z is less than 1.65, the null hypothesis is not rejected. The evidence is not sufficient to show that, among long-term patients, men are confined longer than women.

Notice that, in using the Wilcoxon signed-rank test, you may number the two populations in either order. However, once you have made a choice, W *must* be the sum of the ranks from the population identified as number 1. If, in this example, the population of women had been identified as population number 1, then the test statistic would be

$$z = \frac{56 - \dfrac{8(8 + 9 + 1)}{2}}{\sqrt{\dfrac{(8)(9)(9 + 8 + 1)}{12}}}$$

$$= \frac{56 - 72}{10.4}$$

$$= -1.54$$

This is the same z value you found earlier, except the sign has changed because the populations were renumbered.

Self-Review 17-5

a. H_0: there is no difference in the distances traveled by Dino and by Maxi

  H_a: there is a difference in the distances traveled by Dino and by Maxi

b. Do not reject H_0 if the computed z is between 1.96 and −1.96 · (From Appendix C); otherwise, reject H_0 and accept H_a.

c. $z = 2.38$
  $n_1 = 8$, the number of observations in the sample.

Dino Distance	Rank	Maxi Distance	Rank
252	4	262	9
263	10	242	2
279	14	256	5
273	13	260	8
271	12	258	7
265	11	243	3
257	6	239	1
280	15		—
Total	85		35

$W = 85$

$$z = \frac{85 - \dfrac{8(8 + 7 + 1)}{2}}{\sqrt{\dfrac{(8)(7)(8 + 7 + 1)}{12}}}$$

$$= \frac{21}{8.64} = 2.43$$

d. Reject H_0; accept H_a. There is a difference in the distances traveled by the two golf balls.

The research director for a golf ball manufacturer is interested in determining if there is any difference in the distance two of their golf balls travel. Eight of their Dino brand and seven of their Maxi brand were hit by an automatic driving machine. The results (distances in yards traveled) were as follows:

Dino 252, 263, 279, 273, 271, 265, 257, 280
Maxi 262, 242, 256, 260, 258, 243, 239

Using the 0.05 level and the Wilcoxon rank-sum test:

a. State the null and alternate hypotheses.
b. State the decision rule.
c. Compute z.
d. What is your decision?

Chapter Exercises

7. A study dealing with the social mobility of women gathered data on two groups of women. One group was composed of women who had never married. The second group consisted of married women. Each group was rated on an upward-mobility scale from 0 to 100, with a higher score denoting greater mobility. At the 0.10 level of significance, is there a difference between these two groups with respect to mobility?

Never married 42, 49, 53, 54, 55, 56, 58, 60, 65, 68, 70, 75
Married   30, 35, 42, 45, 55, 60, 60, 70

8. Samples of eight socially active students and ten socially nonactive students were randomly selected to determine if there was a difference between the two groups in a measure of social prestige. Each student was evaluated on a 15-point scale in which a low score denotes lack of prestige. The findings are recorded in the following table. Test the null hypothesis of no difference between the two populations, using a 0.10 level of significance.

Active students   11, 13, 14, 15, 8, 10, 5, 6
Nonactive students 8, 9, 10, 16, 17, 1, 7, 2, 5, 4

Analysis of Variance by Ranks (the Kruskal-Wallis Test)

As noted at the beginning of Chapter 12, in order to apply the ANOVA method to test if the means of several populations are equal, it is assumed that (1) the populations are normally distributed, and (2) the standard deviations of the populations are equal. If these assumptions cannot be met—or when the level of measurement is of ordinal scale (ranked), an analysis of variance technique developed in 1952 by W. H. Kruskal and W. A. Wallis can be applied. It is often referred to as the **Kruskal-Wallis one-way analysis of variance by ranks.**

The procedure for the Kruskal-Wallis test is to substitute the *rankings* of the items in each sample for the actual values. Also, instead of working with the means of each treatment—as in ANOVA—the sum of the ranks for each treatment is used. The test statistic, designated as *H*, is

$$H = \frac{12}{n(n+1)}\left(\frac{S_1^2}{n_1} + \frac{S_2^2}{n_2} + \cdots + \frac{S_k^2}{n_k}\right) - 3(n+1)$$

where

S_1 is the sum of the ranks for the sample designated 1, S_2 is the sum of the ranks for the sample designated 2, and so on.

n_1 is the number in the sample designated 1, n_2 is the number in the sample designated 2, and so on.

n is the combined number of observations for all samples.

k is the number of populations.

In practice, this distribution is well approximated by the chi-square distribution with $k - 1$ degrees of freedom. This approximation is accurate whenever each sample consists of at least five observations. Should any one of the samples include fewer than five observations, a special table of critical values for the analysis of variance by ranks (Kruskal-Wallis test) will have to be consulted. Many advanced texts contain such tables; for our purposes, an analysis of samples equal to, or greater than, five will suffice.

Problem

Five persons from each of three different ethnic minorities were randomly sampled and their perceptions of their own social mobility were recorded. The groups are identified as Black, Oriental, and Hispanic. Each person's response was recorded on a scale where 100 indicates extremely mobile and 0 indicates no mobility. The actual responses by group were

Black   73, 51, 55, 64, 71
Oriental 63, 74, 53, 92, 84
Hispanic 81, 93, 84, 91, 84

The hypothesis to be tested is whether the mean perception scores of the three groups are the same.

Solution

The responses were ranked without regard to ethnic background and the sum of each of the three ranks determined (see Table 17-9).

Note there was a tie for the eleventh rank. There were three responses of 84. In order to break the tie and still have the entire sample ordered from 1 through 15, the averaged rank ($10 + 11 + 12 \div 3 = 11$) was assigned to each of the three tied values.

The sums of the ranks for the samples are: $S_1 = 22$, $S_2 = 39$, and $S_3 = 59$. These values are placed into the formula and the value of H is computed.

$$H = \frac{12}{n(n+1)} \left(\frac{S_1}{n_1} + \frac{S_2}{n_2} + \frac{S_3}{n_3} \right) - 3(n+1)$$

$$= \frac{12}{15(15+1)} \left(\frac{(22)^2}{5} + \frac{(39)^2}{5} + \frac{(59)^2}{5} \right) - 3(15+1)$$

$$= 54.86 - 48 = 6.86$$

The critical value of H is located in Appendix F. Recall there are $k - 1$ degrees of freedom where k stands for the number of populations. There are three populations (Black, Oriental, and Hispanic). So, $k - 1 = 3 - 1 = 2$ degrees of freedom. The critical value for 2 degrees of freedom at the 0.05 significance level, as found in Appendix F, is 5.991. We conclude that at least one population preceives its social mobility in a manner significantly different from that of the others.

Table 17-9

Ranking of Minorities' Perceived Mobility

Black (S_1)		Oriental (S_2)		Hispanic (S_3)	
Score	Rank	Score	Rank	Score	Rank
73	7	63	4	81	9
51	1	74	8	93	15
55	3	53	2	84	11
64	5	92	14	91	13
71	6	84	11	84	11
	22		39		59

A food market is interested in studying the effect of packaging on sales of fresh produce. In order to do so, the 24 supermarkets in a nearby metropolitan area were randomly assigned to one of three groups. The first group of stores sell their produce in bulk; customers select their own produce and place them into paper bags. The second group of stores place produce in a mesh bag, visible to the customer. The third group's produce are prepackaged in kraft bags, where they cannot be seen by the customer. The amount of produce (in pounds) sold in one week by each group is shown in the table that follows.

Weekly Sales of Produce

Kraft Bags	Mesh Bags	Paper Bags
576	464	272
640	480	356
752	720	224
784	756	335
368	208	279
596	512	304
608	288	192
448	496	
	672	

Use the Kruskal-Wallis test and the 0.05 significance level to determine if the different packaging methods result in the same weekly produce sales.

Self-Review 17-6

H_0: the sale of produce is the same

H_a: the sale of produce is not the same

Weekly Sales of Produce Rankings

Kraft Rank	Mesh Rank	Paper Rank
16	12	4
19	13	9
22	21	3
24	23	8
10	2	5
17	15	7
18	6	1
11	14	
	20	
137	126	37

$$H = \frac{12}{24(24+1)} \left[\frac{(137)^2}{8} + \frac{(126)^2}{9} + \frac{(37)^2}{7} \right] - 3(24+1)$$

$$= 86.11 - 75 = 11.11$$

Since the computed value of H is greater than 5.991, H_0 is rejected and H is accepted. The conclusion is that the sale of produce is not the same under the various packaging conditions.

Chapter Exercises

9. The Orange County Sheriff's Department is considering four different car models for its fleet of patrol cars. Samples of all four makes were driven 10,000 miles each and separate records were kept of the expenditures for repairs. Use the Kruskal-Wallis test and the 0.05 significance level to determine if there is a significant difference in the amounts spent on repairs for the four makes of cars.

Model A	Model B	Model C	Model D
$220	$120	$180	$168
290	250	215	247
256	173	231	241
273	235	200	210
265	179	190	276
	249	258	193
	236		273
	147		

10. An insurance company is preparing a new advertising campaign aimed at professionals and is interested in determining the amount of term insurance carried by lawyers, physicians, veterinarians, and college professors. The amount of term insurance carried for a sample taken from each of these four categories is shown in the following table. Is there a difference in the amount of insurance carried by the four groups? Use the 0.01 significance level and the Kruskal-Wallis test.

Lawyers	Physicians	Veterinarians	College Professors
$200,000	$120,000	$ 20,000	$ 30,000
120,000	190,000	105,000	25,000
75,000	200,000	90,000	120,000
65,000	195,000	70,000	190,000
140,000	130,000	200,000	45,000
135,000		125,000	
		175,000	

Summary

This chapter presents four of the many nonparametric tests in existence: (1) the sign test, (2) the Wilcoxon signed-rank test, (3) the Wilcoxon rank-sum test, and (4) the Kruskal-Wallis analysis of variance by ranks. The four tests covered in this chapter illustrate the general nature of "distribution-free" hypothesis tests. With nonparametric tests you need not make the assumption that the populations in question are normally distributed. Usually, such tests are also easier to apply than the corresponding parametric tests and have wider applications because they require only ordinal-level measurements.

The sign test, which employs the sign of the difference of each paired observation, is used to test a hypothesis about two sets of data. The sign of the difference is either positive (+) or negative (−). The sign test actually builds on the binomial distribution because of this "success-or-failure" approach for each outcome.

The Wilcoxon signed-rank test can be used for two independent samples. It considers not only the sign, but also the magnitude of the differences. The test is based on the sum of the ranks; critical values are found in special tables.

In order to test two or more unrelated or independent samples, we have presented the Wilcoxon rank-sum test and the Kruskal-Wallis analysis of variance by ranks. Both these tests rank all the data together as if the observations had come from a common population. Then, in order to perform the actual test, the sum of the ranks for a particular group is computed. The Wilcoxon rank-sum test can be applied only to two populations at one time. When the null hypothesis is true, its test statistic is approximately normal. The Kruskal-Wallis analysis of variance by ranks, on the other hand, can apply to any number of populations. The distribution of the test statistic for the null hypothesis in this case approximates that of chi-square.

Nonparametric Methods: Analysis of Ranked Data

Chapter Outline

I. Characteristics of Nonparametric Tests
  A. Objective. To perform distribution-free hypothesis tests— that is, to compare samples that may come from nonnormal populations.
  B. They are based on either ordinal-level or interval-level measurements.
  C. The operations are usually simple, easy to apply.
  D. No assumptions about the distribution of the population, such as normality, are made.
  E. They are less powerful than corresponding tests such as z or t tests.

II. The Sign Test
  A. Objective. To make comparisons of data where the "sign" of the difference (+ and/or −) is sufficient.
  B. The test statistic follows the binomial probability distribution.
  C. Often, the normal approximation to the binomial is used:

$$z = \frac{X - np}{\sqrt{np(1 - p)}}$$

III. The Wilcoxon Signed-Rank Test
A. Objective. To make comparisons where not only the "sign" of the difference, but also the magnitude of the difference is important.
B. Used for paired samples.
C. The magnitude of the paired differences is ranked without regard to sign.
D. The test statistic is the sum of the positive ranks or the sum of the negative ranks, whichever is smaller.
E. Critical values can be found in a special table.

IV. The Wilcoxon Rank-Sum Test
A. Objective. To make comparisons between two independent samples.
B. All data are ranked together as if in a single sample.
C. The ranks for one of the samples are obtained.
D. The test statistic is approximately normal when both samples are at least of size eight. It is computed by

$$z = \frac{W - \dfrac{n_1(n_1 + n_2 + 1)}{2}}{\sqrt{\dfrac{n_1 n_2 (n_1 + n_2 + 1)}{12}}}$$

V. The Kruskal-Wallis Analysis of Variance by Ranks
A. Objective. To compare two or more samples for equality.
B. Populations are independent.
C. All data are ranked as if they were a single sample.
D. The test statistic is computed by

$$H = \frac{12}{n(n + 1)} \left(\frac{S_1^2}{n_1} + \frac{S_2^2}{n_2} + \cdots + \frac{S_k^2}{n_k} \right) - 3(n + 1)$$

Under the null hypothesis, this statistic is distributed approximately as the chi-square distribution with $k - 1$ degrees of freedom.

Chapter Exercises

11. The work-methods department of an electronics manufacturer has suggested that soft background music might increase productivity. A stereo system has been set up, and the productivity

of a random sample of eight workers is checked against their productivity before music was piped in. Can it be concluded that the workers are more productive with the background music on? Use the 0.05 significance level and the sign test.

Worker	With Music	Without Music
Smith	34	32
Clark	34	37
Craig	41	28
Berry	34	33
Roberts	31	28
Nardelli	36	33
Palmer	29	27
Tang	39	38

12. The number of traffic citations given by two policemen are to be compared. The following is a 25-day sample of citations. At the 0.01 level of significance, can it be concluded that policeman A gives more citations than B? Use the sign test.

Date	A	B	Date	A	B	Date	A	B	Date	A	B	Date	A	B
5 - 1	10	7	5 - 8	7	3	5 - 15	11	8	5 - 22	6	5	5 - 29	10	8
5 - 2	11	5	5 - 9	4	2	5 - 16	12	10	5 - 23	7	9	5 - 30	9	7
5 - 3	7	4	5 - 10	5	8	5 - 17	11	7	5 - 24	9	3	5 - 31	8	4
5 - 4	8	8	5 - 11	8	1	5 - 18	7	2	5 - 25	12	8	6 - 1	5	2
5 - 5	8	9	5 - 12	7	2	5 - 19	5	6	5 - 26	10	7	6 - 2	6	7

13. A study conducted by the National Weather Service on the change in temperature following a thunderstorm found that after 40 storms the temperature increased on 5 occasions, stayed the same on 10 occasions, and decreased on 25 occasions. Is this sufficient evidence, at the 0.05 significance level, to conclude that there is a significant decrease in temperature after a thunderstorm?

14. Two major-league-baseball fans were comparing the relative strengths of the National and American Leagues. One contended that the National League home-run leader had more home runs than the American League leader. A 17-year record of home runs hit by the leader in each of the two leagues follows.
   a. Use the sign test to determine whether the difference found is significant at the 0.05 level.

American	32	36	32	32	37	33	44	49	44	44	49	32	49	45	48	61	40
National	38	38	36	44	40	48	45	45	36	39	44	52	47	44	49	46	41

b. Compare the results using the Wilcoxon signed-rank test on the same data.

15. Seventeen randomly selected married men with incomes greater than $30,000 have been selected and have been asked whether they consider their marriages to be satisfactory. Each man has also been asked to report the length of his marriage. The results of the study follow. At the 0.01 significance level, use the Wilcoxon signed-rank test to determine if there is a difference between the two groups.

Length of Marriage

Satisfied	Not Satisfied
15	8
3	5
9	9
12	12
23	10
29	13
14	4
6	7
16	

16. Two judges normally preside over the family court where all divorce hearings are held. Ms. Kerger, a young lawyer new to the area, represents the husband in an upcoming case. She gathers the following data on the monthly alimony payments (in dollars per month) recently awarded by the two judges. At the 0.01 significance level, use the Wilcoxon rank-sum test to determine if there is a difference in the alimony settlements awarded by the two judges. The samples for each group have been arranged from low to high.

Judge Cain	Judge Stevens
80	120
160	140
220	192
520	204
580	252
640	440
840	500
920	560
1,200	
1,360	
2,000	

17. A study has been made of the murder rates (per 1,000 population) for a sample of medium-sized cities in the Northeast,

South, and Far West. Use the 0.05 significance level to determine if there is a difference in the murder rates of different geographical areas. The sample for each region has been arranged from low to high.

Northeast	South	Far West
2.3	1.6	1.7
4.5	1.9	3.7
6.7	3.0	3.8
7.8	6.5	4.3
9.5	7.2	5.9
12.7	11.6	6.2
13.1	13.1	7.9
	14.5	8.4
	15.1	

First answer all the questions. Then check your answers against those given in the Answer section of the book.

Chapter Achievement Test

I. Multiple-Choice Questions. Select the response that best answers each of the questions that follow (4 points each).

1. Which of the following is *not* a nonparametric test
   a. sign test
   b. *t* test
   c. Wilcoxon signed-rank test
   d. Wilcoxon rank-sum test
   e. all are nonparametric tests

2. The sign test should be employed instead of the paired *t* test when
   a. the sample is greater than 30
   b. the population from which the sample is drawn is known to be nonnormal
   c. both np and $n(1 - p)$ are less than 5
   d. all of the above are correct
   e. none of the above are correct

3. When the sign test is employed, the null hypothesis is that the probability of a positive difference in a paired observation is 0.5 and the sampling distribution for testing follows a binomial distribution.
   a. true
   b. false

4. In comparing the sign test and the Wilcoxon signed-rank test, the Wilcoxon signed-rank test
   a. considers the magnitude of the difference between paired observations

b. considers only the sign of the difference
c. is actually the same as the sign test
d. assumes a normal distribution
e. must always have 30 or more observations

5. In comparing the sign test and the Wilcoxon signed-rank test to the Wilcoxon ranked-sum test, the Wilcoxon ranked-sum test assumes
   a. a normal distribution
   b. a positive difference
   c. dependent samples
   d. independent samples
   e. all of the above are correct

6. If the Wilcoxon signed-rank test is being employed on a set of data with a sample size of 15 and if a two-tailed test with a 0.01 significance level is desired, then the null hypothesis will be rejected if
   a. R^+ is greater than 19
   b. R^- is greater than 19
   c. The smaller of R^+ and R^- is less than or equal to 15
   d. The smaller of R^+ and R^- is less than or equal to 25
   e. none of the above are correct

7. In the Wilcoxon signed-rank test, if the absolute differences between paired observations are equal (tied) and greater than 0, the suggested procedure is
   a. discard the tied observations
   b. use the smaller value of the corresponding ranks
   c. use the larger of the corresponding ranks
   d. assign the average value of the corresponding ranks
   e. none of the above are correct

8. In the Wilcoxon rank-sum test, the differences between pairs of scores are ranked.
   a. true
   b. false

9. A principal advantage of the nonparametric tests is the fact that the underlying assumptions are often less restrictive.
   a. true
   b. false

10. The region of rejection for the Wilcoxon rank-sum test can be
    a. in the upper tail only
    b. in the lower tail only
    c. either the lower tail or the upper tail or both tails
    d. none of the above are correct

II. Computation Problems (20 points each).

11. A medical researcher wishes to determine if the lung capacity of nonsmokers is greater than that of persons who smoke. A random sample of 15 matched pairs has been selected and tested. The lung capacity of the nonsmokers has been subtracted from that of the smoking partners. A total of 13 positive differences were obtained (2 were negative). Is this sufficient evidence to show that the lung capacity of nonsmokers is greater? Use a 0.01 level of significance and the sign test.

12. Several married couples were asked to estimate the age at which their partner achieved emotional maturity. Use the Wilcoxon signed-rank test at the 1% level of significance to test for a difference between the perceptions of husbands and wives.

Husbands	Wives
43	31
28	32
36	34
39	39
43	35
28	22
31	28
23	24
29	34
39	39
26	19
34	29
35	35

13. A study compared the achievement scores of a group of randomly selected inner-city sixth-graders to the scores of a similar group of students from a rural area within the same school district. Use a 0.05 significance level and the Wilcoxon rank-sum test to determine if a difference exists between the two groups. Note that the data have been ordered.

Inter-city students	95, 100, 100, 106, 108, 115, 120, 125, 130
Rural students	108, 114, 118, 122, 124, 126, 130, 136, 142, 150

HIGHLIGHTS
From Chapters 16 and 17

These last two chapters presented some of the basic concepts of hypothesis testing when nominal and ordinal level of measurement are involved. We discussed several tests including the chi-square goodness-of-fit test, the sign test, Wilcoxon's signed-rank test, Wilcoxon's rank-sum test, and the Kruskal-Wallis analysis of variance by ranks.

Key Concepts

1. **Nominal level of measurement.** The nominal level of measurement is considered the "lowest" level of measurement. Data on this level can only be classified into categories; there is no particular order for the categories; and the categories are mutually exclusive. Recall that mutually exclusive means that a respondent can be placed in one and only one category.

2. **Ordinal level of measurement** is the next-higher level of measurement. For ordinal-scaled data, one category is rated higher than the previous one. The categories are also mutually exclusive.

3. **Nonparametric (distribution-free) tests.** Hypothesis tests applied to nominal-level and ordinal-level data are often referred to as nonparametric tests, or distribution-free tests. These tests do not make any assumptions about the distribution of the population from which the sample, or samples, are selected. That is, to apply these nonparametric tests it is not required that the population be normally distributed.

4. **Chi-square goodness-of-fit test.** This test requires only nominal-scaled data. It is concerned with a single trait—the sex of the respondent, for example. The purpose of the test is to find out how well an observed set of data compares to an expected set of data.

Key Concepts (*continued*)

5. **Contingency tables.** This application of the chi-square statistic is concerned with the simultaneous classification of two traits. That is, are the two traits related or not? The chi-square statistic compares observed frequencies with expected frequencies.

6. The **chi-square distribution** has the following major characteristics:
   a. Its value is nonnegative
   b. There is a different chi-square distribution for each number of degrees of freedom.
   c. It is positively skewed, but approaches a symmetric distribution as the number of degrees of freedom increases.

7. The **sign test** is used to investigate changes in paired or related observations. The sign of the difference may be either positive or negative, and the binomial distribution is used as the test statistic.

8. The **Wilcoxon signed-rank test** is an extension of the sign test. It considers not only the sign of the difference in related observations, but also the magnitude of the differences.

9. The **Wilcoxon rank-sum test** concerns random samples taken from two independent populations. It is an alternative to the t test, but the assumption of a normal population is not required. If both samples are of at least size eight, the normal distribution may be used as the test statistic.

10. The **Kruskal-Wallis test** concerns random samples obtained from more than two independent samples. It is a nonparametric alternative to the ANOVA test, but the normality assumption is not required. The chi-square distribution is used as the test statistic.

Key Terms

Nonparametric test
Distribution-free tests
Chi-square distribution
Observed frequency
Expected frequency
Goodness-of-fit test
Contingency table

Nominal level of data
Ordinal level of data
Sign test
Wilcoxon signed-rank test
Wilcoxon rank-sum test
Kruskal-Wallis analysis of variance by ranks

Key Symbols

f_e Frequency expected.
f_o Frequency observed.
k Number of categories.
χ^2 Chi-square.
R^+ Sum of the positive signed differences.

R^- Sum of the negative signed differences.
W Sum of ranks for one population.
H Test statistic for the Kruskal-Wallis test.

Review Problems

1. A traffic engineer is studying the traffic pattern on a four-lane inbound expressway. For an hour he counts the number of vehicles using each lane. The results are as follows:

Lane	Frequency
1	100
2	80
3	60
4	60
	300

At the 0.05 significance level, test the hypothesis that the traffic uses all four lanes equally.

2. Historically, 25.2% of the used cars purchased in a particular geographical area are subcompacts, 52.9% are mid-sized, and 21.9% are full-sized cars. A sample of 500 sales during the first quarter revealed that 129 of the used cars sold were subcompacts, 258 were mid-sized, and 113 were full-sized. At the 0.05 level, test the hypothesis that there has been no change in the buying habits with respect to used cars.

3. There is interest in finding out if there is a relationship between the political affiliation of registered voters and their reaction to a proposed ban on the buildup of nuclear armament. The responses to a questionnaire were tabulated and are shown in the following table:

Political Affiliation	Favor Ban	Oppose Ban	No Opinion
Democrat	295	82	73
Republican	191	69	62
Independent	99	17	41
All others	45	12	14

Test the hypothesis, at the 0.05 significance level, that political affiliation is independent of the reaction to a ban on nuclear armament buildup.

4. A sample of 200 cola drinkers is obtained. They are classified by sex and whether they prefer regular or diet cola.

Cola	Male	Female	Total
Regular	45	40	85
Diet	55	60	115
	100	100	200

At the 0.01 significance level, can we conclude there is a difference in the cola preferences of men and women?

5. You want to use a simple "before-and-after" test to find out whether or not the experience of playing softball on a team made up of both boys and girls is successful in reducing sexism among boys. You test the boys for sexism both before league play begins and at the end of the season. The scores are

	Score	
Boy	Before Season	After Season
Carter	22	20
Jones	30	26
Ford	28	28
Yamamoto	38	30
Cosell	12	11
West	40	41
Archer	37	32
Oreon	32	29
Giles	50	40
Raggi	42	36
Nice	36	29
Cork	33	21
Mueller	36	21

Use the 0.05 significance level to test whether the softball experience has been effective in reducing sexism among boys.

6. An oil company advertises that customers will obtain higher gas mileage by using "Super" lead-free gasoline than using their regular lead-free. To investigate this claim, a sample of 14 cars of various makes and models is obtained. These 14 cars are driven the same distances over the same roads with both the "Super" lead-free and the regular gasolines. The results follow. At the 0.01 significance level, can it be concluded that the lead-free mileage is greater? Use the sign test.

Car	Super	Regular
1	21.5	20.9
2	32.5	32.9
3	29.7	30.9
4	35.4	35.7
5	26.2	26.0
6	40.4	41.7
7	25.9	25.3
8	25.1	25.0
9	24.0	23.8
10	18.9	18.5
11	23.4	23.2
12	30.2	28.5
13	26.8	26.2
14	27.6	27.0

7. Refer to Problem 5. Use the Wilcoxon signed-rank test to investigate whether the softball experience reduced sexism among boys. Use the 0.05 significance level.

8. Refer to Problem 6. Use the Wilcoxon signed-rank test to investigate whether Super lead-free gasoline yields higher mileage. Use the 0.01 significance level.

9. The amount of money spent for lunches by working men and women is being compared. The amount spent by a sample of 10 men and 12 women on a particular day is tabulated; the results are shown in the table that follows. Can it be concluded that men spend more than women? Use the Wilcoxon rank-sum test and the 0.01 significance level.

Men	Women
$3.73	$3.85
4.40	4.05
4.72	4.10
3.95	4.15
5.10	3.90
3.95	3.65
4.00	3.55
4.35	3.20
4.45	3.05
3.85	3.55
	2.90
	4.62

10. A study of the amount of overpayment to welfare recipients is made in three regions. Use the Kruskal-Wallis test to determine if there is a difference in the amount of overpayment in the three regions. Use the 0.05 level of significance.

West	Northwest	Midwest
$20	$24	$33
18	27	32
26	26	40
14	29	23
25	30	28
19	31	35
22	21	38
17	28	30

Case Analysis

(Data for these two cases can be found in the first Chapter Highlights, pp. 126-130).

The McCoy's Market Case

1. Use the 0.05 level of significance to test whether the typical number of items purchased by men is different from that purchased by women. Do *not* assume that the distributions follow a normal distribution.

2. At the 0.01 level of significance, is the sex of the shopper independent of the time of the week in which that person shops?

3. Group the amounts spent into three categories: one for those who spent less than $20, one for those who spent more than $20 but less than $40, and one for those who spent more than $40. Then test at the 0.10 level of significance to determine whether the distribution among these groups is independent of their members' sex.

The St. Mary's Emergency Room Case

1. At the 0.10 level of significance, test to determine whether the rates of patient admissions are the same for each shift.

2. Match as many pairs of patients as you can on the basis of (a) number of staff required to service them, and (b) shift during which they were admitted. Then test at the 0.01 level of significance to determine whether the older member of each pair was charged more than the younger one.

3. Use the Kruskal-Wallis test at the 0.05 level of significance to see whether the age distribution of the patients varies among the three shifts.

Answers

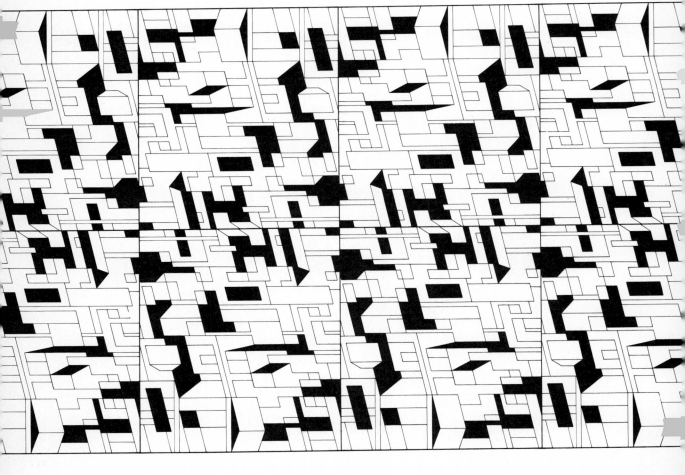

Answers to Chapter Achievement Tests

1. b **2.** b **3.** d **4.** b **5.** a **6.** b **7.** a

8. d **9.** a **10.** c

11.

Time		No. of Patients
1 - 5	///	3
6 - 10	⟊⟊⟊ /	6
11 - 15	⟊⟊⟊	5
16 - 20	⟊⟊⟊ /	6
21 - 25	⟊⟊⟊	5
		25

12.

Depth	Cumulative Frequency
1	1
3	4
5	11
7	19
9	29
11	44
13	53
15	58
17	60

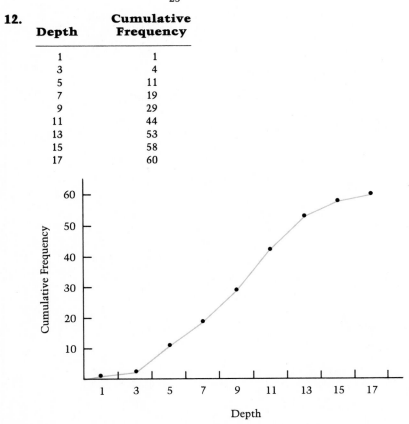

Nineteen out of 60, or 32%, have less than 7/32″ tread depth;
40% have less than 8/32 = 1/4″ tread depth.

13.

Expense	Amount	Percent of Total
Housing	$ 400	38%
Utilities	140	13
Medical	25	2
Food	190	18
Transportation	150	14
Clothing	50	5
Savings	50	5
Miscellaneous	50	5
	$1,055	

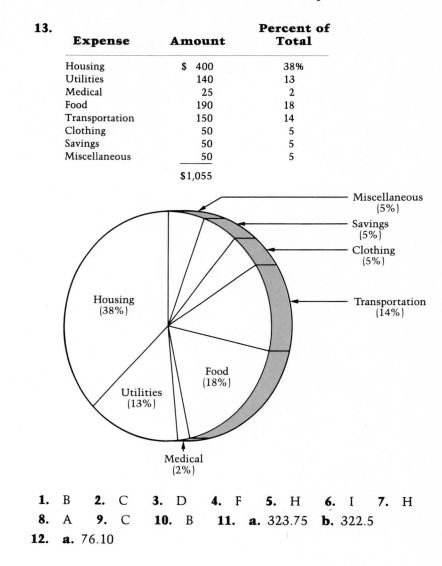

Miscellaneous (5%)
Savings (5%)
Clothing (5%)
Transportation (14%)
Housing (38%)
Food (18%)
Utilities (13%)
Medical (2%)

**CHAPTER 3
ACHIEVEMENT
TEST ANSWERS**

1. B   **2.** C   **3.** D   **4.** F   **5.** H   **6.** I   **7.** H
8. A   **9.** C   **10.** B   **11. a.** 323.75   **b.** 322.5
12. a. 76.10

13.

Pounds	f	X	fX	CF
0.0 to 0.4	3	0.2	0.6	3
0.5 to 0.9	8	0.7	5.6	11
1.0 to 1.4	20	1.2	24.0	31
1.5 to 1.9	10	1.7	17.0	41
2.0 to 2.4	6	2.2	13.2	47
2.5 to 2.9	3	2.7	8.1	50
	50		68.5	

Mean = 68.5/50 = 1.37 pounds.

14. Median = $0.95 + [(25 - 11)/20]0.5 = 1.30$ pounds.

15. Mode = 1.20 pounds. It is the midpoint of the 1.0 to 1.4 class.

1. F; omit "half."   **2.** T   **3.** T   **4.** T   **5.** T

6. F; negatively skewed.   **7.** T   **8.** F; skewness is zero.   **9.** F; CV A = 5%; CV B = 3%   **10.** T   **11.** T

12. T   **13.** T   **14.** T   **15.** T   **16.** T

17. $6 - 2 = 4$

18. $\overline{X} = 4$; $MD = (2 + 2 + 1 + 0 + 1)/5 = 1.2$

19. $s^2 = 2.5$

20. $\sqrt{2.5} = 1.58$

21. $79 - 20 = 59$

22. $Q_1 = 39.5 + [(30 - 23)/20]10 = 43.0$
$Q_3 = 49.5 + [(90 - 43)/50]10 = 58.9$
$QD = (58.9 - 43.0)/2 = 7.95$

23. $s^2 = [341,800 - (6,220)^2/120]/119 = 163$
$s = \sqrt{163} = 12.8$

24. CV(Jefferson) = 12.8/51.8 = 0.25
CV(Samford) = 9/87 = 0.10

25. sk(Jefferson) = 3(51.8 − 53.4)/12.8 = −0.38
sk(Samford) = 3(87 − 84)/9 = 1.00

CHAPTER 4 ACHIEVEMENT TEST ANSWERS

1. T   **2.** T   **3.** F; $P(A$ and $B)$.   **4.** T   **5.** F

6. F; independent.   **7.** T   **8.** F   **9.** T   **10.** F

11. F; permutation.

12. $(10,907 - 2,438)/29,355 = 0.45$

13. $(2,438 + 1,642)/29,355 = 0.14$

CHAPTER 5 ACHIEVEMENT TEST ANSWERS

14. $(0.75)^4 = 0.3164$

15. $(10)^3 = 1,000$

16. $P(\text{History or Math}) = 0.7 + 0.6 - 0.5 = 0.8$

**CHAPTER 6
ACHIEVEMENT
TEST ANSWERS**

1. b **2.** c **3.** d **4.** d **5.** e **6.** a **7.** a

8. NT; the probability of a success changes from trial to trial.

9. T **10.** T

11. **a.** $P(8) = 0.231$
   b. $0 + 0 + 0 + 0.001 + 0.008 = 0.009$
   c. $0.168 + 0.071 + 0.014 = 0.253$

12. **a.** $p = 0.30$, $n = 20$
      $P(8) = 0.114$
   b. $0.001 + 0.007 + 0.028 + 0.072 + 0.130 + 0.179 = 0.417$
   c. $0.031 + 0.012 + 0.004 + 0.001 = 0.048$

**CHAPTER 7
ACHIEVEMENT
TEST ANSWERS**

1. c **2.** a **3.** c **4.** c **5.** d **6.** c **7.** d

8. d **9.** d **10.** a

11. **a.** $(50 - 39.52)/6.29 = 1.67$
   b. $(25 - 39.52)/6.29 = -2.31$
   c. $0.5000 - 0.4525 = 0.0475$
   d. $0.5000 - 0.4896 = 0.0104$
   e. $1.28 = (X - 39.52)/6.29$
      $X = 47.57$

12. **a.** $z = (150 - 160)/5.66 = -1.77$
      $0.4616 + 0.5000 = 0.9616$
   b. $z = (148 - 160)/5.66 = -2.12$
      $0.5000 - 0.4830 = 0.0170$
   c. 0.4616
   d. $-1.28 = (X - 160)/5.66$
      $X = 160 - (1.28)(5.66) = 153$ letters

1. F; population. **2.** F; probability. **3.** T **4.** T
5. F; not necessarily. **6.** T
7. F; sampling distribution. **8.** F; decrease.
9. F; it approaches normal as n increases. **10.** T
11. **a.** $\mu = (1 + 4 + 12 + 11 + 9 + 8)/6 = 9.0$

b. $_nC_r = \dfrac{6!}{4!2!} = \dfrac{6 \cdot 5}{2} = 15$

Sample	Items	Means
1	10, 4, 12, 11	9.25
2	10, 4, 12, 9	8.75
3	10, 4, 12, 8	8.50
4	10, 12, 11, 9	10.50
5	10, 12, 11, 8	10.25
6	10, 12, 9, 8	9.75
7	10, 11, 9, 8	9.50
8	4, 12, 11, 9	9.00
9	4, 12, 11, 8	8.75
10	4, 11, 9, 8	8.00
11	12, 11, 9, 8	10.00
12	10, 4, 11, 9	8.50
13	10, 4, 11, 8	8.25
14	10, 4, 9, 8	7.75
15	4, 12, 9, 8	8.25

c.

Mean	Frequency	$\overline{X}f$
7.75	1	7.75
8.00	1	8.00
8.25	2	16.50
8.50	2	17.00
8.75	2	17.50
9.00	1	9.00
9.25	1	9.25
9.50	1	9.50
9.75	1	9.75
10.00	1	10.00
10.25	1	10.25
10.50	1	10.50
	15	135.00

$\overline{\overline{X}} = 135/15 = 9.0$

d.

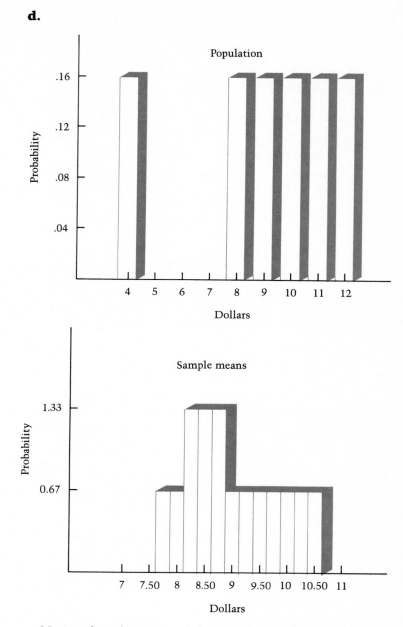

e. Notice that the mean of the sampling distribution (9.00) is exactly equal to the population mean. Notice also that there is less variability in the sampling distribution than in the population. Finally, notice how the shape of the distribution has changed.

12. The standard error of the mean is 3.33, found by $10/\sqrt{9}$.

13. $96.2 \pm (1.96)(9.7/\sqrt{817}) = 96.2 \pm 0.665$
    The interval is from 95.535 to 96.865. About 95% of the similarly constructed intervals will include the population mean.

14. $35 \pm (2.58)(6.3/\sqrt{40}) = 35 \pm 2.57$
    The interval is from 32.43 to 37.57. About 99% of the similarly constructed intervals will include the population mean.

15. $n = \left[\dfrac{1.96(3,000)}{200} \right]^2 = 865$

CHAPTER 9 ACHIEVEMENT TEST ANSWERS

1. c   2. a   3. d   4. b   5. a   6. e   7. d
8. e   9. d; the correct answer is 5.89.   10. d
11. $z = (32.9 - 29.6)/\sqrt{[(5.7)^2/36] + [(5.5)^2/49]} = 3.3/1.23$
    $= 2.67$
    H_0 is rejected. There is a significant difference between the ages of the women in the Atlantic Coast and the Midwest.
12. $z = (34 - 32)/(4.5/\sqrt{49}) = 2/0.643 = 3.11$
    Since $3.11 > 1.65$, H_0 is rejected. The people in the South devoted significantly more time to volunteer activities.

CHAPTER 10 ACHIEVEMENT TEST ANSWERS

1. b   2. d   3. b   4. b   5. c   6. b   7. c
8. a; $z = -0.53$
9. $\bar{p} = (82 + 40)/(100 + 60) = 0.7625$
    $z = 0.15/0.0695 = 2.16$
    Since 2.16 is less than the critical value of 2.33, H_0 is not rejected.
10. $H_0: p \geq 0.85$, $H_a: p < 0.85$
    $z = (0.80 - 0.85)/\sqrt{[(0.85)(1 - 0.85)]/150} = -1.72$
    Since -1.72 does not fall in the rejection region beyond -2.33, the null hypothesis is not rejected.

CHAPTER 11 ACHIEVEMENT TEST ANSWERS

1. d   2. d   3. b   4. c   5. b   6. b   7. d
8. a   9. b   10. e   11. c
12. $H_0: \mu \geq 98.6°F$, $H_a: \mu < 98.6°F$, df $= 25 - 1 = 24$
    Reject H_0 if t is beyond -1.711.
    $t = -2.34$
    Reject H_0 and accept H_a.
13. $H_0: D \leq 0$, $H_a: D > 0$, df $= 7 - 1 = 6$
    Reject H_0 if t is greater than 3.143.
    $\bar{d} = 50/7 = 7.14$, $t = 7.14/(9.15/\sqrt{7}) = 2.06$
    H_0 is not rejected.

14. $H_0: \mu_1 = \mu_2$, $H_a: \mu_1 \neq \mu_2$, df $= 5 + 8 - 2 = 11$
Critical values for this two-tailed test are $+2.201$ and -2.201.
$t = (35.6 - 26)/16.76\sqrt{(1/5) + (1/8)} = 1.00$
H_0 is not rejected.

CHAPTER 12
ACHIEVEMENT
TEST ANSWERS

1. a   **2.** c   **3.** c   **4.** c   **5.** c   **6.** a   **7.** a
8. d   **9.** b   **10.** a

11. $SST = [(18)^2/3 + (31)^2/5 + (26)^2/4] - (75)^2/12 = 0.45$
$SSE = 495 - [(18)^2/3 + (31)^2/5 + (26)^2/4] = 25.8$

$$F = \frac{0.45/2}{25.8/9} = 0.08$$

$H_0: \mu_1 = \mu_2 = \mu_3$, H_a: at least one pair of means is different

12.

Source	Sum of Squares	df	Mean Square
Treatments	0.45	2	0.225
Error	25.80	9	2.867
	26.25		

13.   The critical value is 4.26.

14. H_0 is not rejected. These data show no significant difference among the mean stopping distances.

CHAPTER 13
ACHIEVEMENT
TEST ANSWERS

1. c   **2.** d   **3.** a   **4.** c   **5.** a   **6.** b   **7.** c
8. c
9. a.

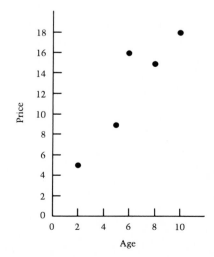

b. $r = \dfrac{5(451) - (31)(63)}{\sqrt{[5(229) - (31)^2][5(911) - (63)^2]}} = \dfrac{302.00}{328.37} = 0.92$

c. The coefficient of determination is $(0.92)^2 = 0.85$. The coefficient of nondetermination is $1 - 0.85 = 0.15$.

d. $H_0: \rho = 0$, $H_a: \rho \neq 0$
Critical values are $+3.182$ and -3.182.
$t = (0.92\sqrt{5 - 2})/\sqrt{0.15} = 4.116$.

e. H_0 is rejected.

10. $r_s = 1 - \dfrac{6(46)}{10[(10)^2 - 1]} = 0.72$

There is a strong correlation between the rankings.

1. a **2.** c **3.** d **4.** a **5.** c **6.** a **7.** b
8. c **9.** e
10. a.

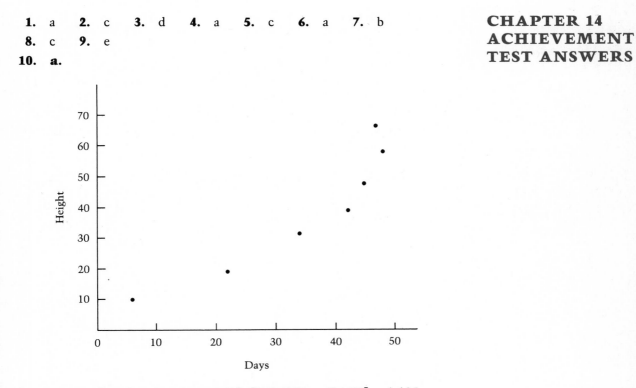

b. $b = [7(11{,}171) - 244(270)]/[7(9{,}978) - (244)^2] = 1.195$
$a = [270 - 1.195(244)]/7 = -3.083$
$Y' = -3.083 + 1.195X$

c. $\sqrt{355.07/5} = 8.43$

d. $Y' = -3.083 + 1.195(25) = 26.792$
$26.792 \pm 7.762 = 19.030$ to 34.554

CHAPTER 15 ACHIEVEMENT TEST ANSWERS

1. d 2. c 3. a 4. b 5. a 6. c 7. b
8. d 9. b 10. b 11. a
12. $Y' = 12.49883 - 0.4690281(\text{VAR3}) + 1.5634668(\text{VAR4})$
13. R SQUARE is 45%.
14. Since the equation explains less than half of the variability, the correlation is not very strong.
15. VAR2 would enter the equation in the next step. It has the largest (in absolute value) correlation with the unexplained variation.
16. There will be three independent variables in the regression equation when STEP 3 is reached.

CHAPTER 16 ACHIEVEMENT TEST ANSWERS

1. c 2. d 3. d 4. c 5. a 6. b 7. a
8. b 9. b 10. a
11. H_0: party affiliation and position are not related
H_a: party affiliation and position are related
df $= (2 - 1)(3 - 1) = 2$
H_0 is rejected if χ^2 is greater than 5.991.

Party Affiliation	Favor f_o	Favor f_e	Oppose f_o	Oppose f_e	Undecided f_o	Undecided f_e	Total f_o	Total f_e
Democrat	22	21	13	14	28	28	63	63
Republican	8	9	7	6	12	12	27	27
	30	30	20	20	40	40	90	90

$$\chi^2 = \frac{(22-21)^2}{21} + \frac{(13-14)^2}{14} + \frac{(28-28)^2}{38} + \frac{(8-9)^2}{9}$$
$$+ \frac{(7-6)^2}{6} + \frac{(12-12)^2}{12} = 0.40$$

H_0 cannot be rejected. Party and position are not related.

12. H_0: distribution is as given
H_a: distribution has changed
df $= 4 - 1 = 3$
H_0 is rejected if χ^2 is greater than 11.345.

Car	f_o	Ratio	f_e	$f_e - f_o$	$(f_e - f_o)^2$	$\dfrac{(f_o - f_e)^2}{f_e}$
Full-sized	38	1	20	18	324	16.20
Medium-sized	62	2	40	22	484	12.10
Compact	41	3	60	−19	361	6.02
Subcompact	59	4	80	−21	441	5.51
	200	10	200			39.83

$\chi^2 = 39.83$

H_0 is rejected. The distribution has changed.

1. b **2.** b **3.** a **4.** a **5.** d **6.** c **7.** d

8. b **9.** a **10.** c

11. H_0: $p \geq 0.5$, H_a: $p < 0.5$
Since $15(0.5)$ is greater than 5.0, the normal approximation is used.
$z = (13 - 7.5)/\sqrt{15(0.5)(0.5)} = 6.60$
H_0 is rejected.

12. H_0: age at emotional maturity is the same
H_a: age at emotional maturity is not the same
The sample size is reduced to ten because of three differences of zero. The critical value is 3. R^+ is 10; R^- is 45.
H_0 cannot be rejected.

13. H_0: achievement scores are the same for the two groups
H_a: achievement scores are not the same for the two groups
The critical values are −1.96 and +1.96.
The sum of ranks for one group is 62.0; for the other 128.0.
$z = -28/\sqrt{150} = -2.29$
H_0 is rejected.

Solutions to Even-Numbered Exercises

2.

Number of Visits	Class Frequencies
0 - 2	10
3 - 5	23
6 - 8	10
9 - 11	3
12 - 14	3
15 - 17	1
Total	50

**Chapter 2
Summarizing Data:
Frequency
Distributions
and Graphic
Presentation**

Based on the frequency distribution, the number of visits to relatives ranges from 0 to 17. The largest concentration is in the 3 - 5 class.

4.

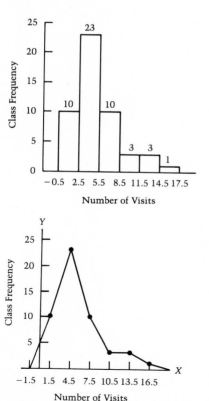

6.

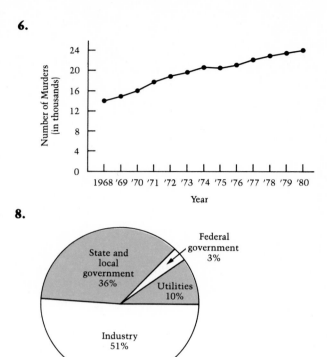

8.

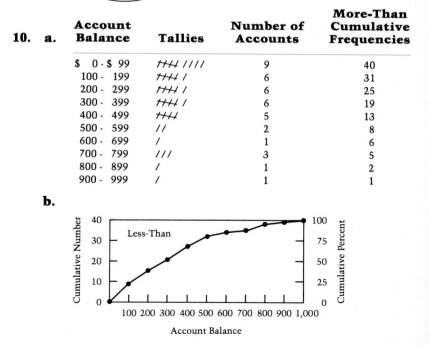

10. **a.**

Account Balance	Tallies	Number of Accounts	More-Than Cumulative Frequencies
$ 0 - $ 99	⫝̸⫝̸ ////	9	40
100 - 199	⫝̸⫝̸ /	6	31
200 - 299	⫝̸⫝̸ /	6	25
300 - 399	⫝̸⫝̸ /	6	19
400 - 499	⫝̸⫝̸	5	13
500 - 599	//	2	8
600 - 699	/	1	6
700 - 799	///	3	5
800 - 899	/	1	2
900 - 999	/	1	1

b.

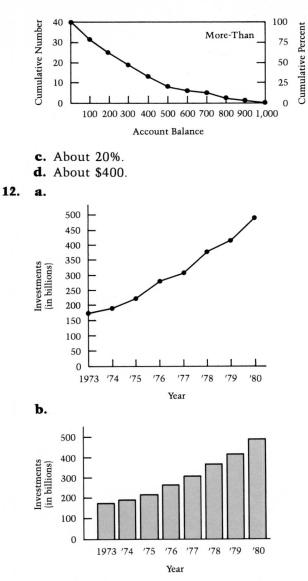

c. About 20%.

d. About $400.

12. a.

b.

14. The sales manager should select KROK because its listening audience is much younger and therefore more likely to purchase a sports car.

2. $\bar{X} = \Sigma X/n = \$163/9 = \$18.11$ (in thousands)

4. $10,000

Chapter 3
Descriptive
Statistics: Measures
of Central Tendency

6.

Miles per Gallon	Class Frequencies f	Midpoints X	d	fX	fd
20 - 24	2	22	−2	44	−4
25 - 29	7	27	−1	189	−7
30 - 34	15	32	0	480	0
35 - 39	8	37	+1	296	8
40 - 44	3	42	+2	126	6
Total	35			1,135	+3

a. 35
b. Using true class limits: $34.5 - 29.5 = 5$. Then, $29.5 + 5/2 = 32$.
c. $\overline{X} = 1,135/35 = 32.43$ miles per gallon
d. $\overline{X} = 32.0 + 5(3/35) = 32.0 + 0.43 = 32.43$ miles per gallon

8.

Number of Hours	Class Frequency	Cumulative Frequency
0	9	9
1	14	23
2	10	33
3	9	42
4	5	47
5 or more	3	50
Total	50	

a. $1.5 + \left(\dfrac{50/2 - 23}{10}\right) 1 = 1.7$ hours

b. Mode is 1 hour.

10.　**a.** $\mu = \Sigma X/N = 1,124,000/5 = 224,800$ prisoners
　　b. Median is 213,000 prisoners.

12. Fifty percent of the candidates take more than 8.5 years, and the other half take less than 8.5 years.

14.　**a.** Mean $= 408/8 = 51.0\%$
　　b. Median $= 42.0\%$

16.　**a.** Median $= \$299,999.50 + \left(\dfrac{1,052/2 - 525}{241}\right)(\$100,000)$

$= \$300,414.43$, or $\$300,414$ rounded

Half of the top officials earn less than $\$300,414$; the other half more than $\$300,414$.

b. The widths of the classes vary, making the mean difficult to compute.

c. Mode is $250,000.

18. **a.** $\bar{X} = \Sigma X/n = \$8.82/7 = \$1.26$
Median $= \$1.27$; mode $= \$1.27$
b. All three averages would be representative.

2. **a.** Pond A: $\bar{X} = 14.2$ inches
Pond B: $\bar{X} = 15.9$ inches
b. Pond A: range $= 16.5 - 12.0 = 4.5$ inches
Pond B: range $= 20.5 - 12.0 = 8.5$ inches
The range in the length of the trout in pond B is significantly greater than those in pond A.
c. Pond A: $MD = 10.4/10 = 1.04$ inches
Pond B: $MD = 27.8/10 = 2.78$ inches
The average amount of variation in the lengths of the trout in pond B is more than twice as great as that in pond A.

Chapter 4
Descriptive Statistics:
Measures of
Dispersion and
Skewness

4. **a.**

Pond A		Pond B	
$X - \bar{X}$	$(X - \bar{X})^2$	$X - \bar{X}$	$(X - \bar{X})^2$
−1.7	2.89	2.1	4.41
−0.2	0.04	4.1	16.81
−0.7	0.49	−3.9	15.21
0.3	0.09	−1.4	1.96
0.8	0.64	3.1	9.61
2.3	5.29	−2.4	5.76
1.8	3.24	−1.4	1.96
−2.2	4.84	−2.9	8.41
−0.2	0.04	4.6	21.16
−0.2	0.04	−1.9	3.61
	17.60		88.90

Pond A: $s^2 = 17.60/(10 - 1) = 1.96$

Pond B: $s^2 = 88.90/(10 - 1) = 9.88$

b. Pond A: $s = \sqrt{1.96} = 1.40$ inches

Pond B: $s = \sqrt{9.88} = 3.14$ inches

c. The variation in the lengths of the trout in pond B (3.14 inches) is greater than for pond A (1.40 inches).

6. Pond A:

$$s^2 = \frac{\Sigma X^2 - (\Sigma X)^2/n}{n-1} = \frac{2,034 - (142)^2/10}{10 - 1} = 1.96$$

$$s = \sqrt{1.96} = 1.40 \text{ inches (same answer)}$$

Pond B:

$$s^2 = \frac{\Sigma X^2 - (\Sigma X)^2/n}{n-1} = \frac{2{,}617 - (159)^2/10}{10-1} = 9.88$$

$$s = \sqrt{9.88} = 3.14 \text{ inches (same answer)}$$

8. **a.** Monday:

$$Q_1 = 11.5 + \frac{18-5}{15}(4) = 14.97$$

$$Q_3 = 19.5 + \frac{54-46}{16}(4) = 21.50$$

$$QD = \frac{21.50 - 14.97}{2} = 3.27$$

Friday:

$$Q_1 = 11.5 + \frac{16-5}{21}(4) = 13.60$$

$$Q_3 = 19.5 + \frac{48-48}{13}(4) = 19.50$$

$$QD = (19.50 - 13.60)/2 = 2.95$$

b. The dispersion between the quartiles is greater on Monday than on Friday (3.27 is greater than 2.95).

10. **a.** Considering these as all the admissions on Monday and Friday (that is, populations):

X	f	Monday fX	fX^2
5.5	1	5.5	30.25
9.5	4	38.0	361.00
13.5	15	202.5	2,733.75
17.5	26	455.0	7,962.50
21.5	16	344.0	7,396.00
25.5	7	178.5	4,551.75
29.5	3	88.5	2,610.75
	72	1,312.0	25,646.00

$$\sigma^2 = \frac{25{,}646 - (1{,}312)^2/72}{72} = 24.15$$

Standard deviation $\sigma = \sqrt{24.15} = 4.91$

Friday: $\sigma^2 = \dfrac{81{,}916 - (1{,}068)^2/64}{64} = 17.09$

Standard deviation $\sigma = \sqrt{17.09} = 4.13$

b. Since 4.91 is greater than 4.13, the dispersion on Monday is greater than on Friday.

12. Salary: $CV = \dfrac{\$3{,}000}{\$21{,}000}(100) = 14.3\%$

Length of employment: $CV = \dfrac{4}{15}(100) = 26.7\%$

There is relatively more dispersion in their lengths of employment.

14. $sk = \dfrac{3(11.5 - 11.95)}{4.5} = -0.3$

This distribution has a slight negative skewness.

16. **a.** Population

b. $390 - 134 = 256$

c. $MD = \dfrac{3.4 + 10.4 + 51.4 + 79.4 + 176.6 + 29.4 + 2.4}{7}$

$= 50.4$

d. $\sigma^2 = \dfrac{359{,}986 - (1{,}494)^2/7}{7} = 5{,}874.82$

e. $\sigma = \sqrt{5{,}874.82} = 76.6$

f. $sk = \dfrac{3(21.3 - 20.3)}{76.6} = 0.41$

18. **a.** Range $= 7 - 0 = 7$

b. $MD = \dfrac{\begin{array}{c}5(2.4) + 11(1.4) + 9(0.4) + 5(0.6) \\ + 5(1.6) + 4(3.6) + 1(4.6)\end{array}}{40}$

$= 61/40 = 1.5$

c. $s^2 = \dfrac{365 - (95)^2/40}{40 - 1} = 3.57$

d. $s = \sqrt{3.57} = 1.89$

e. $QD = \dfrac{3.5 - 1}{2} = 1.25$

f. $sk = \dfrac{3(2.4 - 2.0)}{1.89} = 0.63$

20.   **a.** Range $= 29 - 0 = 29$

b.

Weight	CF
0 - 4	10
5 - 9	47
10 - 14	97
15 - 19	165
20 - 24	195
25 - 29	200

$$Q_1 = 9.5 + \frac{50 - 47}{50}(5) = 9.8$$

$$Q_3 = 14.5 + \frac{150 - 97}{68}(5) = 18.4$$

$$QD = \frac{18.4 - 9.8}{2} = 4.3$$

c.

X	f	fX	fX²
2	10	20	40
7	37	259	1,813
12	50	600	7,200
17	68	1,139	19,363
22	30	660	14,520
27	5	135	3,645
	200	2,813	46,581

Considering this as all persons enrolled (a population):

$$\sigma^2 = \frac{46,581 - (2,813)^2/200}{200} = \frac{46,581 - 39,564.845}{200}$$

$$= 35.08$$

$$\sigma = \sqrt{35.08} = 5.92$$

d. For weights: $CV = \dfrac{5.92}{14.07}(100) = 42\%$

For times: $CV = \dfrac{0.8}{6.0}(100) = 13\%$

There is more relative dispersion in weights because 42 is greater than 13.

e. For weights: $sk = \dfrac{3(14.07 - 15.22)}{5.92} = -0.58$

For times: $sk = \dfrac{3(6.0 - 5.6)}{0.8} = 1.50$

22. a. American League: range $= 374 - 150 = 224$

$$Q_1 = 224.5 + \frac{33 - 17}{22}(25) = 242.68$$

$$Q_3 = 274.5 + \frac{99 - 75}{28}(25) = 295.93$$

$$QD = \frac{295.93 - 242.68}{2} = 26.63$$

$$MD = \frac{\begin{array}{c} 2(93) + 10(68) + 17(43) + 28(18) + 36(7) \\ + 22(32) + 12(57) + 1(82) + 4(107) \end{array}}{132}$$

$$= 32.2$$

$$\sigma^2 = \frac{9{,}745{,}133 - (35{,}459)^2 / 132}{132} = 1{,}665.34$$

$$\sigma = \sqrt{1{,}665.34} = 40.8$$

National League: range $= 349 - 175 = 174$

$$Q_1 = 224.5 + \frac{24.25 - 14}{17}(25) = 239.57$$

$$Q_3 = 274.5 + \frac{72.75 - 58}{25}(25) = 289.25$$

$$QD = \frac{289.25 - 239.57}{2} = 24.84$$

$$MD = \frac{\begin{array}{c} 5(73) + 9(48) + 25(23) + 27(2) \\ + 17(27) + 10(52) + 4(77) \end{array}}{97}$$

$$= 28.0$$

$$\sigma^2 = \frac{6{,}900{,}743 - (25{,}639)^2 / 97}{97} = 1{,}276.84$$

$$\sigma = \sqrt{1{,}276.84} = 35.7$$

b. The distribution of the American League batting averages is more spread out than the distribution for the National League (40.8 is greater than 35.7).

Chapter 5
An Introduction
to Probability

2. **a.** The election
   b. Democrat, Republican, Independent
   c. Male or Female

4. **a.** P(two major) = P(lung) + P(prostate)
$$= 12{,}226/37{,}555 + 10{,}835/37{,}555 = 0.614$$
   b. P(stomach or pancreas) = P(stomach) + P(pancreas)
$$= 3{,}037/37{,}555 + 3{,}031/37{,}555$$
$$= 0.162$$
   c. P(not lung) = P(prostate) + P(colon) + P(stomach)
$$+ P(\text{pancreas})$$
$$= 10{,}835/37{,}555 + 8{,}426/37{,}555$$
$$+ 3{,}037/37{,}555 + 3{,}031/37{,}555$$
$$= 25{,}329/37{,}555 = 0.674$$

6. Let W = women members
U = university persons
$$P(W \text{ or } U) = P(W) + P(U) - P(W \text{ and } U)$$
$$= 0.20 + 0.05 - 0.02$$
$$= 0.23$$

8. **a.** $50/200 = 0.25$
   b. $50/200 = 0.25$
   c.

50/130 Low	130/200 × 50/130	= 0.25
30/130 Avg.	130/200 × 30/130	= 0.15
50/130 High	130/200 × 50/130	= 0.25
20/70 Low	70/200 × 20/70	= 0.10
30/70 Avg.	70/200 × 30/70	= 0.15
20/70 High	70/200 × 20/70	= 0.10

10. **a.** Yes
   b. P(both) = P(first breaks down) $\cdot$ P(second breaks down)
$$= (0.05)(0.10) = 0.005$$
   c. P(neither) = P(first does not break down) $\cdot$ P(second does
not break down)
$$= (0.95)(0.90) = 0.855$$

12. $10 \cdot 10 \cdot 10 \cdot 10 = 10{,}000$ possible plates

14. $_8P_5 = \dfrac{8!}{(8-5)!} = \dfrac{8!}{3!} = \dfrac{8 \cdot 7 \cdot 6 \cdot 5 \cdot 4 \cdot 3 \cdot 2 \cdot 1}{3 \cdot 2 \cdot 1} = 6{,}720$

16. $_{20}C_8 = \dfrac{20!}{(20-8)!8!} = \dfrac{20!}{12!8!}$

$$= \dfrac{20 \cdot 19 \cdot 18 \cdot 17 \cdot 16 \cdot 15 \cdot 14 \cdot 13}{8 \cdot 7 \cdot 6 \cdot 5 \cdot 4 \cdot 3 \cdot 2 \cdot 1}$$

$$= 125{,}970$$

18. Type of car is the experiment. The possible outcomes are: selling a subcompact, selling a compact, selling a luxury car. The possible outcomes with respect to doors are: two-door or four-door.

20. $P(\text{all}) = 0.25 + 0.20 + 0.10 = 0.55$

$$P(\text{sporting event}) = P(\text{basketball}) + P(\text{horse race})$$
$$= 0.25 + 0.10$$
$$= 0.35$$

22. Let $D =$ DisneyWorld and $B =$ Busch Gardens

$$P(D \text{ or } B) = P(D) + P(B) - P(D \text{ and } B)$$
$$= 0.70 + 0.50 - 0.40$$
$$= 0.80$$

24. Let $S =$ snow and $P =$ profit

$$P(S \text{ and } P) = P(S) \cdot P(P|S)$$
$$= (0.80) \cdot (0.85)$$
$$= 0.68$$

26. $P(H \text{ and } W) = P(H) \cdot P(W) = (0.5)(0.7) = 0.35$

28. $10 \cdot 10 \cdot 10 \cdot 10 = 10,000$

30. $_8C_3 = \dfrac{8!}{(8-3)!3!} = \dfrac{8!}{5!3!} = \dfrac{8 \cdot 7 \cdot 6}{3 \cdot 2} = 56$

32. $P(\text{all three}) = P(\text{first}) \cdot P(\text{second}) \cdot P(\text{third})$
$$= (0.90)(0.90)(0.90)$$
$$= 0.729$$

Let $D =$ detected and $ND =$ not detected

$$P(\text{two out of three detected}) = P(D) \cdot P(D) \cdot P(ND) + P(D)$$
$$\cdot P(ND) \cdot P(D) + P(ND)$$
$$\cdot P(D) \cdot P(D)$$
$$= (0.9)(0.9)(0.1)$$
$$+ (0.9)(0.1)(0.9)$$
$$+ (0.1)(0.9)(0.9)$$
$$= 0.243$$

Chapter 6
Probability
Distributions

2. $P(0) = \dfrac{5!}{0!5!} (0.9)^5 (0.1)^0$

$$= (1)(0.59049)$$
$$= 0.59049$$

4. Let $X =$ the number of girls

a. $P(3) = \dfrac{5!}{3!(5-3)!} (1/2)^3 (1 - 1/2)^2$

$$= 10(1/8)(1/4) = 10/32$$
$$= 0.313$$

b. $P(5) = \dfrac{5!}{5!(5-5)!} \, (1/2)^5$

$= 1/32$

$= 0.031$

c. $P \text{ (no girls)} = P(0) = \dfrac{5!}{0!(5-0)!} \, (1/2)^0 (1/2)^5$

$= 1/32$

$= 0.031$

$P \text{ (at least one girl)} = 1 - P \text{ (no girls)}$

$= 1 - 0.031$

$= 0.969$

6. $p = 0.3, n = 5$
a. $P(2) = 0.309$
b. $P(X \le 2) = P(0) + P(1)$
$= 0.168 + 0.360$
$= 0.528$
c. $P(X \ge 1) = 1 - P(0)$
$= 1 - 0.168$
$= 0.832$

8. $p = 0.6, n = 15$
$P(X \ge 10) = P(10) + P(11) + \cdots + P(15)$
$= 0.186 + 0.127 + 0.063 + 0.022 + 0.005 + 0.000$
$= 0.403$

10. $p = 0.4, n = 6$
$P(5) = 0.037$
$P(\text{between two and five}) = P(2) + P(3) + P(4) + P(5)$
$= 0.311 + 0.276 + 0.138 + 0.037$
$= 0.762$

Chapter 7
The Normal
Probability
Distribution

2. $\mu = 90, \sigma = 10$
a. $\mu \pm 2\sigma$ includes about 95% of the values.
$90 \pm 2(10) = 90 \pm 20;$
Time interval: 70 up to 110.
b. $\mu \pm 3\sigma$ includes 99.7% of the values.
$90 \pm 3(10) = 90 \pm 30;$
Time interval: 60 up to 120.

4. Plumbers: Carpenters:
$\mu = \$15, \sigma = \$1.75$ $\mu = \$10, \sigma = \$1.25$
Joe earns \$14/hour   Neil earns \$12/hour
$z = (\$14 - \$15)/\$1.75$ $z = (\$12 - \$10)/\$1.25$
   $= -0.57$ $= 1.60$

While Joe earns more than Neil. Joe is -0.57 standard units below the mean for his trade whereas Neil is 1.6 standard units above average for his trade.

6. $\mu = 10, \sigma = 2$

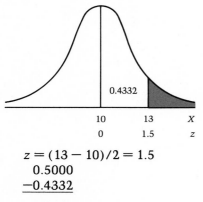

$z = (13 - 10)/2 = 1.5$
   0.5000
-0.4332

   0.0668

The probability that a tree produces more than 13 gallons is 0.0668.

b.

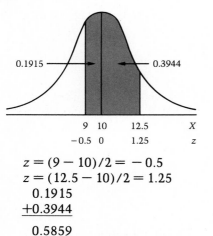

$z = (9 - 10)/2 = -0.5$
$z = (12.5 - 10)/2 = 1.25$
   0.1915
$+0.3944$

   0.5859

The probability that a tree produces between 9.0 and 12.5 gallons is 0.5859.

8. $\mu = 40,000$, $\sigma = 3,000$

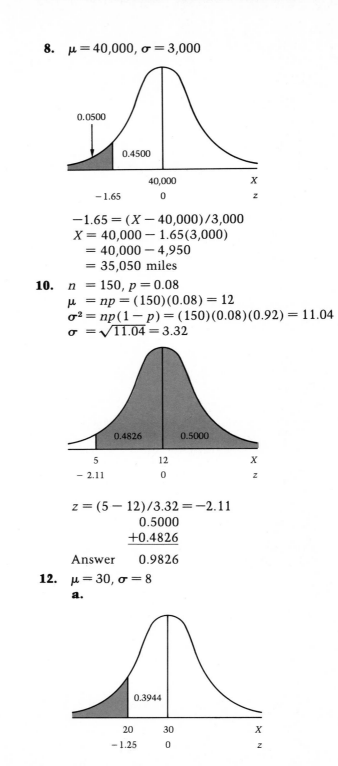

$-1.65 = (X - 40,000)/3,000$
$X = 40,000 - 1.65(3,000)$
$\quad = 40,000 - 4,950$
$\quad = 35,050$ miles

10. $n = 150$, $p = 0.08$
$\mu = np = (150)(0.08) = 12$
$\sigma^2 = np(1 - p) = (150)(0.08)(0.92) = 11.04$
$\sigma = \sqrt{11.04} = 3.32$

$z = (5 - 12)/3.32 = -2.11$
$\qquad 0.5000$
$\underline{+0.4826}$

Answer $\quad 0.9826$

12. $\mu = 30$, $\sigma = 8$
a.

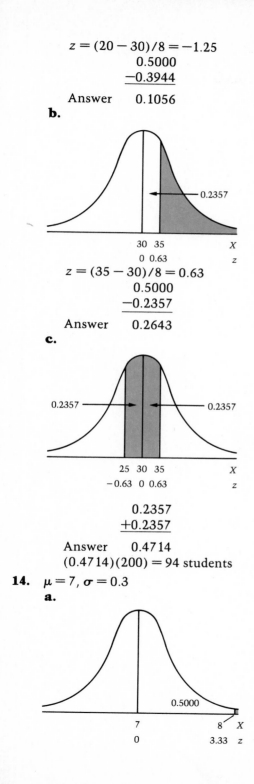

$z = (20 - 30)/8 = -1.25$

$$0.5000$$
$$\underline{-0.3944}$$

Answer 0.1056

b.

0.2357

30 35 X
0 0.63 z

$z = (35 - 30)/8 = 0.63$

$$0.5000$$
$$\underline{-0.2357}$$

Answer 0.2643

c.

0.2357   0.2357

25 30 35 X
-0.63 0 0.63 z

$$0.2357$$
$$\underline{+0.2357}$$

Answer 0.4714

$(0.4714)(200) = 94$ students

14. $\mu = 7,\ \sigma = 0.3$

a.

0.5000

7   8 X
0   3.33 z

$$z = (8 - 7)/0.3 = 3.33$$
$$0.5000$$
$$\underline{-0.5000}$$

Answer   0.0000

They will virtually never overflow.

b.

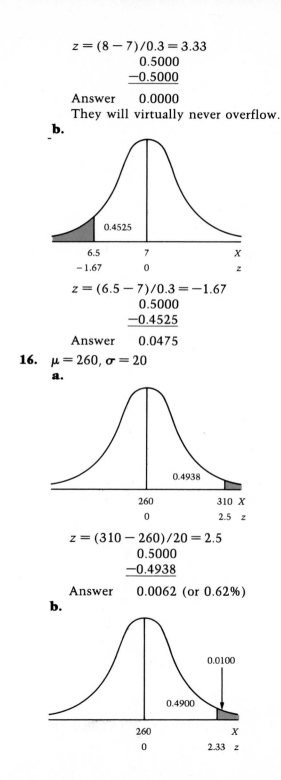

$$z = (6.5 - 7)/0.3 = -1.67$$
$$0.5000$$
$$\underline{-0.4525}$$

Answer   0.0475

16. $\mu = 260, \sigma = 20$

a.

$$z = (310 - 260)/20 = 2.5$$
$$0.5000$$
$$\underline{-0.4938}$$

Answer   0.0062 (or 0.62%)

b.

$$2.33 = (X - 260)/20$$
$$X = 260 + 2.33(20)$$
$$= 260 + 46.6$$
$$= 306.6 \text{ pounds}$$

18. $p = 0.4$, $n = 50$
$\mu = np = (50)(0.4) = 20$
$\sigma^2 = np(1 - p) = (50)(0.4)(0.6) = 12$
$\sigma = \sqrt{12} = 3.46$

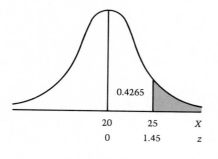

$z = (25 - 20)/3.46 = 1.45$
$\quad\quad 0.5000$
$\quad\quad \underline{-0.4265}$

Answer $\quad 0.0735$

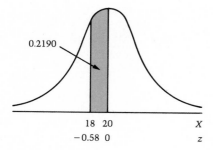

$z = (18 - 20)/3.46$
$\quad = -0.58$
Answer $\quad 0.2190$

2. $\$4.44 \pm (1.65)\left(\dfrac{\$1.27}{\sqrt{300}}\right) = \$4.44 \pm \$0.12$

We are 90% sure that the mean hourly wage is between $4.32 and $4.56.

Chapter 8
Sampling
Distributions and
the Central Limit
Theorem

4. $n = \left[\dfrac{z \cdot s}{E}\right]^2 = \left[\dfrac{(1.96)(\$2.50)}{\$0.50}\right]^2 = 96.04$

Use a sample of 97.

6. Obtain a list of all voters. With a table of random numbers, select the desired number of voters. Contact these voters and ask their opinion of the legislation. If personal interviews are needed, it might be possible to use cluster sampling to avoid travel costs.

8. One possibility is to make an observation each hour. A table of random numbers can be used to select the number of minutes after the hour (for example, 8:10, 9:37, 10:08) that the observation is made. Prepare a list of possible duties. Then, make an inspection each hour for one week and, at that time, record the activity performed.

10. $5.9 \pm (2.58)\left(\dfrac{2}{\sqrt{100}}\right) = 5.9 \pm 0.52$

From 5.38 to 6.42 years.

12. $25.6 \pm (1.65)\left(\dfrac{5.1}{\sqrt{200}}\right) = 25.6 \pm 0.60$

From 25.00 to 26.20 days.

14. Yes, a sample can be representative without being randomly selected. The sample mean, for example, might be exactly equal to the population mean. However, no assessment of variance or uncertainty of the estimate can be given.

16. $n = \left[\dfrac{z \cdot s}{E}\right]^2 = \left[\dfrac{(1.96)(75)}{20}\right]^2 = (7.35)^2 = 54.02$

Use a sample of 55.

Chapter 9
Hypothesis Tests:
Large-Sample
Methods

2. **a.** H_0: $\mu = \$1,010$
   H_a: $\mu \neq \$1,010$
   b. Reject H_0 if z is (not) between -2.58 and 2.58.

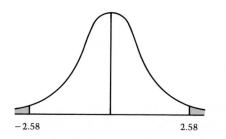

-2.58 2.58

c. $z = \dfrac{\$1,090 - \$1,010}{\$300/\sqrt{50}} = \dfrac{\$80}{\$42.43} = 1.88$

The evidence is insufficient to warrant concluding that there is a significant increase.

4. H_0: $\mu \le \$16,000$
H_a: $\mu > \$16,000$
$\alpha = 0.01$
Reject H_0 if $z \ge 2.33$.

$z = \dfrac{\$17,000 - \$16,000}{\$3,000/\sqrt{75}} = \dfrac{\$1,000}{\$346} = 2.89$

Reject H_0. The population mean is significantly greater than $16,000.

6. **a.** H_0: $\mu_1 = \mu_2$
H_a: $\mu_1 \ne \mu_2$
b. Reject H_0 if z is not between -2.58 and 2.58.

c. $z = \dfrac{202.6 - 200.0}{\sqrt{\dfrac{(3.3)^2}{40} + \dfrac{(2.0)^2}{50}}} = \dfrac{2.6}{\sqrt{.272 + .08}} = \dfrac{2.6}{.594} = 4.38$

Reject H_0. The means are different.

8. H_0: $\mu_1 \ge \mu_2$
H_a: $\mu_1 < \mu_2$
$\alpha = 0.01$
Reject H_0 if $z < -2.33$.

$z = \dfrac{265 - 268}{\sqrt{\dfrac{(60)^2}{800} + \dfrac{(50)^2}{600}}} = \dfrac{-3}{2.94} = -1.02$

We fail to reject H_0. The evidence does *not* support the claim.

10. Population 1 is from the windowless schools.
H_0: $\mu_1 \le \mu_2$
H_a: $\mu_1 > \mu_2$
$\alpha = 0.01$
Reject H_0 if $z > 2.33$.

$z = \dfrac{94 - 90}{\sqrt{\dfrac{8^2}{100} + \dfrac{10^2}{80}}} = \dfrac{4}{1.375} = 2.91$

Reject H_0. There is significantly more anxiety in windowless schools.

12. H_0: $\mu \geq 30$
H_a: $\mu < 30$
$\alpha = 0.02$
Reject H_0 if $z < -2.05$.

$$z = \frac{28.5 - 30}{5/\sqrt{40}} = \frac{-1.5}{0.79} = -1.90$$

The evidence is insufficient to reject H_0. There is no significant improvement in the processing time.

Chapter 10
Testing a Hypothesis
About Proportions

2. H_0: $p \geq 0.20$
H_a: $p < 0.20$
Accept H_0 if z is greater than -2.33.

$$z = \frac{\frac{60}{400} - 0.20}{\sqrt{\frac{(0.20)(1 - 0.20)}{400}}} = \frac{0.15 - 0.20}{0.02} = \frac{-0.05}{0.02} = -2.5$$

Fewer than 20% of those receiving welfare payments are ineligible.

4. H_0: $p_1 \leq p_2$
H_a: $p_1 > p_2$
p_1 refers to young drivers. Accept H_0 if it is greater than or equal to -1.65.

$$\bar{p} = \frac{30 + 55}{100 + 200} = \frac{85}{300} = 0.283$$

$$z = \frac{30 - 0.275}{\sqrt{(0.283)(0.717)\left(\frac{1}{100} + \frac{1}{200}\right)}} = \frac{0.025}{0.055} = 0.45$$

There is no difference in the proportion of drivers who take risks.

6. H_0: $p_1 = p_2$
H_a: $p_1 \neq p_2$
Accept H_0 if z is between -1.96 and 1.96.

$$\bar{p} = \frac{200 + 110}{1,000 + 500} = \frac{310}{1,500} = 0.207$$

$$z = \frac{0.20 - 0.22}{\sqrt{(0.207)(1 - 0.207)\left(\frac{1}{1,000} + \frac{1}{500}\right)}} = \frac{-0.02}{0.022} = -0.90$$

The difference is not significant.

8. $H_0: p_1 \leq p_2$
$H_a: p_1 > P_2$
p_1 = second period. Accept H_0 if z is greater than 2.33.

$$\bar{p} = \frac{83}{420} + \frac{146}{423} = \frac{229}{843} = 0.272$$

$$z = \frac{0.345 - 0.198}{\sqrt{0.272(1 - 0.272)\left(\frac{1}{42} + \frac{1}{42}\right)}}$$

$$z = \frac{0.148}{0.031} = 4.74$$

There has been a significant increase in car pooling.

2. $H_0: \mu \leq 4.3$ years
$H_a: \mu > 4.3$ years
$\alpha = 0.01$
$df = n - 1 = 20 - 1 = 19$
Reject H_0 if t exceeds 2.539; otherwise, fail to reject H_0.

$$t = \frac{\bar{X} - \mu}{\frac{s}{\sqrt{n}}} = \frac{4.6 - 4.3}{\frac{1.2}{\sqrt{20}}} = 1.12$$

The evidence is insufficient to support the new method.

4. $H_0: \mu \leq 1.6$
$H_a: \mu > 1.6$
$\alpha = 0.01$
$df = n - 1 = 5 - 1 = 4$
Reject H_0 if t is less than 3.747; otherwise, fail to reject H_0.

$$\bar{X} = \frac{1.2 + 2.5 + 1.9 + 3.0 + 2.4}{5} = 2.2$$

$$s = \sqrt{\frac{26.06 - \frac{(11)}{5}}{5 - 1}} = 0.68$$

**Chapter 11
Hypothesis Tests:
Small-Sample
Methods**

$$t = \frac{\overline{X} - \mu}{\frac{s}{\sqrt{n}}} = \frac{2.2 - 1.6}{\frac{0.68}{\sqrt{5}}} = 1.97$$

The data agree with the assumption that the mean antibody strength is ≤ 1.6. The null hypothesis cannot be rejected.

6. $H_0: \mu_1 \geq \mu_2$
$H_a: \mu_1 < \mu_2$
$\alpha = 0.01$
$df = 10 + 10 - 2 = 18$
Reject H_0 if t is less than -2.552; otherwise, fail to reject H_0.

$$s_p = \sqrt{\frac{9(0.8)^2 + 9(1.0)^2}{10 + 10 - 2}} = 0.906$$

$$t = \frac{2.5 - 3.1}{0.906 \sqrt{\frac{1}{10} + \frac{1}{10}}} = -1.48$$

The data do not support the claim that the isotonic method is more effective.

8.

LH	RH	d	d^2
140	138	2	4
90	87	3	9
125	110	15	225
130	132	-2	4
95	96	-1	1
121	120	1	1
85	86	-1	1
97	90	7	49
131	129	2	4
110	100	10	100
		36	398

$\overline{d} = 36/10 = 3.6$

$$s_d = \sqrt{\frac{398 - \frac{(36)^2}{10}}{10 - 1}} = 5.46$$

$H_0: D \leq 0$
$H_a: D > 0$
$\alpha = 0.01$
$n = 10$
Reject H_0 if t exceeds 2.821; otherwise, fail to reject H_0.

$$t = \frac{3.6}{\frac{5.46}{\sqrt{10}}} = 2.085$$

The evidence is insufficient to prove they have greater strength in their left hand.

10. H_0: $\mu \geq 20$ pounds overweight
    H_a: $\mu < 20$ pounds overweight
    $\alpha = 0.05$
    df $= 15 - 1 = 14$
    Reject H_0 if t is less than -1.761; otherwise, fail to reject H_0.

$$t = \frac{\overline{X} - \mu}{\frac{s}{\sqrt{n}}} = \frac{18 - 20}{\frac{5}{\sqrt{15}}} = -1.55$$

The evidence is insufficient to doubt the claim in the newspaper.

12. H_0: $\mu \geq 26.0$ miles per gallon
    H_a: $\mu < 26.0$ miles per gallon
    $\alpha = 0.01$
    df $= 6 - 1 = 5$
    Reject H_0 if t is less than -3.365; otherwise, fail to reject H_0.

$$t = \frac{\overline{X} - \mu}{\frac{s}{\sqrt{n}}} = \frac{25.03 - 26.0}{\frac{0.468}{\sqrt{6}}} = -5.08$$

$$\overline{X} = \frac{24.3 + 25.2 + 24.9 + 24.8 + 25.6 + 25.4}{6} = \frac{150.2}{6} = 25.03$$

$$s = \sqrt{\frac{\Sigma X^2 - \frac{(\Sigma X)^2}{n}}{n - 1}} = \sqrt{\frac{3{,}761.1 - \frac{(150.2)^2}{6}}{6 - 1}} = 0.468$$

The car does not meet specifications.

14. H_0: $D = 0$
    H_a: $D \neq 0$
    $\alpha = 0.10$
    df $= 16 - 1 = 15$
    Reject H_0 if t is less than -1.753 or more than 1.753; otherwise, fail to reject H_0.

Temperature		d	d^2
22	25	−3	9
26	25	1	1
28	25	3	9
24	25	−1	1
27	25	2	4
20	25	−5	25
29	25	4	16
32	25	7	49
28	25	3	9
21	25	−4	16
25	25	0	0
27	25	2	4
26	25	1	1
28	25	3	9
30	25	5	25
22	25	−3	9
		15	187

$$\bar{d} = \frac{15}{16} = 0.9375$$

$$s_d = \sqrt{\frac{\Sigma d^2 - \frac{(\Sigma d)^2}{n}}{n-1}}$$

$$= \sqrt{\frac{187 - \frac{(15)^2}{16}}{16-1}}$$

$$s_d = 3.3955$$

$$t = \frac{0.9375}{\frac{3.3955}{\sqrt{16}}} = 1.10$$

The claim of the National Weather Service cannot be refuted.

16. $H_0: D \leq 0$

$H_a: D > 0$

$\alpha = 0.01$

$\text{df} = 12 - 1 = 11$

Reject H_0 if t exceeds 2.718; otherwise, fail to reject H_0.

With	Without	d	d^2
230	217	13	169
225	198	27	729
223	208	15	225
216	222	− 6	36
229	223	6	36
201	214	−13	169
205	187	18	324
193	187	6	36
177	178	− 1	1
201	195	6	36
178	169	9	81
207	194	13	169
		93	2,011

$$\bar{d} = \frac{93}{12} = 7.75$$

$$s_d = \sqrt{\frac{\Sigma d^2 - \frac{(\Sigma d)^2}{n}}{n-1}}$$

$$= \sqrt{\frac{2,011 - \frac{(93)^2}{12}}{12-1}}$$

$$s_d = 10.83$$

$$t = \frac{7.75}{\frac{10.83}{\sqrt{12}}} = 2.48$$

The evidence does not support the manufacturer's claim.

18. $H_0: D \leq 0$
$H_a: D > 0$
df $= 8 - 1 = 7$
Reject H_0 if t is greater than 1.895; otherwise, fail to reject H_0.

After	Before	d	d²
14.0	13.5	0.5	0.25
10.7	11.4	−0.7	0.49
12.4	10.7	1.7	2.89
11.1	11.1	0.0	0.00
10.9	9.8	1.1	1.21
10.5	9.6	0.9	0.81
10.8	10.7	0.1	0.01
13.0	11.7	1.3	1.69
		4.9	7.35

$$\bar{d} = \frac{4.9}{8} = 0.61$$

$$s_d = \sqrt{\frac{7.35 - \frac{(4.9)^2}{8}}{8 - 1}}$$

$$= 0.788$$

$$t = \frac{0.61}{\frac{0.788}{\sqrt{8}}} = 2.19$$

H_0 is rejected and H_a accepted. The sales training program is effective.

**Chapter 12
Analysis of Variance**

2. $H_0: \mu_1 = \mu_2 = \mu_3 = \mu_4$
$H_a:$ not all equal
Reject H_0 if the computed value of F is greater than 5.01.

$$SS \text{ total} = 9{,}089 - \frac{(407)^2}{20} = 806.55$$

$$SST = \frac{(80)^2}{5} + \frac{(107)^2}{6} + \frac{(139)^2}{5} + \frac{(81)^2}{4} - \frac{(407)^2}{20} = 410.17$$

$$SSE = 806.55 - 410.17 = 396.38$$

Source	Sum of Squares	df	Mean Square
Treatment	410.17	3	136.72
Within	396.38	16	24.77
Total	806.55		

$$F = \frac{136.72}{24.77} = 5.52$$

H_0 is rejected and H_a accepted. The ethical behavior of attorneys varies by region.

4. $H_0: \mu_1 = \mu_2 = \mu_3 = \mu_4$
$H_a:$ not all equal
Reject H_0 if F is greater than 3.24; otherwise, do not reject.

$$SS \text{ total} = 16{,}257 - \frac{(555)^2}{20} = 855.75$$

$$SST = \frac{(161)^2}{5} + \frac{(122)^2}{5} + \frac{(120)^2}{5} + \frac{(152)^2}{5} - \frac{(555)^2}{20} = 260.55$$

$$SSE = 855.75 - 260.55 = 595.20$$

Source	Sum of Squares	df	Mean Square
Treatment	260.55	3	86.85
Within	595.20	16	37.20
Total	855.75		

$$F = \frac{86.85}{37.20} = 2.33.$$

H_0 cannot be rejected. The evidence does not suggest any differences in reading assignments.

6. $H_0: \mu_1 = \mu_2 = \mu_3 = \mu_4$
$H_a:$ not all equal
Reject H_0 if F is greater than 3.41; otherwise, do not reject.

$$SS \text{ total} = 3{,}370.58 - \frac{(216.6)^2}{17} = 610.84$$

$$SST = \frac{(48.1)^2}{4} + \frac{(51.9)^2}{5} + \frac{(33.1)^2}{3} + \frac{(83.5)^2}{5} - \frac{(216.6)^2}{17}$$
$$= 117.04$$

$$SSE = 610.84 - 117.04 = 493.80$$

Source	Sum of Squares	df	Mean Square
Treatment	117.04	3	39.01
Within	493.80	13	37.985
Total	610.84	16	

$$F = \frac{39.01}{37.985} = 1.03$$

Do not reject H_0. There is no difference in the time to solve the puzzle.

8. H_0: $\mu_1 = \mu_2 = \mu_3$
H_a: not all equal
Reject H_0 if the computed F is greater than 4.56.

$$SS \text{ total} = 8,173 - \frac{(379)^2}{21} = 1,332.95$$

$$SST = \frac{(210)^2}{8} + \frac{(103)^2}{7} + \frac{(66)^2}{6} - \frac{(379)^2}{21} = 914.03$$

$$SSE = 1,332.95 - 914.03 = 418.92$$

Source	Sum of Squares	df	Mean Square
Treatment	914.03	2	457.02
Within	418.92	18	23.27
Total	1,332.95		

$$F = \frac{457.02}{23.27} = 19.64$$

H_0 is rejected and H_a accepted. There is a difference among the three methods.

10. H_0: $\mu_1 = \mu_2 = \mu_3 = \mu_4$
H_a: not all equal
Reject H_0 if F is greater than 2.86.

Source	Sum of Squares	df	Mean Square
Treatment	3,055,700,000	3	1,018,523,000
Within	5,793,700,000	39	148,556,000
Total	8,849,400,000	42	

$$F = \frac{1,018,523,000}{148,556,000} = 6.86$$

H_0 is rejected. The attendance was different depending on the starting pitcher.

Chapter 13
Correlation Analysis

2.

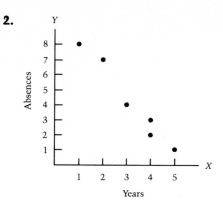

Absences decrease as years of employment increase.

4.

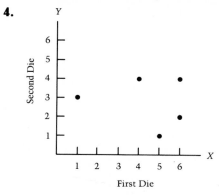

The two dice are independent; there is no pattern to the scatter diagram.

6.

X	Y	XY	X²	Y²
1	8	8	1	64
5	1	5	25	1
2	7	14	4	49
4	3	12	16	9
4	2	8	16	4
3	4	12	9	16
19	25	59	71	143

$$r = \frac{6(59) - (19)(25)}{\sqrt{[6(71) - (19)^2][6(143) - (25)^2]}}$$

$$= \frac{-121}{\sqrt{(65)(233)}} = -0.98$$

8. a.

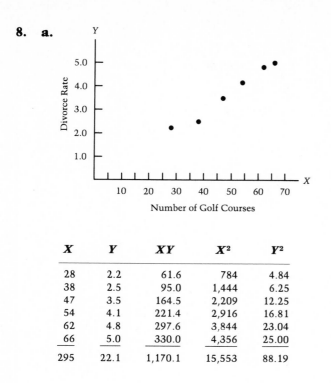

X	Y	XY	X²	Y²
28	2.2	61.6	784	4.84
38	2.5	95.0	1,444	6.25
47	3.5	164.5	2,209	12.25
54	4.1	221.4	2,916	16.81
62	4.8	297.6	3,844	23.04
66	5.0	330.0	4,356	25.00
295	22.1	1,170.1	15,553	88.19

b. $r = \dfrac{6(1,170.1) - (295)(22.1)}{\sqrt{[6(15,553) - (295)^2][6(88.19) - (22.1)^2]}}$

$\quad = \dfrac{501.1}{\sqrt{(6,293)(40.73)}} = 0.99$

c. The divorce rate definitely increases as the number of golf courses increases.

10. a. $(0.961)^2 = 0.924$
  b. $1 - 0.924 = 0.076$
  c. 92.4% of the variation in the number of people on the beach is accounted for by the variation in the high temperature.

12. a. $(-0.683)^2 = 0.466$
  b. $1 - 0.466 - 0.534$
  c. 53.4% of the variation in suspensions is *not* accounted for by the variation in the number of activities offered.

14. $t = \dfrac{0.40\sqrt{20 - 2}}{\sqrt{1 - (0.4)^2}} = \dfrac{1.697}{0.917} = 1.85$

The critical value is 2.8784; therefore, the population correlation could be zero.

16. $t = \dfrac{0.21\sqrt{20 - 2}}{\sqrt{1 - (0.21)^2}} = 0.91$

The null hypothesis of zero correlation is rejected if t is not between -2.101 and $+2.101$; therefore, the population coefficient of correlation is not significantly different from zero.

18.

d	d^2
1	1.00
-4	16.00
-0.5	0.25
1	1.00
-2	4.00
-1.5	2.25
2	4.00
3	9.00
2	4.00
-1	1.00
0	42.50

$$r_s = 1 - \frac{6(42.5)}{10(10^2 - 1)} = 0.74$$

20. a.

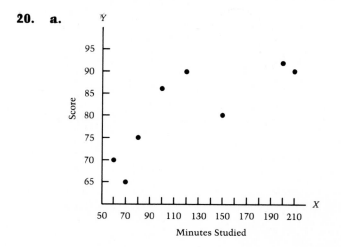

b.

X	Y	XY	X²	y²
60	70	4,200	3,600	4,900
70	65	4,550	4,900	4,225
80	75	6,000	6,400	5,625
100	86	8,600	10,000	7,396
120	90	10,800	14,400	8,100
150	80	12,000	22,500	6,400
200	92	18,400	40,000	8,464
210	90	18,900	44,100	8,100
990	648	83,450	145,900	53,210

$$r = \frac{8(83,450) - (990)(648)}{\sqrt{[8(145,900) - (990)^2][8(53,210) - (648)^2]}}$$

$$= \frac{26,080}{\sqrt{[187,100][5,776]}} = 0.793$$

$$r^2 = (0.793)^2 = 0.629$$

$$1 - r^2 = 1 - 0.629 = 0.371$$

c. $t = \dfrac{0.793\sqrt{8-2}}{\sqrt{1-(0.793)^2}} = \dfrac{1.942}{0.609} = 3.19$

At the 0.05 significance level with 6 degrees of freedom, the critical value of t is 2.447. So the population correlation is significantly different from zero.

d. This correlation is not due to sampling error. As the time spent studying increased, so did the score. Sixty-three percent of the variation in score is explained by the variation in time spent studying.

22. a. $n = 11$, $\Sigma X = 666$, $\Sigma Y = 32.0$, $\Sigma XY = 4,067.11$, $\Sigma X^2 = 71,413.32$, and $\Sigma Y^2 = 252.64$

$$r = \frac{11(4,067.11) - (666)(32.0)}{\sqrt{[11(71,413.32) - (666)^2][11(252.64) - (32.0)^2]}}$$
$$= 0.96$$

$$r^2 = (0.96)^2 = 0.92$$

$$1 - r^2 = 1 - 0.92 = 0.08$$

b. $t = \dfrac{0.96\sqrt{11-2}}{\sqrt{1-(0.96)^2}} = 10.29$

The critical value of t with 9 degrees of freedom at the 0.05 level of significance is 2.262. Since t is larger than the critical value, we reject the hypothesis of no correlation.

c. There appears to be a very close relationship between the number of crimes and the number of law enforcement personnel. Ninety-two percent of the variation in one explains the variation in the other variable.

24. **a.**

Completion Rank	Grade Rank	d	d^2
1	3.5	−2.5	6.25
2	3.5	−1.5	2.25
3	6	−3	9.00
4	2	2	4.00
5	1	4	16.00
6	5	1	1.00
7	7	0	0.00
8	10	−2	4.00
9	8	1	1.00
10	9	1	1.00
			44.50

b.

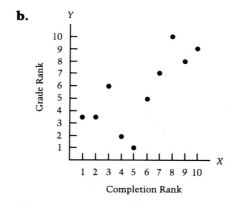

c. $r_s = 1 - \dfrac{6(44.5)}{10[10^2 - 1]} = 0.73$

d. There does appear to be a strong direct relationship between order of completion and grade rank.

Chapter 14
Regression Analysis

2. **a.** $Y' = a + bX$
b. Approximately −100 employees.
c. Approximately 0.20 minorities per nonminority, estimated

by using the points (1,000, 100) and (2,000, 300) [(300 − 100)/(2,000 − 1,000) = 0.20].

d. $Y' = -100 + 0.2(3,000) = 500$ minority employees

e. The negative value for **a** indicates that smaller firms have no minority employees. The slope b means they generally hire 2 minorities for every 10 nonminorities.

4. **a.** $Y' = 39 - 0.002(7,000) = 25$

b. The Y-intercept is 39.

c. The slope is −0.002.

d. For each $1,000 increase in median annual income, the birthrate *decreases* by 2 per thousand.

6. **a.**

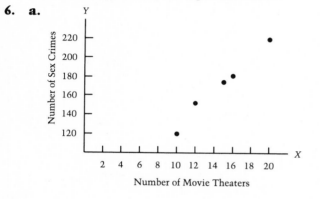

Number of Movie Theaters

b.

X	Y	XY	X²
15	175	2,625	225
20	220	4,400	400
10	120	1,200	100
12	152	1,824	144
16	181	2,896	256
73	848	12,945	1,125

$$b = \frac{5(12,945) - (73)(848)}{5(1,125) - (73)^2} = \frac{2,821}{296} = 9.53$$

$$a = \frac{848 - 9.53(73)}{5} = 30.46$$

$$Y' = 30.46 + 9.53X$$

c. If there were no theaters there would be 30 crimes. Each additional theater adds 9.53 crimes.

d. $Y' = 30.46 + 9.53(18) = 202$

8. a.

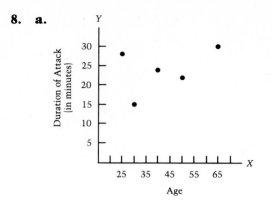

b.

X	Y	XY	X²
30	15	450	900
25	28	700	625
65	30	1,950	4,225
50	22	1,100	2,500
40	24	960	1,600
210	119	5,160	9,850

$$b = \frac{5(5,160) - (210)(119)}{5(9,850) - (210)^2}$$

$$= \frac{810}{5,150} = 0.16$$

$$a = \frac{119 - 0.16(210)}{5} = 17.08$$

$$Y' = 17.08 + 0.16X$$

c. Ten years of age appears to add 1.6 minutes to an attack's duration. The average attack is at least 17.08 minutes.

d. $Y' - 17.08 + 0.16(42) = 23.8$

For a 42-year-old person we would expect an attack to last 23.8 minutes, on the average.

10. $s_{Y \cdot X} = \sqrt{\dfrac{13,213 - (-25.7)(259) - (1.04)(18,992)}{6 - 2}}$

$$= \sqrt{\frac{117.62}{4}} = 5.42$$

12. **a.**

X	Y	XY	X^2	Y^2
5	1,110	5,550	25	1,232,100
2	490	980	4	240,100
12	2,500	30,000	144	6,250,000
4	880	3,520	16	774,400
7	1,530	10,710	49	2,340,900
1	270	270	1	72,900
31	6,780	51,030	239	10,910,400

$$b = \frac{6(51,030) - (31)(6,780)}{6(239) - (31)^2} = \frac{96,000}{473} = 202.96$$

$$a = \frac{6,780 - 202.96(31)}{6} = 81.37$$

$$Y' = 81.37 + 202.96X$$

b. $$s_{Y \cdot X} = \sqrt{\frac{10,910,400 - 81.37(6,780) - 202.96(51,030)}{6 - 2}}$$

$$= \sqrt{\frac{1,662.6}{4}} = 20.39$$

14. **a.** $$57.8 \pm (2.776)(5.42)\sqrt{\frac{1}{6} + \frac{(80 - 66)^2}{27,954 - (396)^2/6}}$$

$$= 57.8 \pm (15.05)(.5253)$$
$$= 57.8 \pm 7.9$$
From 49.9 to 65.7.

b. $$57.8 \pm (2.776)(5.42)\sqrt{1 + \frac{1}{6} + \frac{(80 - 66)^2}{27,954 - (396)^2/6}}$$

$$= 57.8 \pm (15.05)(1.1289)$$
$$= 57.8 \pm 17.0$$
From 40.8 to 74.8.

16. **a.** $$Y' = 81.37 + 202.96(6) = 1,299.13$$

$$1,299.13 \pm (2.132)(20.39)\sqrt{\frac{1}{6} + \frac{(6 - 5.17)^2}{239 - (31)^2/6}}$$

$$= 1,299.13 \pm (43.47)(0.4188)$$
$$= 1,299.13 \pm 18.21$$

We are 90% confident that the mean cost is between $1,280.92 and $1,317.34.

b. $1,299.13 \pm (43.47)(1.0842)$

$= 1,299.13 \pm 47.13$

We are 90% confident that the cost for the individual is between $1,252.00 and $1,346.26.

18. a.

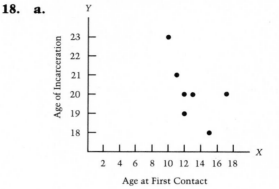

b.

X	Y	XY	X²	Y²
11	21	231	121	441
17	20	340	289	400
13	20	260	169	400
12	19	228	144	361
15	18	270	225	324
10	23	230	100	529
12	20	240	144	400
90	141	1,799	1,192	2,855

$$b = \frac{7(1,799) - (90)(141)}{7(1,192) - (90)^2} = \frac{-97}{244} = -0.40$$

$$a = \frac{141 - (-0.40)(90)}{7} = 25.25$$

$$Y' = 25.25 - 0.40X$$

c. $s_{Y \cdot X} = \sqrt{\dfrac{2,855 - 25.25(141) - (-0.40)(1,799)}{7 - 2}} = 1.37$

d. $Y' = 25.25 - 0.40(14) = 19.65$

$$19.65 \pm (2.571)(1.37)\sqrt{1 + \frac{1}{7} + \frac{(14 - 12.86)^2}{1,192 - (90)^2/7}}$$

$= 19.65 \pm 3.83$

From 15.82 to 23.48.

20. a.

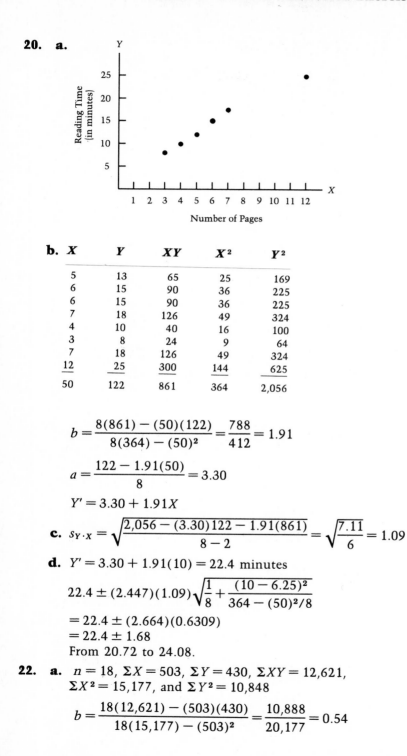

b.

X	Y	XY	X²	Y²
5	13	65	25	169
6	15	90	36	225
6	15	90	36	225
7	18	126	49	324
4	10	40	16	100
3	8	24	9	64
7	18	126	49	324
12	25	300	144	625
50	122	861	364	2,056

$$b = \frac{8(861) - (50)(122)}{8(364) - (50)^2} = \frac{788}{412} = 1.91$$

$$a = \frac{122 - 1.91(50)}{8} = 3.30$$

$$Y' = 3.30 + 1.91X$$

c. $s_{Y \cdot X} = \sqrt{\dfrac{2{,}056 - (3.30)122 - 1.91(861)}{8 - 2}} = \sqrt{\dfrac{7.11}{6}} = 1.09$

d. $Y' = 3.30 + 1.91(10) = 22.4$ minutes

$$22.4 \pm (2.447)(1.09)\sqrt{\frac{1}{8} + \frac{(10 - 6.25)^2}{364 - (50)^2/8}}$$

$$= 22.4 \pm (2.664)(0.6309)$$
$$= 22.4 \pm 1.68$$

From 20.72 to 24.08.

22. a. $n = 18$, $\Sigma X = 503$, $\Sigma Y = 430$, $\Sigma XY = 12{,}621$, $\Sigma X^2 = 15{,}177$, and $\Sigma Y^2 = 10{,}848$

$$b = \frac{18(12{,}621) - (503)(430)}{18(15{,}177) - (503)^2} = \frac{10{,}888}{20{,}177} = 0.54$$

$$a = \frac{430 - 0.54(503)}{18} = 8.81$$

$$Y' = 8.81 + 0.54X$$

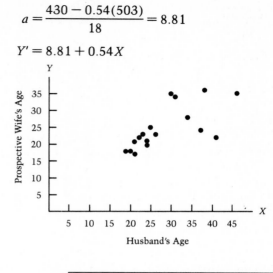

b. $s_{Y \cdot X} = \sqrt{\dfrac{10{,}848 - 8.81(430) - 0.54(12{,}621)}{18 - 2}}$

$$= \sqrt{\frac{249.37}{16}} = 3.95$$

c. $Y' = 8.81 + 0.54(40) = 30.39$

The predicted age of the prospective wife is 30.39 years.

$$30.39 \pm 1.746(3.95) \sqrt{\frac{1}{18} + \frac{(40 - 27.9)^2}{15{,}177 - \dfrac{(503)^2}{18}}}$$

$$= 30.39 \pm 2.98$$

The confidence interval is 27.41 to 33.37. We are 90% sure that the wife's mean age is between 27.41 and 33.37 years.

24. a. $n = 14$, $\Sigma X = 964.3$, $\Sigma Y = 2{,}086$, $\Sigma XY = 144{,}231$, $\Sigma X^2 = 66{,}553.57$, and $\Sigma Y^2 = 316{,}730$

$$b = \frac{14(144{,}231) - (964.3)(2{,}086)}{14(66{,}553.57) - (964.3)^2} = \frac{7{,}704.2}{1{,}875.49} = 4.11$$

$$a = \frac{2{,}086 - 4.11(964.3)}{14} = -133.94$$

$$Y' = -133.94 + 4.11X$$

b. $s_{Y \cdot X} = \sqrt{\dfrac{316{,}730 - (-133.94)(2{,}086) - 4.11(144{,}231)}{14 - 2}}$

$$= \sqrt{\frac{3{,}655.46}{12}} = 17.45$$

c. $Y' = -133.94 + 4.11(72) = 161.82$

$$161.82 \pm (2.179)(17.45)\sqrt{\frac{1}{14} + \frac{(72 - 68.88)^2}{66{,}553.57 - (964.3)^2/14}}$$

$= 161.82 \pm (38.024)(0.3797)$

161.82 ± 14.44 for all men who are six feet tall.
We are 95% sure their average weight is between 147.38 pounds and 176.26 pounds.

d. $161.82 \pm (38.024)(1.0697) = 161.82 \pm 40.67$
The 95% confidence interval for John Kuk is from 121.15 pounds to 202.49 pounds.

2. a. $Y' = 30 + 1.05(3) + 6.5(8) + 0.005(12{,}000) = 145.15$
b. Another year without a child adds $6.5(1) = 6.5$. An additional $5,000 in income adds $0.005(5{,}000) = 25$, so the income contributes more to satisfaction.

4. a. $Y' = 6.0 + 0.06(150) + 0.05(310) - 2.75(3) = 22.25$
b. A turnover costs a team an average of 2.75 points.

6. a. $Y' = 2.43 + 0.32(50) + 8.89(3) + 0.14(40) = 50.7$, or $50,700
b. $8,890
c. $0.32(\$10{,}000) = \$3{,}200$

Chapter 15
Multiple Regression
and Correlation
Analysis

8. a.

Column	C1	C2	C3
Count	14	14	14
Row			
1	70.	7.	21.
2	20.	14.	9.
3	40.	10.	16.
4	70.	8.	18.
5	30.	12.	9.
6	50.	9.	19.
7	60.	6.	17.
8	10.	12.	7.
9	20.	11.	12.
10	65.	15.	10.
11	40.	10.	5.
12	60.	4.	20.
13	50.	8.	8.
14	50.	6.	12.

	C1	C2
C2	−0.537	
C3	0.632	−0.617

The manual dexterity (C3) score enters first because it has a stronger correlation with bonus (C1).

b. The multiple regression equation is
$Y' = 35.83 - 1.46(X_1) + 1.78(X_2)$
where $X_1 =$ years of experience
and $X_2 =$ dexterity score
The coefficient of multiple determination is 0.435.
The regression equation is: $Y = 35.8 - 1.46X_1 + 1.78X_2$

Column		Coefficient	Standard Deviation of Coefficient	T-Ratio = Coefficient/ Standard Deviation
	—	35.83	27.69	1.29
X1	C2	−1.455	1.762	−0.82
X2	C3	1.778	1.054	1.68

The standard deviation of Y about regression line is $S = 15.90$ with $(14 - 3) = 11$ degrees of freedom.

10. The answers will vary depending on the data selected by the student.

Chapter 16
Analysis of Nominal-Level Data: The Chi-Square Distribution

2. H_0: the die is fair
H_a: the die is not fair
df $= 6 - 1 = 5$
Reject H_0 if χ^2 is greater than 11.070.

Number of Spots	f_o	f_e	$f_o - f_e$	$(f_o - f_e)^2$	$\dfrac{(f_o - f_e)^2}{f_e}$
1	15	20	−5	25	1.25
2	29	20	9	81	4.05
3	14	20	−6	36	1.80
4	17	20	−3	9	0.45
5	28	20	8	64	3.20
6	17	20	−3	9	0.45
			0		11.20

H_0 is rejected. The die is not fair.

4. H_0: there is no contradiction
H_a: there is a contradiction
df $= 4 - 1 = 3$
Reject H_0 if χ^2 is greater than 7.815.

Type of Skin	f_o	f_e	$f_o - f_e$	$(f_o - f_e)^2$	$\dfrac{(f_o - f_e)^2}{f_e}$
Round yellow	334	342	−8	64	0.19
Wrinkled yellow	123	114	9	81	0.71
Round green	120	114	6	36	0.32
Wrinkled green	31	38	−1	49	1.29
	608		0		2.51

H_0 cannot be rejected. The results are not contradictory to what is expected.

6. H_0: there is no relationship between rating and performance
H_a: there is a relationship between rating and performance
df $= (4 - 1)(4 - 1) = 9$
Reject H_0 if χ^2 is greater than 21.666.

	Ratings								
	Below Average		Average		Above Average		Outstanding		
Performance	f_o	f_e	f_o	f_e	f_o	f_e	f_o	f_e	Total
Poor	11	4.80	9	6.40	7	9.60	5	11.20	32
Fair	8	7.05	18	9.40	11	14.10	10	16.45	47
Good	6	9.60	8	12.80	28	19.20	22	22.40	64
Superior	5	8.55	5	11.40	14	17.10	33	19.95	57
Total	30		40		60		70		200

$$\chi^2 = \frac{(11 - 4.80)^2}{4.80} + \frac{(9 - 6.40)^2}{6.40} + \frac{(7 - 9.60)^2}{9.60} + \frac{(5 - 11.2)^2}{11.2}$$

$$+ \frac{(8 - 7.05)^2}{7.05} + \frac{(18 - 9.40)^2}{9.40} + \frac{(11 - 14.10)^2}{14.10}$$

$$+ \frac{(10 - 16.45)^2}{16.45} + \frac{(6 - 9.60)^2}{9.60} + \frac{(8 - 12.80)^2}{12.80}$$

$$+ \frac{(28 - 19.20)^2}{19.20} + \frac{(22 - 22.4)^2}{22.4} + \frac{(5 - 8.55)^2}{8.55}$$

$$+ \frac{(5 - 11.40)^2}{11.40} + \frac{(14 - 17.1)^2}{17.1} + \frac{(33 - 19.95)^2}{19.95} = 45.76$$

H_0 is rejected. There is a relationship between rating and performance.

8. H_0: there is no difference with respect to social class
H_a: there is a difference with respect to social class
df $= 5 - 1 = 4$
Reject H_0 if χ^2 is greater than 9.488.

Social Class	f_o	f_e	$f_o - f_e$	$(f_o - f_e)^2$	$\dfrac{(f_o - f_e)^2}{f_e}$
Upper	18	77	−59	3,481	45.21
Upper-middle	31	77	−46	2,116	27.48
Middle	46	77	−31	961	12.48
Lower-middle	126	77	49	2,401	31.18
Lower	164	77	87	7,569	98.30
	385	385			214.65

H_0 is rejected. There is a difference with respect to social class.

10. H_0: years of education follows the expected frequency
H_a: years of education does not follow the expected frequency
df = 4 − 1 = 3
Reject H_0 if χ^2 is greater than 7.815.

Years of Education	f_o	f_e	$f_o - f_e$	$(f_o - f_e)^2$	$\dfrac{(f_o - f_e)^2}{f_e}$
1 - 8	77	100	−23	529	5.29
9 - 12	198	150	48	2,304	15.36
12 - 16	178	200	−22	484	2.42
Over 16	47	50	− 3	9	0.18
					23.25

H_0 is rejected. The observed frequency differs from the expected frequency.

12. H_0: age and income are not related
H_a: age and income are related
df = (3 − 1)(3 − 1) = 4
Reject H_0 if χ^2 is greater than 13.277.

Age	Less than $100,000 f_o	Less than $100,000 f_e	$100,000 to $399,999 f_o	$100,000 to $399,999 f_e	$400,000 or More f_o	$400,000 or More f_e	Total
Under 40	6	6.67	9	7.62	5	5.71	20
40 to 54	18	15.00	19	17.14	8	12.86	45
55 or older	11	13.33	12	15.24	17	11.43	40
Total	35		40		30		105

$$\chi^2 = \frac{(6-6.67)^2}{6.67} + \frac{(9-7.62)^2}{7.62} + \frac{(5-5.71)^2}{5.71}$$

$$+ \frac{(18-15)^2}{15} + \frac{(19-17.14)^2}{17.14} + \frac{(8-12.86)^2}{12.86}$$

$$+ \frac{(11-13.33)^2}{13.33} + \frac{(12-15.24)^2}{15.24} + \frac{(17-11.43)^2}{11.43} = 6.85$$

H_0 cannot be rejected. There is no relationship shown between age and income.

14. H_0: the distribution is binomial
    H_a: the distribution is not binomial
    df $= 5 - 1 = 4$
    Reject H_0 if χ^2 is greater than 13.277.
    With $p = 0.5$ and $n = 4$, Appendix A is used to obtain the probability for the various number of heads. These probabilities are then multiplied by the number of trials, which is 100.

Number of Heads	Observed f_o	Expected f_e		$f_o - f_e$	$(f_o - f_e)^2$	$\dfrac{(f_o - f_e)^2}{f_e}$
0	8	100(0.063) =	6.3	1.7	2.89	0.46
1	30	100(0.250) =	25.0	5.0	25.00	1.00
2	29	100(0.375) =	37.5	−8.5	72.25	1.93
3	23	100(0.250) =	25.0	−2.0	4.00	0.16
4	10	100(0.063) =	6.3	+3.7	13.69	2.17
	100		100.1			5.72

H_0 is not rejected. The possibility that the distribution is binomial cannot be dismissed.

2. H_0: $p = 0.50$
   H_a: $p \neq 0.50$
   $n = 6$
   $P(X = 0) + P(X = 6) = 0.016 + 0.016 = 0.032$
   H_0 is rejected if the signed differences are all +, or all −. Since there are five instances where there are more + signs, H_0 cannot be rejected. The evidence does not suggest a salary difference.

4. H_0: $p = 0.50$
   H_a: $p \neq 0.50$
   Since both np and $n(1 - p)$ are greater than five, the normal approximation is appropriate. H_0 is accepted if z is in the interval −1.96 to 1.96.

Chapter 17
Nonparametric Methods: Analysis of Ranked Data

$$z = \frac{20 - 14}{\sqrt{28(0.5)(0.5)}} = 2.27$$

H_0 is rejected and H_a accepted. The attitudes of parents and offspring are different.

6.

Pair	Difference	Rank	Signed Rank R^+	Signed Rank R^-
1	+ 8	4	4	
2	− 9	5.5		5.5
3	+21	13	13	
4	+13	9.5	9.5	
5	− 5	3		3
6	−22	14.5		14.5
7	− 3	1		1
8	+19	12	12	
9	−10	7		7
10	−13	9.5		9.5
11	−22	14.5		14.5
12	+ 4	2	2	
13	−11	8		8
14	+ 9	5.5	5.5	
15	−16	11		11
			46.0	74.0

H_0: sum of the ranks is the same
H_a: sum of the ranks is different
For a two-tailed test, $\alpha = 0.05$ and $n = 15$, the decision rule is to reject H_0 if the smaller of the sums is less than or equal to 25. Since $R^+ = 46$, H_0 cannot be rejected.

8. H_0: there is no difference
H_a: there is a difference

Active Students Score	Active Students Rank	Nonactive Students Score	Nonactive Students Rank
5	4.5	1	1
6	6	2	2
8	8.5	4	3
10	11.5	5	4.5
11	13	7	7
13	14	8	8.5
14	15	9	10
15	16	10	11.5
		16	17
		17	15
	88.5		82.5

Do not reject H_0 if z is between -1.645 and 1.645.

$$z = \frac{W - \dfrac{n_1(n_1 + n_2 + 1)}{2}}{\sqrt{\dfrac{n_1(n_2)(n_1 + n_2 + 1)}{12}}}$$

$$= \frac{88.5 - \dfrac{8(8 + 10 + 1)}{2}}{\sqrt{\dfrac{8(10)(8 + 10 + 1)}{12}}} = \frac{88.5 - 76}{\sqrt{11.25}} = 1.11$$

Do not reject H_0.

10. H_0: the populations are the same
H_a: the populations are different

Lawyers		Physicians		Veterinarians		College Professors	
Amount	Rank	Amount	Rank	Amount	Rank	Amount	Rank
$200,000	22	$120,000	11	$ 20,000	1	$ 30,000	3
120,000	11	190,000	18.5	105,000	9	25,000	2
75,000	7	200,000	22	90,000	8	120,000	11
65,000	5	195,000	20	70,000	6	190,000	18.5
140,000	16	130,000	14	200,000	22	45,000	4
135,000	15			125,000	13		
				175,000	17		
	76.0		85.5		76		38.5

The decision rule is: Reject H_0 if H exceeds 11.345.

$$H = \frac{12}{23(24)}\left(\frac{(76)^2}{6} + \frac{(85.5)^2}{5} + \frac{(76)^2}{7} + \frac{(38.5)^2}{5}\right) - 3(23 + 1)$$
$$= 5.09$$

Since the computed value of H is less than 11.345, H_0 cannot be rejected. There is no significant difference in the amount of insurance carried by each group.

12. H_0: $p \leq 0.50$
H_a: $p > 0.50$
There is one tie. Both gave eight tickets on 5-4, hence the sample size is reduced to 24. Both np and $n(1 - p)$ are greater than five; therefore, the normal approximation may be used [$24(0.5) = 12$ and $24(1 - 0.5) = 12$]. The null hypothesis is not rejected if z is less than 2.33.

$$z = \frac{19 - 12}{\sqrt{6}} = 2.86$$

Since the computed value of z is greater than the critical value, H_0 is rejected and H_a accepted. We conclude that policeman A gives more tickets than B.

14. **a.** H_0: $p \leq 0.5$

H_a: $p > 0.5$

The number of home runs hit by the American League leader is subtracted from the National League leader's total. Hence the expected sign of the difference is $+$ if H_0 is false. There are 10 $+$ signs and since np and $n(1 - p)$ are both greater than five, the normal approximation to the binomial may be used. The null hypothesis is not rejected if the computed value of z is less than or equal to 1.645.

$$z = \frac{X - np}{\sqrt{np(1 - p)}} = \frac{10 - 8.5}{\sqrt{17(0.5)(0.5)}} = 0.73$$

The null hypothesis cannot be rejected.

b. National	American	Difference	Rank	Sign of Rank R^+	R^-
38	32	+ 6	12	12	
38	36	+ 2	5.5	5.5	
36	32	+ 4	8.5	8.5	
44	32	+12	14	14	
40	37	+ 3	7	7	
48	33	+15	15.5	15.5	
45	44	+ 1	2.5	2.5	
45	49	− 4	8.5		8.5
36	44	− 8	13		13
39	44	− 5	10.5		10.5
44	49	− 5	10.5		10.5
52	32	+20	17	17	
47	49	− 2	5.5		5.5
44	45	− 1	2.5		2.5
49	48	+ 1	2.5	2.5	
46	61	−15	15.5		15.5
41	40	+ 1	2.5	2.5	
				87.0	66.0

From Appendix G, H_0 is rejected if the smaller of the two sums is less than or equal to 41. Since 66 is greater than 41, H_0 cannot be rejected.

16.	Judge Cain Payments	Rank	Judge Stevens Payments	Rank
	20	1	30	2
	40	4	35	3
	55	7	48	5
	130	11	51	6
	145	13	63	8
	160	14	110	9
	210	15	125	10
	230	16	140	12
	300	17		
	340	18		
	500	19	—	
		135		55

H_0: amounts awarded by the judges are the same
H_a: amounts awarded by the judges are not the same
H_0 is not rejected if it is in the interval -2.58 to 2.58.

$$z = \frac{W - \dfrac{n_1(n_1 + n_2 + 1)}{2}}{\sqrt{\dfrac{n_1 \cdot n_2(n_1 + n_2 + 1)}{12}}} = \frac{55 - \dfrac{8(11 + 8 + 1)}{2}}{\sqrt{\dfrac{11 \cdot 8(11 + 8 + 1)}{12}}}$$

$$= \frac{55 - 80}{\sqrt{146.67}} = -2.06$$

The null hypothesis cannot be rejected.

Appendix Tables

A—The Binomial Probability Distribution

X	.05	.1	.2	.3	.4	.5	.6	.7	.8	.9	.95
0	.950	.900	.800	.700	.600	.500	.400	.300	.200	.100	.050
1	.050	.100	.200	.300	.400	.500	.600	.700	.800	.900	.950

X	.05	.1	.2	.3	.4	.5	.6	.7	.8	.9	.95
0	.903	.810	.640	.490	.360	.250	.160	.090	.040	.010	.003
1	.095	.180	.320	.420	.480	.500	.480	.420	.320	.180	.095
2	.003	.010	.040	.090	.160	.250	.360	.490	.640	.810	.903

X	.05	.1	.2	.3	.4	.5	.6	.7	.8	.9	.95
0	.857	.729	.512	.343	.216	.125	.064	.027	.008	.001	.000
1	.135	.243	.384	.441	.432	.375	.288	.189	.096	.027	.007
2	.007	.027	.096	.189	.288	.375	.432	.441	.384	.243	.135
3	.000	.001	.008	.027	.064	.125	.216	.343	.512	.729	.857

N= 4
PROBABILITY

X	.05	.1	.2	.3	.4	.5	.6	.7	.8	.9	.95
0	.815	.656	.410	.240	.130	.063	.026	.008	.002	.000	.000
1	.171	.292	.410	.412	.346	.250	.154	.076	.026	.004	.000
2	.014	.049	.154	.265	.346	.375	.346	.265	.154	.049	.014
3	.000	.004	.026	.076	.154	.250	.346	.412	.410	.292	.171
4	.000	.000	.002	.008	.026	.063	.130	.240	.410	.656	.815

N= 5
PROBABILITY

X	P=.05	.1	.2	.3	.4	.5	.6	.7	.8	.9	.95
0	.774	.590	.328	.168	.078	.031	.010	.002	.000	.000	.000
1	.204	.328	.410	.360	.259	.156	.077	.028	.006	.000	.000
2	.021	.073	.205	.309	.346	.313	.230	.132	.051	.008	.001
3	.001	.008	.051	.132	.230	.313	.346	.309	.205	.073	.021
4	.000	.000	.006	.028	.077	.156	.259	.360	.410	.328	.204
5	.000	.000	.000	.002	.010	.031	.078	.168	.328	.590	.774

.969

N= 6
PROBABILITY

X	.05	.1	.2	.3	.4	.5	.6	.7	.8	.9	.95
0	.735	.531	.262	.118	.047	.016	.004	.001	.000	.000	.000
1	.232	.354	.393	.303	.187	.094	.037	.010	.002	.000	.000
2	.031	.098	.246	.324	.311	.234	.138	.060	.015	.001	.000
3	.002	.015	.082	.185	.276	.313	.276	.185	.082	.015	.002
4	.000	.001	.015	.060	.138	.234	.311	.324	.246	.098	.031
5	.000	.000	.002	.010	.037	.094	.187	.303	.393	.354	.232
6	.000	.000	.000	.001	.004	.016	.047	.118	.262	.531	.735

N= 7
PROBABILITY

X	.05	.1	.2	.3	.4	.5	.6	.7	.8	.9	.95
0	.698	.478	.210	.082	.028	.008	.002	.000	.000	.000	.000
1	.257	.372	.367	.247	.131	.055	.017	.004	.000	.000	.000
2	.041	.124	.275	.318	.261	.164	.077	.025	.004	.000	.000
3	.004	.023	.115	.227	.290	.273	.194	.097	.029	.003	.000
4	.000	.003	.029	.097	.194	.273	.290	.227	.115	.023	.004
5	.000	.000	.004	.025	.077	.164	.261	.318	.275	.124	.041
6	.000	.000	.000	.004	.017	.055	.131	.247	.367	.372	.257
7	.000	.000	.000	.000	.002	.008	.028	.082	.210	.478	.698

N= 8
PROBABILITY

X	.05	.1	.2	.3	.4	.5	.6	.7	.8	.9	.95
0	.663	.430	.168	.058	.017	.004	.001	.000	.000	.000	.000
1	.279	.383	.336	.198	.090	.031	.008	.001	.000	.000	.000
2	.051	.149	.294	.296	.209	.109	.041	.010	.001	.000	.000
3	.005	.033	.147	.254	.279	.219	.124	.047	.009	.000	.000
4	.000	.005	.046	.136	.232	.273	.232	.136	.046	.005	.000
5	.000	.000	.009	.047	.124	.219	.279	.254	.147	.033	.005
6	.000	.000	.001	.010	.041	.109	.209	.296	.294	.149	.051
7	.000	.000	.000	.001	.008	.031	.090	.198	.336	.383	.279
8	.000	.000	.000	.000	.001	.004	.017	.058	.168	.430	.663

N= 9
PROBABILITY

X	.05	.1	.2	.3	.4	.5	.6	.7	.8	.9	.95
0	.630	.387	.134	.040	.010	.002	.000	.000	.000	.000	.000
1	.299	.387	.302	.156	.060	.018	.004	.000	.000	.000	.000
2	.063	.172	.302	.267	.161	.070	.021	.004	.000	.000	.000
3	.008	.045	.176	.267	.251	.164	.074	.021	.003	.000	.000
4	.001	.007	.066	.172	.251	.246	.167	.074	.017	.001	.000
5	.000	.001	.017	.074	.167	.246	.251	.172	.066	.007	.001
6	.000	.000	.003	.021	.074	.164	.251	.267	.176	.045	.008
7	.000	.000	.000	.004	.021	.070	.161	.267	.302	.172	.063
8	.000	.000	.000	.000	.004	.018	.060	.156	.302	.387	.299
9	.000	.000	.000	.000	.000	.002	.010	.040	.134	.387	.630

N= 10
PROBABILITY

X	.05	.1	.2	.3	.4	.5	.6	.7	.8	.9	.95
0	.599	.349	.107	.028	.006	.001	.000	.000	.000	.000	.000
1	.315	.387	.268	.121	.040	.010	.002	.000	.000	.000	.000
2	.075	.194	.302	.233	.121	.044	.011	.001	.000	.000	.000
3	.010	.057	.201	.267	.215	.117	.042	.009	.001	.000	.000
4	.001	.011	.088	.200	.251	.205	.111	.037	.006	.000	.000
5	.000	.001	.026	.103	.201	.246	.201	.103	.026	.001	.000
6	.000	.000	.006	.037	.111	.205	.251	.200	.088	.011	.001
7	.000	.000	.001	.009	.042	.117	.215	.267	.201	.057	.010
8	.000	.000	.000	.001	.011	.044	.121	.233	.302	.194	.075
9	.000	.000	.000	.000	.002	.010	.040	.121	.268	.387	.315
10	.000	.000	.000	.000	.000	.001	.006	.028	.107	.349	.599

N= 11
PROBABILITY

X	.05	.1	.2	.3	.4	.5	.6	.7	.8	.9	.95
0	.569	.314	.086	.020	.004	.000	.000	.000	.000	.000	.000
1	.329	.384	.236	.093	.027	.005	.001	.000	.000	.000	.000
2	.087	.213	.295	.200	.089	.027	.005	.001	.000	.000	.000
3	.014	.071	.221	.257	.177	.081	.023	.004	.000	.000	.000
4	.001	.016	.111	.220	.236	.161	.070	.017	.002	.000	.000
5	.000	.002	.039	.132	.221	.226	.147	.057	.010	.000	.000
6	.000	.000	.010	.057	.147	.226	.221	.132	.039	.002	.000
7	.000	.000	.002	.017	.070	.161	.236	.220	.111	.016	.001
8	.000	.000	.000	.004	.023	.081	.177	.257	.221	.071	.014
9	.000	.000	.000	.001	.005	.027	.089	.200	.295	.213	.087
10	.000	.000	.000	.000	.001	.005	.027	.093	.236	.384	.329
11	.000	.000	.000	.000	.000	.000	.004	.020	.086	.314	.569

N= 12
PROBABILITY

X	.05	.1	.2	.3	.4	.5	.6	.7	.8	.9	.95
0	.540	.282	.069	.014	.002	.000	.000	.000	.000	.000	.000
1	.341	.377	.206	.071	.017	.003	.000	.000	.000	.000	.000
2	.099	.230	.283	.168	.064	.016	.002	.000	.000	.000	.000
3	.017	.085	.236	.240	.142	.054	.012	.001	.000	.000	.000
4	.002	.021	.133	.231	.213	.121	.042	.008	.001	.000	.000
5	.000	.004	.053	.158	.227	.193	.101	.029	.003	.000	.000
6	.000	.000	.016	.079	.177	.226	.177	.079	.016	.000	.000
7	.000	.000	.003	.029	.101	.193	.227	.158	.053	.004	.000
8	.000	.000	.001	.008	.042	.121	.213	.231	.133	.021	.002
9	.000	.000	.000	.001	.012	.054	.142	.240	.236	.085	.017
10	.000	.000	.000	.000	.002	.016	.064	.168	.283	.230	.099
11	.000	.000	.000	.000	.000	.003	.017	.071	.206	.377	.341
12	.000	.000	.000	.000	.000	.000	.002	.014	.069	.282	.540

N= 13
PROBABILITY

X	.05	.1	.2	.3	.4	.5	.6	.7	.8	.9	.95
0	.513	.254	.055	.010	.001	.000	.000	.000	.000	.000	.000
1	.351	.367	.179	.054	.011	.002	.000	.000	.000	.000	.000
2	.111	.245	.268	.139	.045	.010	.001	.000	.000	.000	.000
3	.021	.100	.246	.218	.111	.035	.006	.001	.000	.000	.000
4	.003	.028	.154	.234	.184	.087	.024	.003	.000	.000	.000
5	.000	.006	.069	.180	.221	.157	.066	.014	.001	.000	.000
6	.000	.001	.023	.103	.197	.209	.131	.044	.006	.000	.000
7	.000	.000	.006	.044	.131	.209	.197	.103	.023	.001	.000
8	.000	.000	.001	.014	.066	.157	.221	.180	.069	.006	.000
9	.000	.000	.000	.003	.024	.087	.184	.234	.154	.028	.003
10	.000	.000	.000	.001	.006	.035	.111	.218	.246	.100	.021
11	.000	.000	.000	.000	.001	.010	.045	.139	.268	.245	.111
12	.000	.000	.000	.000	.000	.002	.011	.054	.179	.367	.351
13	.000	.000	.000	.000	.000	.000	.001	.010	.055	.254	.513

N= 14
PROBABILITY

X	.05	.1	.2	.3	.4	.5	.6	.7	.8	.9	.95
0	.488	.229	.044	.007	.001	.000	.000	.000	.000	.000	.000
1	.359	.356	.154	.041	.007	.001	.000	.000	.000	.000	.000
2	.123	.257	.250	.113	.032	.006	.001	.000	.000	.000	.000
3	.026	.114	.250	.194	.085	.022	.003	.000	.000	.000	.000
4	.004	.035	.172	.229	.155	.061	.014	.001	.000	.000	.000
5	.000	.008	.086	.196	.207	.122	.041	.007	.000	.000	.000
6	.000	.001	.032	.126	.207	.183	.092	.023	.002	.000	.000
7	.000	.000	.009	.062	.157	.209	.157	.062	.009	.000	.000
8	.000	.000	.002	.023	.092	.183	.207	.126	.032	.001	.000
9	.000	.000	.000	.007	.041	.122	.207	.196	.086	.008	.000
10	.000	.000	.000	.001	.014	.061	.155	.229	.172	.035	.004
11	.000	.000	.000	.000	.003	.022	.085	.194	.250	.114	.026
12	.000	.000	.000	.000	.001	.006	.032	.113	.250	.257	.123
13	.000	.000	.000	.000	.000	.001	.007	.041	.154	.356	.359
14	.000	.000	.000	.000	.000	.000	.001	.007	.044	.229	.488

N= 15
PROBABILITY

X	.05	.1	.2	.3	.4	.5	.6	.7	.8	.9	.95
0	.463	.206	.035	.005	.000	.000	.000	.000	.000	.000	.000
1	.366	.343	.132	.031	.005	.000	.000	.000	.000	.000	.000
2	.135	.267	.231	.092	.022	.003	.000	.000	.000	.000	.000
3	.031	.129	.250	.170	.063	.014	.002	.000	.000	.000	.000
4	.005	.043	.188	.219	.127	.042	.007	.001	.000	.000	.000
5	.001	.010	.103	.206	.186	.092	.024	.003	.000	.000	.000
6	.000	.002	.043	.147	.207	.153	.061	.012	.001	.000	.000
7	.000	.000	.014	.081	.177	.196	.118	.035	.003	.000	.000
8	.000	.000	.003	.035	.118	.196	.177	.081	.014	.000	.000
9	.000	.000	.001	.012	.061	.153	.207	.147	.043	.002	.000
10	.000	.000	.000	.003	.024	.092	.186	.206	.103	.010	.001
11	.000	.000	.000	.001	.007	.042	.127	.219	.188	.043	.005
12	.000	.000	.000	.000	.002	.014	.063	.170	.250	.129	.031
13	.000	.000	.000	.000	.000	.003	.022	.092	.231	.267	.135
14	.000	.000	.000	.000	.000	.000	.005	.031	.132	.343	.366
15	.000	.000	.000	.000	.000	.000	.000	.005	.035	.206	.463

N= 16
PROBABILITY

X	.05	.1	.2	.3	.4	.5	.6	.7	.8	.9	.95
0	.440	.185	.028	.003	.000	.000	.000	.000	.000	.000	.000
1	.371	.329	.113	.023	.003	.000	.000	.000	.000	.000	.000
2	.146	.275	.211	.073	.015	.002	.000	.000	.000	.000	.000
3	.036	.142	.246	.146	.047	.009	.001	.000	.000	.000	.000
4	.006	.051	.200	.204	.101	.028	.004	.000	.000	.000	.000
5	.001	.014	.120	.210	.162	.067	.014	.001	.000	.000	.000
6	.000	.003	.055	.165	.198	.122	.039	.006	.000	.000	.000
7	.000	.000	.020	.101	.189	.175	.084	.019	.001	.000	.000
8	.000	.000	.006	.049	.142	.196	.142	.049	.006	.000	.000
9	.000	.000	.001	.019	.084	.175	.189	.101	.020	.000	.000
10	.000	.000	.000	.006	.039	.122	.198	.165	.055	.003	.000
11	.000	.000	.000	.001	.014	.067	.162	.210	.120	.014	.001
12	.000	.000	.000	.000	.004	.028	.101	.204	.200	.051	.006
13	.000	.000	.000	.000	.001	.009	.047	.146	.246	.142	.036
14	.000	.000	.000	.000	.000	.002	.015	.073	.211	.275	.146
15	.000	.000	.000	.000	.000	.000	.003	.023	.113	.329	.371
16	.000	.000	.000	.000	.000	.000	.000	.003	.028	.185	.440

N= 17
PROBABILITY

X	.05	.1	.2	.3	.4	.5	.6	.7	.8	.9	.95
0	.418	.167	.023	.002	.000	.000	.000	.000	.000	.000	.000
1	.374	.315	.096	.017	.002	.000	.000	.000	.000	.000	.000
2	.158	.280	.191	.058	.010	.001	.000	.000	.000	.000	.000
3	.041	.156	.239	.125	.034	.005	.000	.000	.000	.000	.000
4	.008	.060	.209	.187	.080	.018	.002	.000	.000	.000	.000
5	.001	.017	.136	.208	.138	.047	.008	.001	.000	.000	.000
6	.000	.004	.068	.178	.184	.094	.024	.003	.000	.000	.000
7	.000	.001	.027	.120	.193	.148	.057	.009	.000	.000	.000
8	.000	.000	.008	.064	.161	.185	.107	.028	.002	.000	.000
9	.000	.000	.002	.028	.107	.185	.161	.064	.008	.000	.000
10	.000	.000	.000	.009	.057	.148	.193	.120	.027	.001	.000
11	.000	.000	.000	.003	.024	.094	.184	.178	.068	.004	.000
12	.000	.000	.000	.001	.008	.047	.138	.208	.136	.017	.001
13	.000	.000	.000	.000	.002	.018	.080	.187	.209	.060	.008
14	.000	.000	.000	.000	.000	.005	.034	.125	.239	.156	.041
15	.000	.000	.000	.000	.000	.001	.010	.058	.191	.280	.158
16	.000	.000	.000	.000	.000	.000	.002	.017	.096	.315	.374
17	.000	.000	.000	.000	.000	.000	.000	.002	.023	.167	.418

N= 18
PROBABILITY

X	.05	.1	.2	.3	.4	.5	.6	.7	.8	.9	.95
0	.397	.150	.018	.002	.000	.000	.000	.000	.000	.000	.000
1	.376	.300	.081	.013	.001	.000	.000	.000	.000	.000	.000
2	.168	.284	.172	.046	.007	.001	.000	.000	.000	.000	.000
3	.047	.168	.230	.105	.025	.003	.000	.000	.000	.000	.000
4	.009	.070	.215	.168	.061	.012	.001	.000	.000	.000	.000
5	.001	.022	.151	.202	.115	.033	.004	.000	.000	.000	.000
6	.000	.005	.082	.187	.166	.071	.015	.001	.000	.000	.000
7	.000	.001	.035	.138	.189	.121	.037	.005	.000	.000	.000
8	.000	.000	.012	.081	.173	.167	.077	.015	.001	.000	.000
9	.000	.000	.003	.039	.128	.185	.128	.039	.003	.000	.000
10	.000	.000	.001	.015	.077	.167	.173	.081	.012	.000	.000
11	.000	.000	.000	.005	.037	.121	.189	.138	.035	.001	.000
12	.000	.000	.000	.001	.015	.071	.166	.187	.082	.005	.000
13	.000	.000	.000	.000	.004	.033	.115	.202	.151	.022	.001
14	.000	.000	.000	.000	.001	.012	.061	.168	.215	.070	.009
15	.000	.000	.000	.000	.000	.003	.025	.105	.230	.168	.047
16	.000	.000	.000	.000	.000	.001	.007	.046	.172	.284	.168
17	.000	.000	.000	.000	.000	.000	.001	.013	.081	.300	.376
18	.000	.000	.000	.000	.000	.000	.000	.002	.018	.150	.397

N= 19
PROBABILITY

X	.05	.1	.2	.3	.4	.5	.6	.7	.8	.9	.95
0	.377	.135	.014	.001	.000	.000	.000	.000	.000	.000	.000
1	.377	.285	.068	.009	.001	.000	.000	.000	.000	.000	.000
2	.179	.285	.154	.036	.005	.000	.000	.000	.000	.000	.000
3	.053	.180	.218	.087	.017	.002	.000	.000	.000	.000	.000
4	.011	.080	.218	.149	.047	.007	.001	.000	.000	.000	.000
5	.002	.027	.164	.192	.093	.022	.002	.000	.000	.000	.000
6	.000	.007	.095	.192	.145	.052	.008	.001	.000	.000	.000
7	.000	.001	.044	.153	.180	.096	.024	.002	.000	.000	.000
8	.000	.000	.017	.098	.180	.144	.053	.008	.000	.000	.000
9	.000	.000	.005	.051	.146	.176	.098	.022	.001	.000	.000
10	.000	.000	.001	.022	.098	.176	.146	.051	.005	.000	.000
11	.000	.000	.000	.008	.053	.144	.180	.098	.017	.000	.000
12	.000	.000	.000	.002	.024	.096	.180	.153	.044	.001	.000
13	.000	.000	.000	.001	.008	.052	.145	.192	.095	.007	.000
14	.000	.000	.000	.000	.002	.022	.093	.192	.164	.027	.002
15	.000	.000	.000	.000	.001	.007	.047	.149	.218	.080	.011
16	.000	.000	.000	.000	.000	.002	.017	.087	.218	.180	.053
17	.000	.000	.000	.000	.000	.000	.005	.036	.154	.285	.179
18	.000	.000	.000	.000	.000	.000	.001	.009	.068	.285	.377
19	.000	.000	.000	.000	.000	.000	.000	.001	.014	.135	.377

N= 20
PROBABILITY

X	.05	.1	.2	.3	.4	.5	.6	.7	.8	.9	.95
0	.358	.122	.012	.001	.000	.000	.000	.000	.000	.000	.000
1	.377	.270	.058	.007	.000	.000	.000	.000	.000	.000	.000
2	.189	.285	.137	.028	.003	.000	.000	.000	.000	.000	.000
3	.060	.190	.205	.072	.012	.001	.000	.000	.000	.000	.000
4	.013	.090	.218	.130	.035	.005	.000	.000	.000	.000	.000
5	.002	.032	.175	.179	.075	.015	.001	.000	.000	.000	.000
6	.000	.009	.109	.192	.124	.037	.005	.000	.000	.000	.000
7	.000	.002	.055	.164	.166	.074	.015	.001	.000	.000	.000
8	.000	.000	.022	.114	.180	.120	.035	.004	.000	.000	.000
9	.000	.000	.007	.065	.160	.160	.071	.012	.000	.000	.000
10	.000	.000	.002	.031	.117	.176	.117	.031	.002	.000	.000
11	.000	.000	.000	.012	.071	.160	.160	.065	.007	.000	.000
12	.000	.000	.000	.004	.035	.120	.180	.114	.022	.000	.000
13	.000	.000	.000	.001	.015	.074	.166	.164	.055	.002	.000
14	.000	.000	.000	.000	.005	.037	.124	.192	.109	.009	.000
15	.000	.000	.000	.000	.001	.015	.075	.179	.175	.032	.002
16	.000	.000	.000	.000	.000	.005	.035	.130	.218	.090	.013
17	.000	.000	.000	.000	.000	.001	.012	.072	.205	.190	.060
18	.000	.000	.000	.000	.000	.000	.003	.028	.137	.285	.189
19	.000	.000	.000	.000	.000	.000	.000	.007	.058	.270	.377
20	.000	.000	.000	.000	.000	.000	.000	.001	.012	.122	.358

B—A Table of Random Numbers

33620	88487	70903	20357	05870	61177	19683	14492	12819	72553
75849	52067	19602	54856	45847	30100	85799	02712	11409	43289
19947	72811	63358	33919	77084	87629	52232	27607	24062	66302
81825	01464	48630	21461	08465	43379	14764	01184	20483	96182
75429	42606	17156	38447	87490	01051	20251	33075	18632	50135
11582	10842	06855	25088	46653	00286	26176	37032	83848	09206
37015	84325	96123	56321	04156	88822	25304	29222	58447	48851
06402	62241	25679	44715	19985	49681	18508	00465	40561	60918
59614	66846	85544	06874	01723	55988	60685	02718	55049	54841
52562	76607	59311	46969	53292	76241	92905	12086	64446	56785
95629	80326	37032	45089	15682	70727	22612	19239	06396	62001
65094	33827	93956	72790	03926	58153	96932	66370	99793	25775
51387	86685	08894	35878	18129	48737	08503	22366	98859	19850
71529	51996	99289	44268	42759	72434	54402	84560	41670	77515
11776	17395	61317	63290	17067	18408	08992	74943	45945	00325
82437	75248	23715	61194	62175	11149	44793	45957	65711	16043
14997	08398	37662	90175	65331	02562	38020	79525	31325	51759
55317	50018	64380	49047	57111	41641	25427	18625	96477	82040
47422	53721	11419	38616	72171	21523	80967	66207	02336	47981
09540	89442	52381	35035	15884	64273	96028	53514	44982	34607
72960	97289	74529	58580	24001	51320	36135	81297	41841	45517
18614	52254	24989	43465	62593	86589	26472	78422	31093	99309
40510	73555	89746	27658	41276	01861	12456	04449	94614	41816
13094	58717	15584	80291	16827	63515	49156	98782	23240	38480
15172	45757	18812	86080	93471	37872	35732	66961	06847	14692
76208	31777	41993	04148	78217	45369	21105	67959	30672	80364
09556	58574	91555	11046	72988	08271	10101	97504	14689	67094
60550	78165	60295	55210	55589	16165	73895	14658	44073	42391
65447	54491	14578	76579	90312	41113	65371	10676	47424	74086
84014	17701	51271	90046	69498	65270	29903	92236	19551	62579
37889	85731	98110	39598	62227	71295	00103	26720	29258	24055
15643	25400	71774	34647	44939	17764	76523	27576	78097	09715
59810	56274	02338	49070	96396	12146	81104	74206	27310	11757
70211	83037	98184	02201	78303	69144	74225	33604	49466	41221
06829	30426	70960	60645	85090	59155	79641	11026	14437	48708
94887	62595	59101	58728	26060	51073	03723	11149	75899	61339
57334	80471	55374	68820	14345	46807	89548	88908	78698	81286
81109	06627	77919	53821	83253	42285	94170	39442	39002	53352
30365	09434	79163	50188	12158	56475	04267	70615	37241	17406
22206	83484	99909	51217	58346	52090	32883	20674	97392	28270
63966	57749	39115	21190	26059	39420	94283	86203	05730	52240
64554	61044	64498	87781	46539	74112	30881	76457	14738	64336
92864	90110	49722	37592	61512	99828	56784	54296	87549	37614
00253	49864	42601	73838	00698	12227	53712	43701	94552	41096
13535	90824	10759	15926	52039	50091	52709	87377	26311	87222
32793	13980	80824	39350	84284	31106	71699	21052	64285	94301
33488	79908	22263	87796	49436	24720	35705	32139	58812	30873
59923	26145	43655	03622	35331	04927	62743	58120	74620	53027
27043	78658	43413	78614	15316	97926	70552	44083	45500	14077
05478	53519	32446	02086	88094	66418	74183	24088	78721	19100

91899	20696	81533	48736	33386	10790	95629	69205	27306	81818
02812	77805	35579	51182	55225	74657	32346	29465	25892	56598
97822	28834	27575	43413	83610	80271	96033	07248	90885	45746
93899	55326	15997	44330	00160	55793	92362	54739	17337	18491
79267	98483	94585	98334	25304	63352	45848	86027	86284	79072
87313	30799	62947	47311	40104	47384	83508	49754	65614	83483
41802	53864	88310	10488	97047	80901	52916	09398	73825	69033
27734	93897	25980	65823	34464	29013	49061	95129	73214	35512
03534	36601	59926	11800	02737	86302	37500	44766	47416	73597
93546	55818	41909	12056	48148	55360	95018	33346	72830	27768
50528	81261	06466	70242	53402	37944	98853	27385	27442	78728
24710	50740	73055	04592	53505	08014	81725	89058	46256	61631
75861	44421	93362	50625	20459	80429	16657	03618	26913	10497
40729	68233	30681	81317	99109	25775	72039	78412	84320	99510
11198	86189	96227	25205	06091	33680	95363	29109	08502	31166
61953	56307	70898	50681	48551	58547	33673	65327	47773	78392
07982	80001	31924	30757	04239	60659	07390	77729	09287	25062
04561	67777	58205	41347	01393	53134	18372	99042	16365	45883
84784	50525	57454	28455	68226	34656	99729	38884	39018	77703
01151	72507	53380	53827	42486	54465	85676	71819	91199	06254
71615	34986	74297	00144	38676	89967	57072	98869	39744	26530
61070	66851	27305	03759	44723	96108	99165	78489	18910	52719
99357	06738	62879	03910	17350	49169	73738	03850	01207	76891
09405	11448	10734	05837	24397	10420	15668	16712	94496	37144
12171	52850	98644	25641	75261	89365	25807	32796	89994	78286
02919	73367	10119	43985	15074	25554	33538	30584	41312	98985
95491	80235	10460	26869	83474	58734	06852	53637	86569	50187
64498	88438	03890	83380	98728	06504	63482	29503	37031	70040
13805	99756	60954	25708	80461	34440	62828	38823	60512	41102
73795	76970	02036	61863	09336	34615	40607	63979	69836	91183
64707	50750	27665	28113	84768	59843	47111	34076	78235	04727
03406	31225	55362	27095	97093	33526	04192	70413	28610	51458
01229	15239	02711	30716	10660	14742	55448	59008	73348	44267
82702	51718	73852	25552	42213	58265	19385	29356	35565	44815
41002	27977	76669	13993	89028	81538	31626	05708	17444	60610
20994	64600	12320	38703	04219	89220	84346	48552	88430	36765
70741	33569	32786	91576	21316	08082	06711	99093	16970	97249
84844	56305	87143	22325	15261	31024	04629	49918	23283	77656
42578	79267	00232	59816	13925	18557	39108	90940	10186	23256
14709	54916	42182	16586	15424	18225	63321	60898	17056	24303
66315	13477	02272	66141	85321	15676	03825	53818	24112	13479
08356	79095	42818	96607	15359	88957	07166	32345	61606	63620
08032	61053	62262	60214	84998	29016	35525	43638	64571	26313
68670	64220	07012	63186	48604	66315	37162	93619	25284	68953
55384	15691	46630	31943	76707	59795	13012	27318	23216	91039
98906	28598	62470	03643	75798	04218	44639	09290	06706	45989
64070	21706	85643	67095	56827	51869	37617	57140	88409	33868
71322	96476	47475	46445	19989	85843	06921	11691	99308	90575
91690	15790	20480	47154	11626	22143	10377	96389	54904	46721
70048	39989	68176	60248	18970	07655	67912	45592	58440	60344

C—The Normal Probability Distribution

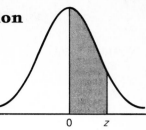

	\multicolumn{10}{c	}{SECOND DECIMAL PLACE OF z}
z	0	1	2	3	4	5	6	7	8	9
0.0	.0000	.0040	.0080	.0120	.0160	.0199	.0239	.0279	.0319	.0359
0.1	.0398	.0438	.0478	.0517	.0557	.0596	.0636	.0675	.0714	.0753
0.2	.0793	.0832	.0871	.0910	.0948	.0987	.1028	.1064	.1103	.1141
0.3	.1179	.1217	.1255	.1293	.1331	.1368	.1406	.1443	.1480	.1517
0.4	.1554	.1591	.1628	.1664	.1700	.1736	.1772	.1808	.1844	.1879
0.5	.1915	.1950	.1985	.2019	.2054	.2088	.2123	.2157	.2190	.2224
0.6	.2257	.2291	.2324	.2357	.2389	.2422	.2454	.2486	.2517	.2549
0.7	.2580	.2611	.2642	.2673	.2704	.2734	.2764	.2794	.2823	.2852
0.8	.2881	.2910	.2939	.2967	.2995	.3023	.3051	.3078	.3106	.3133
0.9	.3159	.3186	.3212	.3238	.3264	.3289	.3315	.3340	.3365	.3389
1.0	.3413	.3438	.3461	.3485	.3508	.3531	.3554	.3577	.3599	.3621
1.1	.3643	.3665	.3686	.3708	.3729	.3749	.3770	.3790	.3810	.3830
1.2	.3849	.3869	.3888	.3907	.3925	.3944	.3962	.3980	.3997	.4015
1.3	.4032	.4049	.4066	.4082	.4009	.4115	.4131	.4147	.4162	.4177
1.4	.4192	.4207	.4222	.4236	.4251	.4265	.4279	.4292	.4306	.4319
1.5	.4332	.4345	.4357	.4370	.4382	.4394	.4406	.4418	.4429	.4441
1.6	.4452	.4463	.4474	.4484	.4495	.4505	.4515	.4525	.4535	.4545
1.7	.4554	.4564	.4573	.4582	.4591	.4599	.4608	.4616	.4625	.4633
1.8	.4641	.4649	.4656	.4664	.4671	.4678	.4686	.4693	.4699	.4706
1.9	.4713	.4719	.4726	.4732	.4738	.4744	.4750	.4756	.4761	.4767
2.0	.4772	.4778	.4783	.4788	.4793	.4798	.4803	.4808	.4812	.4817
2.1	.4821	.4826	.4830	.4834	.4838	.4842	.4846	.4850	.4854	.4857
2.2	.4861	.4864	.4868	.4871	.4875	.4878	.4881	.4884	.4887	.4890
2.3	.4893	.4896	.4898	.4901	.4904	.4906	.4909	.4911	.4913	.4916
2.4	.4918	.4920	.4922	.4925	.4927	.4929	.4931	.4932	.4934	.4936
2.5	.4938	.4940	.4941	.4943	.4945	.4946	.4948	.4949	.4951	.4952
2.6	.4953	.4955	.4956	.4957	.4959	.4960	.4961	.4962	.4963	.4964
2.7	.4965	.4966	.4967	.4968	.4969	.4970	.4971	.4972	.4973	.4974
2.8	.4974	.4975	.4976	.4977	.4977	.4978	.4979	.4979	.4980	.4981
2.9	.4981	.4982	.4982	.4983	.4984	.4984	.4985	.4985	.4986	.4986
3.0	.4987									
3.5	.4998									
4.0	.49997									
4.5	.499997									
5.0	.4999997									

The entries in this table are the proportion of the observations from a *normal* distribution that have z scores between 0 and z. This proportion is represented by the shaded area under the curve in the figure.

Reprinted with permission from *CRC Standard Mathematical Tables*, Fifteenth Edition (West Palm Beach, Fla.: Copyright The Chemical Rubber Co., CRC Press, Inc.).

D—Critical Values of Student's *t* Distribution

One-tailed value

Two-tailed value

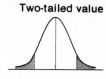

DEGREES OF FREEDOM	ONE-TAILED VALUE					
	0.25	0.10	0.05	0.025	0.01	0.005
	TWO-TAILED VALUE					
	0.50	0.20	0.10	0.05	0.02	0.01
1	1.000	3.078	6.314	12.706	31.821	63.657
2	0.816	1.886	2.920	4.303	6.965	9.925
3	.765	1.638	2.353	3.182	4.541	5.841
4	.741	1.533	2.132	2.776	3.747	4.604
5	.727	1.476	2.015	2.571	3.365	4.032
6	.718	1.440	1.943	2.447	3.143	3.707
7	.711	1.415	1.895	2.365	2.998	3.499
8	.706	1.397	1.860	2.306	2.896	3.355
9	.703	1.383	1.833	2.262	2.821	3.250
10	.700	1.372	1.812	2.228	2.764	3.169
11	.697	1.363	1.796	2.201	2.718	3.106
12	.695	1.356	1.782	2.179	2.681	3.055
13	.694	1.350	1.771	2.160	2.650	3.012
14	.692	1.345	1.761	2.145	2.626	2.977
15	.691	1.341	1.753	2.131	2.602	2.947
16	.690	1.337	1.746	2.120	2.583	2.921
17	.689	1.333	1.740	2.110	2.567	2.898
18	.688	1.330	1.734	2.101	2.552	2.878
19	.688	1.328	1.729	2.093	2.539	2.861
20	.687	1.325	1.725	2.086	2.528	2.845
21	.686	1.323	1.721	2.080	2.518	2.831
22	.686	1.321	1.717	2.074	2.508	2.819
23	.685	1.319	1.714	2.069	2.500	2.807
24	.685	1.318	1.711	2.064	2.492	2.797
25	.684	1.316	1.708	2.060	2.485	2.787
26	.684	1.315	1.706	2.056	2.479	2.779
27	.684	1.314	1.703	2.052	2.473	2.771
28	.683	1.313	1.701	2.048	2.467	2.763
29	.683	1.311	1.699	2.045	2.462	2.756
30	.683	1.310	1.697	2.042	2.457	2.750
35	.682	1.306	1.690	2.030	2.438	2.724
40	.681	1.303	1.684	2.021	2.423	2.704
45	.680	1.301	1.680	2.014	2.412	2.690
50	.680	1.299	1.676	2.008	2.403	2.678
55	.679	1.297	1.673	2.004	2.396	2.669
60	.679	1.296	1.671	2.000	2.390	2.660
70	.678	1.294	1.667	1.994	2.381	2.648
80	.678	1.293	1.665	1.989	2.374	2.638
90	.678	1.291	1.662	1.986	2.368	2.631
100	.677	1.290	1.661	1.982	2.364	2.625
120	.677	1.289	1.658	1.980	2.358	2.617
∞	.674	1.282	1.645	1.960	2.326	2.576

From *The Ways and Means of Statistics* by Leonard J. Tashman and Kathleen R. Lamborn © 1979 by Harcourt Brace Jovanovich, Inc. Reprinted by permission of the publisher.

E—Critical Values of the *F* Statistic
(0.05 level of significance)

Degrees of Freedom in Numerator

	1	2	3	4	5	6	7	8	9	10	12	15	20	24	30	40	60	120	∞
1	161	200	216	225	230	234	237	239	241	242	244	246	248	249	250	251	252	253	254
2	18.5	19.0	19.2	19.2	19.3	19.3	19.4	19.4	19.4	19.4	19.4	19.4	19.4	19.5	19.5	19.5	19.5	19.5	19.5
3	10.1	9.55	9.28	9.12	9.01	8.94	8.89	8.85	8.81	8.79	8.74	8.70	8.66	8.64	8.62	8.59	8.57	8.55	8.53
4	7.71	6.94	6.59	6.39	6.26	6.16	6.09	6.04	6.00	5.96	5.91	5.86	5.80	5.77	5.75	5.72	5.69	5.66	5.63
5	6.61	5.79	5.41	5.19	5.05	4.95	4.88	4.82	4.77	4.74	4.68	4.62	4.56	4.53	4.50	4.46	4.43	4.40	4.37
6	5.99	5.14	4.76	4.53	4.39	4.28	4.21	4.15	4.10	4.06	4.00	3.94	3.87	3.84	3.81	3.77	3.74	3.70	3.67
7	5.59	4.74	4.35	4.12	3.97	3.87	3.79	3.73	3.68	3.64	3.57	3.51	3.44	3.41	3.38	3.34	3.30	3.27	3.23
8	5.32	4.46	4.07	3.84	3.69	3.58	3.50	3.44	3.39	3.35	3.28	3.22	3.15	3.12	3.08	3.04	3.01	2.97	2.93
9	5.12	4.26	3.86	3.63	3.48	3.37	3.29	3.23	3.18	3.14	3.07	3.01	2.94	2.90	2.86	2.83	2.79	2.75	2.71
10	4.96	4.10	3.71	3.48	3.33	3.22	3.14	3.07	3.02	2.98	2.91	2.85	2.77	2.74	2.70	2.66	2.62	2.58	2.54
11	4.84	3.98	3.59	3.36	3.20	3.09	3.01	2.95	2.90	2.85	2.79	2.72	2.65	2.61	2.57	2.53	2.49	2.45	2.40
12	4.75	3.89	3.49	3.26	3.11	3.00	2.91	2.85	2.80	2.75	2.69	2.62	2.54	2.51	2.47	2.43	2.38	2.34	2.30
13	4.67	3.81	3.41	3.18	3.03	2.92	2.83	2.77	2.71	2.67	2.60	2.53	2.46	2.42	2.38	2.34	2.30	2.25	2.21
14	4.60	3.74	3.34	3.11	2.96	2.85	2.76	2.70	2.65	2.60	2.53	2.46	2.39	2.35	2.31	2.27	2.22	2.18	2.13
15	4.54	3.68	3.29	3.06	2.90	2.79	2.71	2.64	2.59	2.54	2.48	2.40	2.33	2.29	2.25	2.20	2.16	2.11	2.07
16	4.49	3.63	3.24	3.01	2.85	2.74	2.66	2.59	2.54	2.49	2.42	2.35	2.28	2.24	2.19	2.15	2.11	2.06	2.01
17	4.45	3.59	3.20	2.96	2.81	2.70	2.61	2.55	2.49	2.45	2.38	2.31	2.23	2.19	2.15	2.10	2.06	2.01	1.96
18	4.41	3.55	3.16	2.93	2.77	2.66	2.58	2.51	2.46	2.41	2.34	2.27	2.19	2.15	2.11	2.06	2.02	1.97	1.92
19	4.38	3.52	3.13	2.90	2.74	2.63	2.54	2.48	2.42	2.38	2.31	2.23	2.16	2.11	2.07	2.03	1.98	1.93	1.88
20	4.35	3.49	3.10	2.87	2.71	2.60	2.51	2.45	2.39	2.35	2.28	2.20	2.12	2.08	2.04	1.99	1.95	1.90	1.84
21	4.32	3.47	3.07	2.84	2.68	2.57	2.49	2.42	2.37	2.32	2.25	2.18	2.10	2.05	2.01	1.96	1.92	1.87	1.81
22	4.30	3.44	3.05	2.82	2.66	2.55	2.46	2.40	2.34	2.30	2.23	2.15	2.07	2.03	1.98	1.94	1.89	1.84	1.78
23	4.28	3.42	3.03	2.80	2.64	2.53	2.44	2.37	2.32	2.27	2.20	2.13	2.05	2.01	1.96	1.91	1.86	1.81	1.76
24	4.26	3.40	3.01	2.78	2.62	2.51	2.42	2.36	2.30	2.25	2.18	2.11	2.03	1.98	1.94	1.89	1.84	1.79	1.73
25	4.24	3.39	2.99	2.76	2.60	2.49	2.40	2.34	2.28	2.24	2.16	2.09	2.01	1.96	1.92	1.87	1.82	1.77	1.71
30	4.17	3.32	2.92	2.69	2.53	2.42	2.33	2.27	2.21	2.16	2.09	2.01	1.93	1.89	1.84	1.79	1.74	1.68	1.62
40	4.08	3.23	2.84	2.61	2.45	2.34	2.25	2.18	2.12	2.08	2.00	1.92	1.84	1.79	1.74	1.69	1.64	1.58	1.51
60	4.00	3.15	2.76	2.53	2.37	2.25	2.17	2.10	2.04	1.99	1.92	1.84	1.75	1.70	1.65	1.59	1.53	1.47	1.39
120	3.92	3.07	2.68	2.45	2.29	2.18	2.09	2.02	1.96	1.91	1.83	1.75	1.66	1.61	1.55	1.50	1.43	1.35	1.25
∞	3.84	3.00	2.60	2.37	2.21	2.10	2.01	1.94	1.88	1.83	1.75	1.67	1.57	1.52	1.46	1.39	1.32	1.22	1.00

Degrees of Freedom in Denominator

Source: This table is reproduced from M. Merrington and C.M. Thompson, "Tables of Percentage Points of the Inverted Beta (*F*) Distribution," *Biometrika*, vol. 33 (1943), by permission of the Biometrika trustees.

E—Critical Values of the F Statistic (0.01 level of significance)

Degrees of Freedom in Numerator

Denominator df	1	2	3	4	5	6	7	8	9	10	12	15	20	24	30	40	60	120	∞
1	4,052	5,000	5,403	5,625	5,764	5,859	5,928	5,982	6,023	6,056	6,106	6,157	6,209	6,235	6,261	6,287	6,313	6,339	6,366
2	98.5	99.0	99.2	99.2	99.3	99.3	99.4	99.4	99.4	99.4	99.4	99.4	99.4	99.5	99.5	99.5	99.5	99.5	99.5
3	34.1	30.8	29.5	28.7	28.2	27.9	27.7	27.5	27.3	27.2	27.1	26.9	26.7	26.6	26.5	26.4	26.3	26.2	26.1
4	21.2	18.0	16.7	16.0	15.5	15.2	15.0	14.8	14.7	14.5	14.4	14.2	14.0	13.9	13.8	13.7	13.7	13.6	13.5
5	16.3	13.3	12.1	11.4	11.0	10.7	10.5	10.3	10.2	10.1	9.89	9.72	9.55	9.47	9.38	9.29	9.20	9.11	9.02
6	13.7	10.9	9.78	9.15	8.75	8.47	8.26	8.10	7.98	7.87	7.72	7.56	7.40	7.31	7.23	7.14	7.06	6.97	6.88
7	12.2	9.55	8.45	7.85	7.46	7.19	6.99	6.84	6.72	6.62	6.47	6.31	6.16	6.07	5.99	5.91	5.82	5.74	5.65
8	11.3	8.65	7.59	7.01	6.63	6.37	6.18	6.03	5.91	5.81	5.67	5.52	5.36	5.28	5.20	5.12	5.03	4.95	4.86
9	10.6	8.02	6.99	6.42	6.06	5.80	5.61	5.47	5.35	5.26	5.11	4.96	4.81	4.73	4.65	4.57	4.48	4.40	4.31
10	10.0	7.56	6.55	5.99	5.64	5.39	5.20	5.06	4.94	4.85	4.71	4.56	4.41	4.33	4.25	4.17	4.08	4.00	3.91
11	9.65	7.21	6.22	5.67	5.32	5.07	4.89	4.74	4.63	4.54	4.40	4.25	4.10	4.02	3.94	3.86	3.78	3.69	3.60
12	9.33	6.93	5.95	5.41	5.06	4.82	4.64	4.50	4.39	4.30	4.16	4.01	3.86	3.78	3.70	3.62	3.54	3.45	3.36
13	9.07	6.70	5.74	5.21	4.86	4.62	4.44	4.30	4.19	4.10	3.96	3.82	3.66	3.59	3.51	3.43	3.34	3.25	3.17
14	8.86	6.51	5.56	5.04	4.70	4.46	4.28	4.14	4.03	3.94	3.80	3.66	3.51	3.43	3.35	3.27	3.18	3.09	3.00
15	8.68	6.36	5.42	4.89	4.56	4.32	4.14	4.00	3.89	3.80	3.67	3.52	3.37	3.29	3.21	3.13	3.05	2.96	2.87
16	8.53	6.23	5.29	4.77	4.44	4.20	4.03	3.89	3.78	3.69	3.55	3.41	3.26	3.18	3.10	3.02	2.93	2.84	2.75
17	8.40	6.11	5.19	4.67	4.34	4.10	3.93	3.79	3.68	3.59	3.46	3.31	3.16	3.08	3.00	2.92	2.83	2.75	2.65
18	8.29	6.01	5.09	4.58	4.25	4.01	3.84	3.71	3.60	3.51	3.37	3.23	3.08	3.00	2.92	2.84	2.75	2.66	2.57
19	8.19	5.93	5.01	4.50	4.17	3.94	3.77	3.63	3.52	3.43	3.30	3.15	3.00	2.92	2.84	2.76	2.67	2.58	2.49
20	8.10	5.85	4.94	4.43	4.10	3.87	3.70	3.56	3.46	3.37	3.23	3.09	2.94	2.86	2.78	2.69	2.61	2.52	2.42
21	8.02	5.78	4.87	4.37	4.04	3.81	3.64	3.51	3.40	3.31	3.17	3.03	2.88	2.80	2.72	2.64	2.55	2.46	2.36
22	7.95	5.72	4.82	4.31	3.99	3.76	3.59	3.45	3.35	3.26	3.12	2.98	2.83	2.75	2.67	2.58	2.50	2.40	2.31
23	7.88	5.66	4.76	4.26	3.94	3.71	3.54	3.41	3.30	3.21	3.07	2.93	2.78	2.70	2.62	2.54	2.45	2.35	2.26
24	7.82	5.61	4.72	4.22	3.90	3.67	3.50	3.36	3.26	3.17	3.03	2.89	2.74	2.66	2.58	2.49	2.40	2.31	2.21
25	7.77	5.57	4.68	4.18	3.86	3.63	3.46	3.32	3.22	3.13	2.99	2.85	2.70	2.62	2.53	2.45	2.36	2.27	2.17
30	7.56	5.39	4.51	4.02	3.70	3.47	3.30	3.17	3.07	2.98	2.84	2.70	2.55	2.47	2.39	2.30	2.21	2.11	2.01
40	7.31	5.18	4.31	3.83	3.51	3.29	3.12	2.99	2.89	2.80	2.66	2.52	2.37	2.29	2.20	2.11	2.02	1.92	1.80
60	7.08	4.98	4.13	3.65	3.34	3.12	2.95	2.82	2.72	2.63	2.50	2.35	2.20	2.12	2.03	1.94	1.84	1.73	1.60
120	6.85	4.79	3.95	3.48	3.17	2.96	2.79	2.66	2.56	2.47	2.34	2.19	2.03	1.95	1.86	1.76	1.66	1.53	1.38
∞	6.63	4.61	3.78	3.32	3.02	2.80	2.64	2.51	2.41	2.32	2.18	2.04	1.88	1.79	1.70	1.59	1.47	1.32	1.00

Degrees of Freedom in Denominator

F—Critical Values of the Chi-Square Statistic

ss of freedom   *Level of Significance*

χ^2

Possible Values of χ^2

$K-1$

DEGREES OF FREEDOM df	RIGHT-TAIL AREA			
	0.10	0.05	0.02	0.01
1	2.706	3.841	5.412	6.635
2	4.605	5.991	7.824	9.210
3	6.251	7.815	9.837	11.345
4	7.779	9.488	11.668	13.277
5	9.236	11.070	13.388	15.086
6	10.645	12.592	15.033	16.812
7	12.017	14.067	16.622	18.475
8	13.362	15.507	18.168	20.090
9	14.684	16.919	19.679	21.666
10	15.987	18.307	21.161	23.209
11	17.275	19.675	22.618	24.725
12	18.549	21.026	24.054	26.217
13	19.812	22.362	25.472	27.688
14	21.064	23.685	26.873	29.141
15	22.307	24.996	28.259	30.578
16	23.542	26.296	29.633	32.000
17	24.769	27.587	30.995	33.409
18	25.989	28.869	32.346	34.805
19	27.204	30.144	33.687	36.191
20	28.412	31.410	35.020	37.566
21	29.615	32.671	36.343	38.932
22	30.813	33.924	37.659	40.289
23	32.007	35.172	38.968	41.638
24	33.196	36.415	40.270	42.980
25	34.382	37.652	41.566	44.314
26	35.563	38.885	42.856	45.642
27	36.741	40.113	44.140	46.963
28	37.916	41.337	45.419	48.278
29	39.087	42.557	46.693	49.588
30	40.256	43.773	47.962	50.892

This table contains the values of χ^2 that correspond to a specific right-tail area and specific numbers of degrees of freedom df.

Source: From Table IV of Fisher & Yates: *Statistical Tables for Biological, Agricultural and Medical Research*, published by Longman Group Ltd., London (previously published by Oliver and Boyd Ltd., Edinburgh), by permission of the authors and publishers.

G—Critical Values of the Wilcoxon Signed-Rank Statistic

ONE-TAILED VALUE	0.10	0.05	0.025	0.01	0.005
TWO-TAILED VALUE	0.20	0.10	0.05	0.02	0.01
n					
5	2	0	—	—	—
6	3	2	0	—	—
7	5	3	2	0	—
8	8	5	3	1	0
9	10	8	5	3	1
10	14	10	8	5	3
11	17	13	10	7	5
12	21	17	13	9	7
13	26	21	17	12	9
14	31	25	21	15	12
15	36	30	25	19	15
16	42	35	29	23	19
17	48	41	34	27	23
18	55	47	40	32	27
19	62	53	46	37	32
20	69	60	52	43	37

Each entry represents the largest value of *W* with a value less than or equal to the value in the column heading.

Adapted from Table II in F. Wilcoxon, S.K. Katti, and Roberta A. Wilcox, *Critical Values and Probability Levels for the Wilcoxon Rank Sum Test and the Wilcoxon Signed Rank Test*, American Cyanamid Company (Lederle Laboratories Division, Pearl River, N.Y.) and The Florida State University (Department of Statistics, Tallahassee, Fla.), August 1963. Used with the permission of American Cyanamid Company and The Florida State University.

Glossary

alternate hypothesis A claim about the population that is accepted if the null hypothesis is rejected.

analysis of variance (ANOVA) A statistical technique used to determine whether more than two populations have the same mean.

arithmetic mean The sum of the values divided by the total number of values.

average A single value that is representative of the set of data.

central limit theorem The distribution of the sample means approaches the normal probability distribution as the sample size increases, regardless of the shape of the population.

chi-square distribution A positively skewed continuous probability distribution based on the number of degrees of freedom. It is nonnegative and approaches a symmetric distribution as the number of degrees of freedom increases.

class An interval within which data are tallied.

class frequency The number of observations, or tallies, that occur in each class.

classical probability The number of favorable outcomes divided by the total number of possible outcomes.

cluster sample A form of stratified sampling. The population is divided into homogeneous groups called clusters, such as geographic regions. Then, samples are randomly obtained from those clusters.

coefficient of correlation A measure of the strength of the association between two interval-scaled variables. It may range from -1.0 to 1.0 with -1.0 and 1.0 indicating perfect correlation and 0 the absence of correlation.

coefficient of determination The proportion of the total variation in one variable that is explained by the variation in the other variable.

coefficient of skewness A measure of the lack of symmetry in a distribution.

coefficient of variation The standard deviation divided by the mean. A measure of the relative dispersion of a data set.

combination One particular group of objects or persons selected from a larger group.

complement rule The probability of an event not happening. It is found by subtracting the probability of it happening from 1.0, written $P(\text{not } A) = 1 - P(A)$.

conditional probability The likelihood an event will occur, assuming that another event has already occurred.

confidence interval A range within which the population parameter is expected to fall for a preselected level of confidence.

contingency table Frequency data from the simultaneous classification of more than one variable or trait of the observed item, often tallied in one table.

continuous probability distribution A probability distribution that assumes an infinite number of values within a specific range of values.

correlation analysis The statistical techniques used to determine the strength of the relationship between two variables. The basic objective of correlation analysis is to determine the degree of correlation (relationship) between variables, from zero (no correlation) to perfect (complete) correlation.

critical value A value or values that separate the region of rejection from the remaining values.

decision rule A statement of the condition or conditions under which the null hypothesis is rejected.

degrees of freedom The number of items in a sample that are free to vary.

dependent samples Samples that are paired or related in some fashion.

dependent variable The variable that is being predicted or estimated.

descriptive statistics The methods used to describe the data that have been collected.

discrete probability distribution A distribution that can assume only certain values. It is usually the result of counting the number of favorable outcomes of an experiment.

event A collection of one or more outcomes of an experiment.

exhaustive Each person, object, or item must be classified in at least one category.

experiment The observation of some activity, or the act of taking some type of measurement.

F distribution A continuous probability distribution where the value of F is always positive. The distribution is always positively skewed.

first quartile The point below which one-fourth of the observations occur.

frequency distribution An arrangement of the data that shows the frequency of occurrence of the values of interest.

frequency polygon A chart using straight lines to graphically portray the frequency distribution.

general rule of addition If two events are combined that are not mutually exclusive, the probability that one or the other will occur is their sum minus the probability of the joint occurrence, written $P(A \text{ or } B) = P(A) + P(B) - P(A \text{ and } B)$.

general rule of multiplication The probability of the events A and B occurring is the probability of event A times the probability of event B, given that event A has occurred, written $P(A \text{ and } B) = P(A) \cdot P(B|A)$.

goodness-of-fit test A test to determine if an observed set of frequencies could have been obtained from a population with a given distribution.

histogram A chart using bars to graphically portray the frequency distribution.

independent events The occurrence of one event does not affect the probability that the other event will take place.

independent samples   Samples that are not related in any way.

independent variable   A variable that provides the basis for estimation. It is the predictor variable and is usually designated X.

inferential statistics   A decision, estimate, prediction, or generalization about a population based on a sample.

interquartile range   The difference between the third quartile and the first quartile.

interval estimate   A range of values within which we have some confidence that the population parameter lies.

interval scale   The distance between numbers is a known, constant size.

joint probability   A measure of the likelihood that two or more events will happen concurrently.

Kruskal-Wallis test   A test based on ranks to determine whether more than two samples come from the same population.

least-squares principle   A method used to determine the regression equation by minimizing the sum of the squares of the distances between the actual Y values and the predicted values of Y.

level of significance   The probability of rejecting the null hypothesis when it is true.

mean deviation   The average of the deviations between all the observations in a set of data and its mean.

median   The midpoint of the values after they have been arranged from the smallest to the largest (or the largest to the smallest).

midpoint   A point that divides a class into two equal parts.

mode   The value of the observation that appears most often.

mutually exclusive   An individual or item that, by virtue of being included in one category, must be excluded from another.

negatively skewed distribution   A distribution having a mean smaller than the median and mode. The long tail of the distribution is to the left, or in a negative direction.

nominal scale   Data that are organized into categories the order of which is not important.

nonparametric tests   Tests where assumptions regarding the shape of the population are not required. Such tests are applicable for nominal or ordinal scale of measurement.

nonprobability sampling   Items whose inclusion in the sample is based on the judgment of the person selecting the sample.

null hypothesis   A claim about the value of a population parameter.

one-tailed test   A hypothesis test in which the rejection region is in one tail of the sampling distribution.

ordinal scale   Data or categories that can be ranked; that is, one category is higher than another.

outcome   A particular result of an experiment.

parameter   One measurable characteristic of a population.

partial correlation   A measure that shows the relationship between the dependent variable and an independent variable not yet considered in the multiple regression equation when the other independent variables in the equation are considered but held constant.

permutation   An ordered arrangement of a group of objects.

point estimate The value, computed from a sample, used to estimate a population parameter.

pooled variance estimate An estimate of the population variance obtained by combining two or more sample variances.

population A collection of all possible members of a set of individuals, objects, or measurements.

positively skewed distribution A distribution having a mean larger than the median or mode. The long tail of the distribution is to the right, or in a positive direction.

probability A fraction that measures the likelihood a particular event will occur.

probability distribution A listing of the outcomes that may occur and of their corresponding probabilities.

probability sampling A method of sampling where each member of the population of interest has a known likelihood of being included in the sample.

proportion A fraction, ratio, percent, or probability that indicates what part of the sample or population has a particular trait.

quartile deviation One-half of the interquartile range. To compute the quartile deviation, the first quartile is subtracted from the third quartile and the difference is divided by two.

random sample A sample chosen so that each member of the population has the same chance of being selected.

random variable A quantity that assumes one and only one numerical value as a result of the outcome of an experiment.

range The difference between the highest and lowest observations in a set of data.

raw data Numerical information presented in an ungrouped form.

regression equation A mathematical equation that defines the relationship between two variables.

relative frequency The number of times a particular event occurred in the past divided by the total number of observations.

sample A part or portion of the population.

sampling distribution of the mean A probability distribution of all possible sample means of a given size selected from a population.

sampling error The difference between the value of the population parameter and its corresponding sample statistic.

scatter diagram A graphic tool that visually portrays the relationship between two variables.

sign test A test based on the sign of the differences in a set of paired observations.

Spearman's rank-order correlation coefficient A measure of the strength of association between two ordinal-scaled sets of variables. It may range from -1.0 to 1.0.

special rule of addition If two mutually exclusive events A and B are combined, the probability that one or the other will happen is their sum, written $P(A \text{ or } B) = P(A) + P(B)$.

special rule of multiplication A rule for combining two or more independent events. The probability of the joint occurrence of two events is $P(A \text{ and } B) = P(A) \cdot P(B)$.

standard deviation The square root of the arithmetic mean of the squared deviations from the mean.

standard error of estimate A measure of the variability of the observed values around the regression line.

standard error of the mean The standard deviation of the sampling distribution of the sample means.

standard normal distribution A

special normal distribution with a mean of zero and a standard deviation of one unit.

stated limits The actual boundaries for a particular class in a frequency distribution.

statistic One measurable characteristic of a sample.

statistics The body of techniques used to facilitate the collection, organization, presentation, analysis, and interpretation of data for the purpose of making better decisions.

stratified random sample After the population of interest is divided into logical strata, a sample is drawn from each stratum or subgroup. The manner in which the sample is gathered may be either nonproportional or proportional to the total number of members in each stratum.

subjective probability The likelihood of an event assigned on the basis of whatever information is available.

symmetric distribution A distribution that has the same shape on both sides of the median.

systematic random sample The members of the population are arranged in some fashion. A random starting point is selected. Then every kth element is chosen for the sample.

t distribution A continuous probability distribution with a mean of zero. It is flatter at the apex and more spread out than the standard normal distribution.

test statistic A quantity, calculated from the sample information, used as a basis for deciding whether or not to reject the null hypothesis.

third quartile The point below which three-fourths of the observations occur.

treatment A specific source or cause of variation in a set of data.

true limits The real boundaries of a class, given continuous data.

two-tailed test A hypothesis test where the rejection region is divided equally between the two tails of the sampling distribution.

Type I error An error that occurs when a true null hypothesis is rejected.

Type II error An error that occurs when a false null hypothesis is not rejected.

weighted mean The values of the observations are weighted by the frequency of occurrence.

Wilcoxon rank-sum test A test based on the sum of the ranks to determine whether two samples came from the same population.

Wilcoxon signed-rank test A test applied to paired data. It is a replacement for the paired t test.

Y-intercept The point where the regression equation crosses the Y-axis.

z value or z score A unit of measure with respect to the standard normal distribution. It measures the distance from the mean of a normal distribution in terms of the number of standard deviations.

Index